TRANSFORMERS AND INDUCTORS FOR POWER ELECTRONICS

TRANSFORMERS AND INDUCTORS FOR POWER ELECTRONICS

THEORY, DESIGN AND APPLICATIONS

W. G. Hurley
National University of Ireland, Galway, Ireland

W. H. Wölfle
Convertec Ltd, Wexford Ireland

WILEY
A John Wiley & Sons, Ltd., Publication

This edition first published 2013

Registered office
John Wiley & Sons Ltd, The Atrium, Southern Gate, Chichester, West Sussex, PO19 8SQ, United Kingdom

For details of our global editorial offices, for customer services and for information about how to apply for permission to reuse the copyright material in this book please see our website at www.wiley.com.

Library of Congress Cataloging-in-Publication Data
Hurley, William G.
Transformers and inductors for power electronics: theory, design and applications / W.G. Hurley, W.H. Wölfle.
pages cm
Includes bibliographical references and index.
ISBN 978-1-119-95057-8 – ISBN 978-1-118-54464-8 – ISBN 978-1-118-54466-2– ISBN 978-1-118-54467-9 – ISBN 978-1-118-54468-6
1. Electric transformers–Design and construction. 2. Electric inductors–Design and construction. I. Wölfle, Werner H. II. Title.
TK2551.H87 2013
621.31′4–dc23

2012039432

ISBN 978-1-119-95057-8

Set in 10/12pt Times-Roman by Thomson Digital, Noida, India
Printed and bound in Malaysia by Vivar Printing Sdn Bhd

To Our Families

Contents

About the Authors

William Gerard Hurley was born in Cork, Ireland. He received the B.E. degree in Electrical Engineering from the National University of Ireland, Cork in 1974, the M.S. degree in Electrical Engineering from the Massachusetts Institute of Technology, Cambridge MA, in 1976 and the PhD degree at the National University of Ireland, Galway in 1988. He was awarded the D.ENG degree by the National University of Ireland in 2011.

He worked for Honeywell Controls in Canada from 1977–1979, and for Ontario Hydro from 1979–1983. He lectured in electronic engineering at the University of Limerick, Ireland from 1983 to 1991 and is currently Professor of Electrical Engineering at the National University of Ireland, Galway. He is the Director of the Power Electronics Research Centre there. He served on the faculty at the Massachusetts Institute of Technology as a Visiting Professor of Electrical Engineering in 1997–1998. Prof. Hurley has given invited presentations on magnetics in Mexico, Japan, Singapore, Spain, the Czech Republic, Hong Kong, China and USA.

His research interests include high frequency magnetics, power quality, and renewable energy systems. He received a Best Paper Prize for the *IEEE Transactions on Power Electronics* in 2000. Prof. Hurley is a Fellow of the IEEE. He has served as a member of the Administrative Committee of the Power Electronics Society of the IEEE and was General Chair of the Power Electronics Specialists Conference in 2000.

Werner Hugo Wölfle was born in Bad Schussenried, Germany. He graduated from the University of Stuttgart in Germany in 1981 as a Diplom-Ingenieur in Electronics. He completed a PhD degree in Electrical Engineering at the National University of Ireland, Galway in 2003.

He worked for Dornier Systems GmbH from 1982–1985 as a Development Engineer for power converters in space craft applications. From 1986–1988 he worked as a Research and Development Manager for industrial AC and DC power. Since 1989 he has been Managing Director of Convertec Ltd. in Wexford, Ireland, a company of the TRACOPOWER Group. Convertec develops high reliability power converters for industrial applications. He is currently an Adjunct Professor in Electrical Engineering at the National University of Ireland, Galway.

Acknowledgements

We would like to acknowledge Prof. John Kassakian, M.I.T. for his continued support for our magnetics work for many years. We are indebted to the numerous staff and students of the National University of Ireland, Galway, past and present who have contributed to this work.

A special thanks to Dr Eugene Gath, University of Limerick for his mathematical input to the optimisation problems. The contributions of Dr Ningning Wang, Tyndall Institute and Dr Jian Liu, Volterra to the planar magnetics material is much appreciated.

A special word of gratitude goes to PhD students Dr Maeve Duffy, Dr John Breslin who contributed to many of the ideas in this text. Their PhD theses form the foundations upon which this book is based.

We appreciate the many insights and ideas that arose in discussions with Joe Madden, Enterprise Ireland; Prof. Dean Patterson, University of Nebraska-Lincoln; Prof. Ron Hui, University of Hong Kong; Prof. Dave Perreault, M.I.T.; Prof. Charles Sullivan, Dartmouth College; Dr Arthur Kelley and Prof Cian Ó'Mathúna, University College Cork.

We acknowledge the reviewers for their thorough efforts: Dr Noel Barry, National Maritime College of Ireland, Cork; Dr Ziwei Ouyang, Danish Technical University; Dr Kwan Lee, Hong Kong University and Jun Zhang, NUI, Galway. The graphics were prepared by Longlong Zhang, Zhejiang University and Francois Lemarchand, University of Nantes. Designs and solutions were provided by Ignacio Lope, University of Zaragoza. The references were assembled by Migle Makelyte, NUI, Galway. The measurements were performed by Slawomir Duda, Convertec Ltd.; Robin Draye, Université Paul Sabatier, Toulouse and Lionel Breuil, University of Nantes. Dr Pádraig Ó'Catháin wrote the equations in Latex. Credit for the cover design goes to Dee Enright and John Breslin.

Two individuals converted diverse notes into a cohesive manuscript and deserve special mention and thanks: Mari Moran who edited the whole document and Francois Lemarchand who completed the graphics, wrote the MATLAB programs and organised the references.

We are grateful for the support of the Wiley staff in Chichester who guided us in the process of preparing the manuscript for publication.

This work was supported by the Grant-in-Aid Publications Fund at the National University of Ireland, Galway and the Scholarly Publication Grants Scheme of the National University of Ireland.

Finally we would like to acknowledge the support of our families: our wives (Kathleen and Ingrid) and sons and daughters (Deirdre, Fergus, Yvonne, Julian and Maureen) who have all inspired our work.

Foreword

It's too big! It's too hot! It's too expensive! And the litany goes on, recognizable to those of us who have designed inductors and transformers, the bane of power electronics. In writing this book, Professor Hurley and Doctor Wölfle have combined their expertise to produce a resource that, while not guaranteeing freedom from pain, at least provides substantial anaesthesia.

Ger Hurley has been engaged in research, teaching and writing about magnetic analysis and design for almost 40 years, since his time as a graduate student at MIT completing his thesis on induction heating under my supervision. And Werner Wölfle brings to this text, in addition to his extensive industrial experience, the benefit of having been Prof. Hurley's student. So, in some very small way, I take some very small credit for this book.

Today's demands on power electronics are unprecedented and, as their application moves ever further into the commodity marketplace (solar PV converters, EV and hybrid drives, home automation, etc.), the emphases placed on cost and efficiency are driving a sharp focus on the high-cost transformers and inductors in these products. As we venture into design domains, where electroquasistatics no longer obtains, and where the contradictory demands of efficiency and size reduction create an engineering confrontation, we need the guidance that this book provides.

While many books have been written to aid the engineer in the design of magnetics, they almost exclusively present design rules and formulas without exposing the underlying physics that governs their use. Hurley and Wölfle, too, provide formulas and rules, but the emphasis is on understanding the fundamental physical phenomena that lead to them. As we move to higher frequencies, new geometries, new materials and new manufacturing technologies, we can no longer simply find an appropriate formula, go to a catalogue to select a pot core, C-core or E-core, and begin winding. An understanding of electromagnetic fundamentals, modelling and analysis is now critically important to successful design – an understanding that Hurley and Wölfle convey most effectively.

With its comprehensive scope and careful organization of topics, covering fundamentals, high-frequency effects, unusual geometries, loss mechanisms, measurements and application examples, this book is a 'must have' reference for the serious power electronics engineer pursuing designs that are not too big, not too hot and not too expensive. Hurley and Wölfle have produced a text that is destined to be a classic on all our shelves, right next to 'The Colonel's' book[1]. A remarkable achievement.

John G. Kassakian
Professor of Electrical Engineering
The Massachusetts Institute of Technology

[1] McLyman, Colonel W.T. (1978) *Transformer and Inductor Design Handbook*. Marcel Dekker, Inc., New York.

Preface

The design of magnetic components such as transformers and inductors has been of interest to electronic and electrical engineers for many years. Traditionally, treatment of the topic has been empirical, and the 'cook-book' approach has prevailed. In the past, this approach has been adequate when conservative design was acceptable. In recent years, however, space and cost have become premium factors in any design, so that the need for tighter designs is greater. The power supply remains one of the biggest components in portable electronic equipment. Power electronics is an enabling technology for power conversion in energy systems. All power electronic converters have magnetic components in the form of transformers for power transfer and inductors for energy storage.

The momentum towards high-density, high-efficiency power supplies continues unabated. The key to reducing the size of power supplies is high-frequency operation, and the bottleneck is the design of the magnetic components. New approaches are required, and concepts that were hitherto unacceptable to the industry are gaining ground, such as planar magnetics, integrated magnetics and matrix configurations.

The design of magnetic components is a compromise between conflicting demands. Conventional design is based on the premise that the losses are equally divided between the core and the winding. Losses increase with frequency, and high-frequency design must take this into account.

Magnetic components are unique, in that off-the-shelf solutions are not generally available. The inductor is to the magnetic field what the capacitor is to the electric field. In the majority of applications, the capacitor is an off-the-shelf component, but there are several reasons for the lack of standardization in inductors and transformers. In terms of duality, the voltage rating is to the capacitor what the current rating is to the inductor. Dielectric materials used in capacitor manufacture can be chosen so that voltage rating greatly exceeds the design specification without incurring extra cost. In this way, a spectrum of voltage ratings can be covered by a single device.

On the other hand, the current flow in an inductor gives rise to heat loss, which contributes to temperature rise, so that the two specifications are interlinked. This, in turn, determines the size of the conductors, with consequential space implications. Magnetic components are usually the most bulky components in a circuit, so proper sizing is very important.

Returning to the duality analogy, the dielectric material in a capacitor is to the electric field what ferromagnetic material in a magnetic component is to the magnetic field. In general, dielectrics are linear over a very large voltage range and over a very wide frequency range. However, ferromagnetic materials are highly non-linear and can be driven into saturation with small deviations from the design specifications. Furthermore, inductance is a frequency-dependent phenomenon. Dielectric loss does not contribute to temperature rise in a critical way, whereas magnetic core loss is a major source of temperature rise in an inductor.

The totality of the above factors means that magnetic component design is both complex and unique to each application. Failure mechanisms in magnetic components are almost always due to excessive temperature rise, which means that the design must be based on both electrical and thermal criteria. A good designer must have a sound knowledge of circuit analysis, electromagnetism and heat transfer. The purpose of this book is to review the fundamentals in all areas of importance to magnetic component design and to establish sound design rules which are straightforward to implement.

The book is divided into four sections, whose sequence was chosen to guide the reader in a logical manner from the fundamentals of magnetics to advanced topics. It thus covers the full spectrum of material by providing a comprehensive reference for students, researchers and practising engineers in transformer and inductor design.

The Introduction covers the fundamental concepts of magnetic components that serve to underpin the later sections. It reviews the basic laws of electromagnetism, as well as giving a historical context to the book. Self and mutual inductance are introduced and some important coil configurations are analyzed; these configurations form the basis of the practical designs that will be studied later on. The concepts of geometric mean distance and geometric mean radius are introduced to link the formulas for filaments to practical coils with finite wires such as litz wires.

In Section I, the design rules for inductor design are established and examples of different types of inductors are given. The single coil inductor, be it in air or with a ferromagnetic core or substrate, is the energy storage device. A special example is the inductor in a flyback converter, since it has more than one coil. This treatment of the inductor leads on to the transformer in Section II, which has multiple coils and its normal function is to transfer energy from one coil to another.

Section II deals with the general design methodology for transformers, and many examples from rectifiers and switched mode power supplies are given. Particular emphasis is placed on modern circuits, where non-sinusoidal waveforms are encountered and power factor calculations for non-sinusoidal waveforms are covered. In a modern power converter, the transformer provides electrical isolation and reduces component stresses where there is a large input/output conversion ratio. The operation of the transformer at high frequency reduces the overall size of the power supply.

There is an inverse relationship between the size of a transformer and its frequency of operation, but losses increase at high frequency. There is skin effect loss and proximity effect loss in the windings due to the non-uniform distribution of the current in the conductors. The core loss increases due to eddy currents circulating in the magnetic core and also due to hysteresis. General rules are established for optimizing the design of windings under various excitation and operating conditions – in particular, the type of waveforms encountered in switching circuits are treated in detail. A simple, straightforward formula is presented to optimize the thickness of a conducting layer in a transformer winding.

Finally, Section III treats some advanced topics of interest to power supply designers. The authors feel that the book would be incomplete without a section on measurements, a topic that is often overlooked. Advances in instrumentation have given new impetus to accurate measurements. Practitioners are well aware of the pitfalls of incorrect measurement techniques when it comes to inductance, because of the non-linear nature of hysteresis. Planar magnetics have now become mainstream. The incorporation of power supplies into integrated circuits is well established in current practice.

This book is of interest to students of electrical engineering and electrical energy systems – graduate students dealing with specialized inductor and transformer design and practising engineers working with power supplies and energy conversion systems. It aims to provide a clear and concise text based on the fundamentals of electromagnetics. It develops a robust methodology for transformer and inductor design, drawing on historical references. It is also a strong resource of reference material for researchers. The book is underpinned by a rigorous approach to the subject matter, with emphasis on the fundamentals, and it incorporates both depth and breadth in the examples and in setting out up-to-date design techniques.

The accompanying website www.wiley.com/go/hurley_transformers contains a full set of instructors' presentations, solutions to end-of-chapter problems, and digital copies of the book's figures.

Prof. W. G. Hurley and Dr Werner Wölfle
National University of Ireland, Galway, Ireland
March 2013

Nomenclature

The following is a list of symbols used in this book, and their meanings.

A	Average or geometric mean radius
A_c	Cross-sectional area of magnetic core
A_g	Cross-sectional area of the gap
A_L	Inductance per turn
A_m	Effective cross-sectional area of magnetic circuit
A_p	Product of window winding area × cross-sectional area
A_t	Surface area of wound transformer
A_w	Bare wire conduction area
a	Transformer turns ratio
a_1, a_2	Inside and outside radii of a coil
$B_{\max}$	Maximum flux density
B_o	Optimum flux density
B_{sat}	Saturation flux density
b	Winding dimension: see Figure 6.4
C_{eff}	Effective capacitance of a transformer
D	Duty cycle
d	Thickness of foil or layer
d_1, d_2	Height of filaments or coil centres above ferromagnetic substrate
Φ	Magnetomotive force, mmf
f	Frequency in hertz
G, g	Maximum and minimum air gap lengths
GMD	Geometric mean distance between coils
$g(x)$	Air-gap length at x
h	Winding dimension: see Figure 2.14
h_c	Coefficient of heat transfer by convection
h_1, h_2	Coil heights in axial direction
$\hat{I}$	Peak value of the current waveform
I_{dc}	Average value of current
I_n	RMS value of the nth harmonic of current
$I_n(x), K_n(x)$	Modified Bessel functions of the first and second kind, respectively
I'_{rms}	RMS value of the derivative of the current waveform
I_{rms}	RMS value of the current waveform
J_o	Current density

$J(r)$	Current density at radius r
$J_0(x)$, $J_1(x)$	Bessel functions of the first kind
K_c	Material parameter
$K(f)$, $E(f)$	Complete elliptic integrals of the first and second kind, respectively
K_i	Current waveform factor
K_t	48.2×10^3
K_v	Voltage waveform factor
k	Coupling coefficient
k_a, k_c, k_w	Dimensionless constants (see Equations 3.25, 3.26 and 3.27)
k_f	Core stacking factor A_m/A_c
k_i	Defined in Figure 7.28
k_p	Power factor
k_{pn}	Ratio of the AC resistance to DC resistance at *n*th harmonic frequency
k_s	Skin-effect factor
k_u	Window utilization factor
L	Self-inductance
L_{eff}	Effective inductance
L_l	Leakage inductance
L_m	Magnetizing inductance
L_s	Additional coil inductance due to ferromagnetic substrate
l_c	Magnetic path length of core
M	Mutual inductance
MLT	Mean length of a turn
m	$\sqrt{(j\omega\mu_0\sigma)}$
N	Number of turns in coil
n	Harmonic number
P_{cu}	Copper or winding loss
P_{fe}	Iron or core loss
P_o	Output power
P_v	Power loss per unit volume
p	Number of layers
R	Average or geometric mean radius
$\mathcal{R}$	Reluctance
R_{ac}	AC resistance of a winding with sinusoidal excitation
R_{dc}	DC resistance of a winding
R_{eff}	Effective AC resistance of a winding, with arbitrary current waveform
R_δ	DC resistance of a winding of thickness δ_0
R_θ	Thermal resistance
r_1, r_2	Inside and outside radii of a coil
r_o	Radius of bare wire
s	Substrate separation in sandwich structure
T	Period of a waveform
T_a	Ambient temperature
T_{max}	Maximum operating temperature
t	Substrate thickness
t_r	Rise time (0–100%)

V_{rms}	RMS value of the voltage waveform
VA	Voltampere rating of winding
V_c	Volume of core
V_o	DC output voltage
V_s	DC input voltage
V_w	Volume of winding
$\langle v \rangle$	Average value of voltage over time τ
W_a	Window winding area of core
W_c	Electrical conduction area
W_m	Stored energy in a magnetic field
w	Winding dimension: see Figure 6.4
Z	Impedance
Z_i	Internal impedance of a conductor
z	Axial separation
α, β	Material constants
α_{20}	Temperature co-efficient of resistivity at 20°C
Δ	Ratio d/δ_0
ΔB	Flux density ripple
ΔT	Temperature rise
ΔV	Output voltage ripple
δ	Skin depth
δ_0	Skin depth at fundamental frequency
δ_n	Skin depth at the *n*th harmonic frequency
ϕ	Flux
$\phi(k)$	Defined in Equation 9.49
ϕ_0	Defined in Equation 9.58
γ	Ratio of iron loss to copper loss
Λ	Defined in Equation 9.36
λ	Flux linkage
μ	Static or absolute permeability
μ_0	Magnetic permeability of free space $4\pi \times 10^{-7}$ H/m
μ_{eff}	Effective relative permeability
μ_i	Initial permeability
μ_{inc}	Incremental permeability
μ_{opt}	Optimum value of effective relative permeability
μ_r	Relative permeability
μ_{rs}	Complex relative permeability
η	Porosity factor
ρ_{20}	Electrical resistivity at 20 °C
ρ_w	Electrical resistivity
σ	Electrical conductivity
τ	Time for flux to go from zero to its maximum value
Ψ	$(5p^2-1)/15$
ω	Angular frequency (rad/s)

1

Introduction

In this chapter, we describe the historical developments that led to the evolution of inductance as a concept in electrical engineering. We introduce the laws of electromagnetism which are used throughout the book. Magnetic materials that are in common use today for inductors and transformers are also discussed.

1.1 Historical Context

In 1820, Oersted discovered that electric current flowing in a conductor produces a magnetic field. Six years later, Ampere quantified the relationship between the current and the magnetic field. In 1831, Faraday discovered that a changing magnetic field causes current to flow in any closed electric circuit linked by the magnetic field, and Lenz showed that there is a relationship between the changing magnetic field and the induced current. Gauss established that magnetic poles cannot exist in isolation. These phenomena established the relationship between electricity and magnetism and became the basis for the science of electromagnetism.

In 1865, Maxwell unified these laws in the celebrated form of Maxwell's equations, which established the basis for modern electrical engineering. He also established the link between phenomena in electromagnetics and electrostatics. Father Nicholas Joseph Callan, who was Professor of Natural Philosophy at the National University of Ireland, Maynooth, in the 1830 s, invented the induction coil. Alexander Anderson was Professor of Natural Philosophy at the National University of Ireland, Galway in the early 1900 s and gave his name to the Anderson Bridge for measuring inductance.

These individuals provide the inspiration for a textbook on magnetic design that focuses on the issues that arise in power electronics. Power electronics is an enabling technology for modern energy conversion systems and inductors and transformers are at the heart of these systems.

Figure 1.1 shows a straight conductor carrying a current, *i*. The presence of the magnetic field is detected by placing a freely-suspended magnet in the vicinity of the conductor. The direction of the magnetic field (a vector) is given by the direction in which the north pole of the search magnet points. It turns out that the magnitude of the magnetic field is constant on any circle concentric with the conductor, and its direction is tangential to that circle, given by

Transformers and Inductors for Power Electronics: Theory, Design and Applications, First Edition.
W. G. Hurley and W. H. Wölfle.

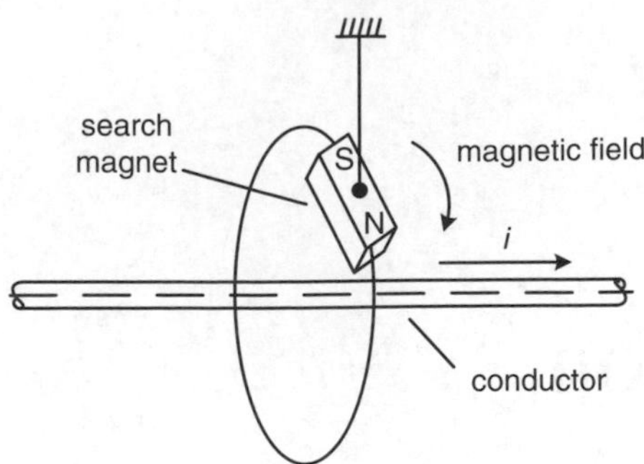

Figure 1.1 Magnetic field created by a current.

the right hand rule – that is, a conventional (right-handed) cork screw will rotate in the direction of the magnetic field if it is driven in the direction of the current flow. It also turns out that the magnitude of the magnetic field is proportional to the current in the conductor and is inversely proportional to the radial distance from the conductor axis.

The magnetic field around a straight conductor is illustrated in Figure 1.2. The direction of the magnetic field as shown complies with the right hand screw rule. An alternative to the right hand screw rule for establishing the direction of the magnetic field created by the current is to point the thumb of your right hand along the conductor in the direction of the current flow, and your fingers will wrap themselves around the conductor in the direction of the magnetic field. The higher density of the lines near the conductor indicates a stronger magnetic field in this area.

The magnetic field around the current carrying conductor is described by two vector quantities: the magnetic flux density **B** and the magnetic field intensity **H**.

The magnetic field intensity **H** is best explained by Ampere's law, which expresses these observations about the current-carrying conductor in their most general form:

$$\oint_C \mathbf{H} \cdot d\mathbf{l} = \int_S \mathbf{J}_f \cdot \mathbf{n} da \tag{1.1}$$

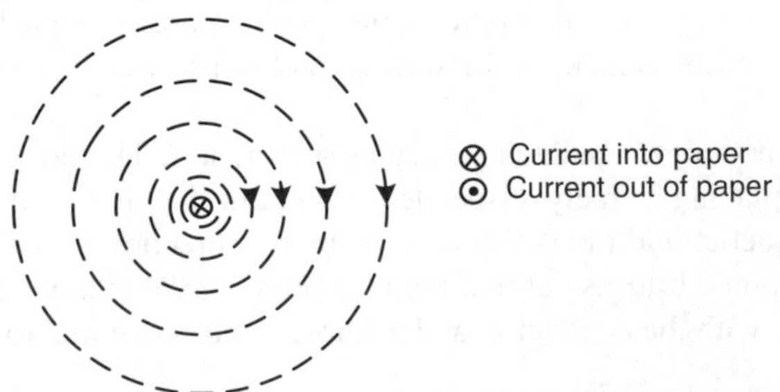

Figure 1.2 Magnetic field around a current-carrying conductor.

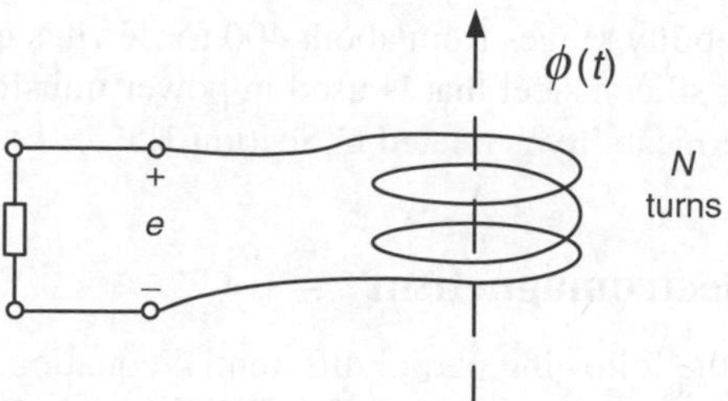

Figure 1.3 Conductor in a changing magnetic field.

The closed contour C, the surface S and the normal vector are defined by convention: S is the surface enclosed by C and $\mathbf{n}$ is the unit vector normal to S. $\mathbf{H}$ is the magnetic field intensity in A/m and $\mathbf{J}_f$ is the current density in A/m^2. The quantity on the right hand side of Equation 1.1 is the current enclosed by the contour.

Figure 1.3 shows a coil with N turns in a magnetic field. The magnetic flux that links each turn of the coil is ϕ and the electromotive force (emf) induced in the coil is given by:

$$e = -N\frac{d\phi}{dt} \tag{1.2}$$

This states that the induced electromotive force (emf) in a coil of N turns is proportional to the rate of change of the magnetic flux that links the coil. The negative sign indicates that current flow in the external circuit will create an opposing magnetic field.

In a more general form, Equation 1.2 may be stated as:

$$e = -\frac{d}{dt}\int_s \mathbf{B}\cdot\mathbf{n}\,da \tag{1.3}$$

The integral in Equation 1.3 represents the flux linking the coil. The surface S and the normal vector are defined as before. The flux density $\mathbf{B}$ in Wb/m^2 or tesla is the flux per unit area inside the coil.

The magnetic field intensity $\mathbf{H}$ gives rise to a magnetic flux density $\mathbf{B}$ in a medium of permeability μ, so that:

$$\mathbf{B} = \mu\mathbf{H} \tag{1.4}$$

The units for permeability are H/m and for free space $\mu_0 = 4\pi \times 10^{-7}$ H/m. For magnetic media, μ could be up to 10 000 times greater than μ_0. The permeability is usually presented as the product of μ_0 and the relative permeability μ_r.

$$\mu = \mu_r\mu_0 \tag{1.5}$$

Typically, relative permeability ranges from about 400 for ferrites used for power electronics applications to 10 000 for silicon steel that is used in power transformers at 50 Hz or 60 Hz. μ_r is taken as 1 for air. Permeability is treated in Section 1.5.

1.2 The Laws of Electromagnetism

In Maxwell's equations, the following partial differential equation relates the magnetic field intensity **H** to the current density $\mathbf{J}_f$ and the electric displacement **D**:

$$\nabla \times \mathbf{H} = \mathbf{J}_f + \frac{\partial \mathbf{D}}{\partial t} \tag{1.6}$$

In general, the laws of electricity and magnetism are broadly divided into quasi-static magnetic field systems and quasi-static electric field systems. In this book, we concern ourselves with quasi-static magnetic field systems and the contribution of the displacement current is considered negligible. The electric field intensity is then:

$$\nabla \times \mathbf{H} = \mathbf{J}_f \tag{1.7}$$

This is Ampere's law in differential form.

1.2.1 Ampere's Magnetic Circuit Law

This law states that the line integral of H around any closed contour is equal to the total current enclosed by that contour, and it may be stated in the integral form of Equation 1.4:

$$\oint_C \mathbf{H} \cdot d\mathbf{l} = i \tag{1.8}$$

The right hand side of Equation 1.1 is simply the current i enclosed by the contour, and corresponds to the right hand side of Equation 1.8.

We have already discovered that magnitude of **H** is constant around a circle concentric with the axis of the conductor of Figure 1.2. Evaluation of the closed integral of Equation 1.8 for the straight conductor of Figure 1.2 gives the magnitude of the magnetic field intensity $H(r)$ at a radius r from the conductor:

$$\int_0^{2\pi} H(r)dr = i \tag{1.9}$$

$$2\pi H(r) = i \tag{1.10}$$

$$H(r) = \frac{i}{2\pi r} \tag{1.11}$$

and the corresponding magnetic flux density in air is from Equation 1.4:

$$B(r) = \mu_0 \frac{i}{2\pi r} \tag{1.12}$$

We will meet further examples of this law later in our study of inductors.

We have seen Ampere's law in the form of a differential equation (Equation 1.7) and in the form of an integral equation (Equation 1.8). In many practical applications, it makes more sense to state the law in discrete form or in the form of a difference equation. Specifically, if there are a limited number of discrete sections with a constant value of H over a length l, then:

$$\sum H \cdot l = Ni \tag{1.13}$$

In this form, H is summed around the loop for discrete lengths, as in the case of the closed core of an inductor or transformer, and the loop encloses a total current corresponding to N turns, each carrying a current i. We will return this topic in Chapter 2.

1.2.2 Faraday's Law of Electromagnetic Induction

In Maxwell's equations, Faraday's law of Electromagnetic Induction takes the form:

$$\nabla \times \mathbf{E} = -\frac{\partial \mathbf{B}}{\partial t} \tag{1.14}$$

In its integral form, this is:

$$\oint_C \mathbf{E} \cdot dl = -\frac{d}{dt}\int_s \mathbf{B} \cdot \mathbf{n}da \tag{1.15}$$

This states that the integral of the electric field intensity $\mathbf{E}$ around a closed loop C is equal to the rate of change of the magnetic flux that crosses the surface S enclosed by C. $\mathbf{E}$ normally includes a velocity term in the form of $\mathbf{v} \times \mathbf{B}$, which takes into account the movement of a conductor in a magnetic field, such as an electric motor or generator. However, for inductors and transformers, this does not arise. The differential form of Equation 1.15 is described by Equation 1.2.

$N\phi$ is called the flux linkage, which is the total flux linking the circuit. A coil with N turns may have a flux ϕ linking each turn, so that the flux linkage is

$$\lambda = N\phi \tag{1.16}$$

The polarity of the induced electromagnetic field (emf) is established by noting that the effect of the current caused by the emf is to oppose the flux creating it; this is Lenz's law. The induced emf opposes the creating flux by generating secondary currents called eddy currents. Eddy currents flow in magnetic materials and in conductors operating at high frequency. These topics arise in later chapters.

We will see examples of Faraday's law and Lenz's law in our study of transformers.

In a simple magnet consisting of a north pole and a south pole, flux emanating from the north pole returns to the south pole and through the magnet back to the north pole. This means that the total flux emanating from a closed surface surrounding the magnet is zero. This is Gauss' law and, in the form of Maxwell's equations, it states that the divergence of the magnetic field is zero:

$$\nabla \cdot \mathbf{B} = 0 \tag{1.17}$$

So for a closed surface S:

$$\oint_S \mathbf{B} \cdot \mathbf{n} da = 0 \tag{1.18}$$

In other words the lines of magnetic flux are continuous and always form closed loops as illustrated in Figure 1.2. Kirchhoff's current law is another example of this and in Maxwell's equations it is expressed as:

$$\nabla \times \mathbf{J}_f = 0 \tag{1.19}$$

with the more recognized integral form of

$$\oint_S \mathbf{J}_f \cdot \mathbf{n} da = 0 \tag{1.20}$$

This means that when a node in an electrical circuit is surrounded by a closed surface, the current into the surface is equal to the current leaving the surface.

Example 1.1

Derive an expression for the magnetic flux density inside a conductor of radius r_o carrying current I that is uniformly distributed over the cross-section.

We have already established the magnetic field outside the conductor in Equation 1.12. The magnetic field inside of the conductor is shown in Figure 1.4 and observes the right hand rule.

Assuming that uniform current density in the conductor, the current inside a loop of radius r is:

$$i(r) = \frac{\pi r^2}{\pi r_o^2} I$$

We can now apply Ampere's law on a closed contour at radius r, yielding:

$$H(r) = \frac{r}{2\pi r_o^2} I$$

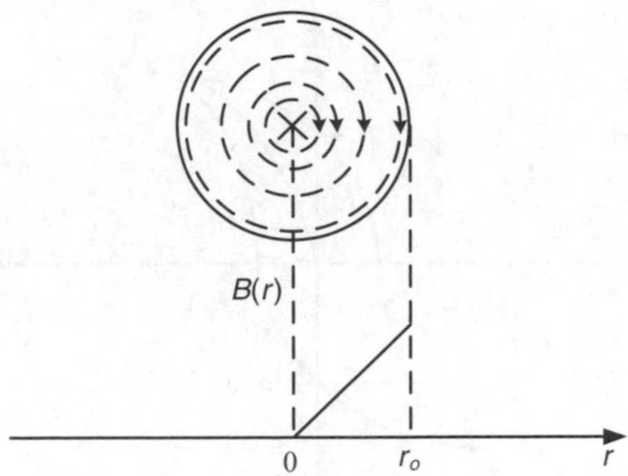

Figure 1.4 Magnetic field inside a current carrying conductor.

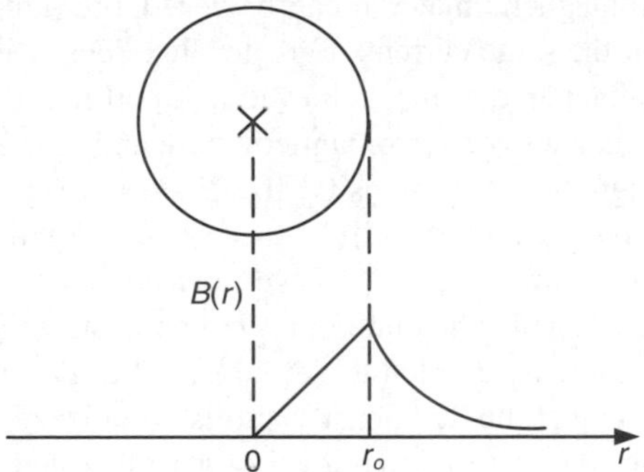

Figure 1.5 Magnetic field inside and outside a current carrying conductor.

and the flux density for a non-magnetic conductor ($\mu_r = 1$) is:

$$B(r) = \frac{\mu_0 r}{2\pi r_o^2} I$$

Combining this result with Equation 1.12, the internal flux increases linearly with radius inside the conductor, and outside the conductor the flux density falls off inversely with radius. The result in shown in Figure 1.5.

1.3 Ferromagnetic Materials

We have already seen that for a current-carrying coil, H is proportional to the current in the coil (Ampere's law) and B is proportional to the integral of the voltage across the coil (Faraday's law). The magnetic flux density is related to the magnetic field intensity by the magnetic permeability μ. In a ferromagnetic material, the magnetic flux density is enhanced or amplified compared to a medium such as air; the amplification factor is the permeability μ.

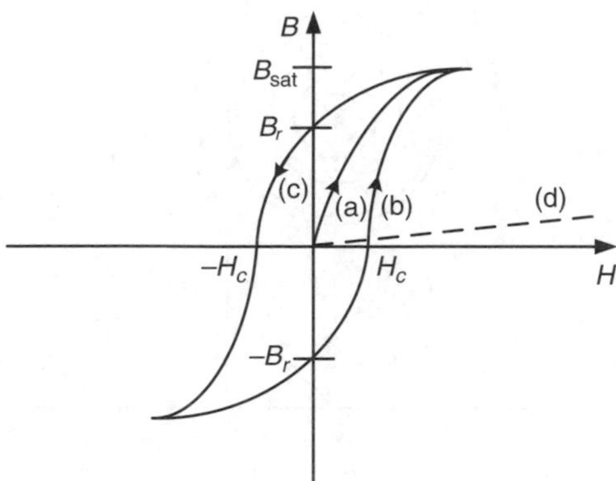

Figure 1.6 *B-H* magnetization curve.

The permeability in a ferromagnetic material can be very large (unless it is limited by saturation), which means that for the same current, a greater flux density is achieved in a core made of ferromagnetic material compared to that achieved in a coil in air.

Taking a completely demagnetized ferromagnetic core and slowly increasing the flux density by increasing the magnetic field intensity, the *B-H* curve will follow the curve (a) in Figure 1.6; the details of this experiment will be described in Section 1.4.2. As the *H* field is increased, the flux density saturates at B_{sat}. If we now decrease H, the flux density B will follow curve (c) in Figure 1.6. This phenomenon is called hysteresis.

When the magnetizing force is returned to zero, a residual flux density exists in the magnetic material. The magnitude of the flux density at this point is called the remanent magnetization B_r. In order to return the material to a level of zero flux density, a negative value of H is required; $-H_c$ is called the coercive force of the material. Further decreases in H will eventually cause the material to saturate at $-B_{sat}$ and a positive coercive force $+H_c$ will again return the material to a state of zero flux.

Increasing H further causes B to follow curve (b). If we continue to vary H in a periodic manner, the *B-H* loop will settle into a fixed loop as illustrated, and the closed loop is called the hysteresis loop. In its most simplified form, the hysteresis loop is characterized by the saturation flux density B_{sat}, the coercive force H_c and the slope of the *B-H* curve μ. If the core material were non-magnetic such as air, then the *B-H* magnetization curve would be linear, as shown by (d) in Figure 1.6. Clearly, the magnetic medium has a much higher flux density for the same magnetizing force.

The relationship in a ferromagnetic medium is not linear, although it is reasonably linear up to a value labelled B_{sat} in Figure 1.6. Beyond B_{sat}, the medium assumes the characteristic of a non-magnetic medium such as air, and the relative permeability μ_r approaches 1; this effect is called saturation. In a practical design, it is customary to set the maximum flux density B_{max} at a value below the saturation flux density B_{sat}.

The explanation of the above phenomena is rooted in the complex area of atomic physics. However, we can explain the macro effects by magnetic domains. An electron spinning around an atomic nucleus produces a magnetic field at right angles to its orbital plane. An electron can also spin about its own axis, giving rise to a magnetic field. These effects combine to form a magnetic moment or dipole. The atoms in ferromagnetic material form

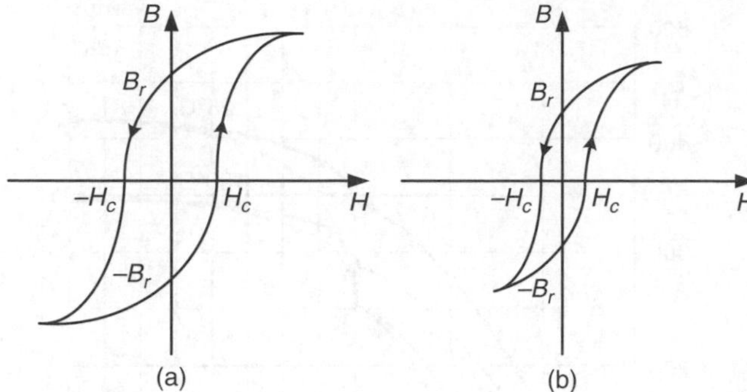

Figure 1.7 *B-H* magnetization curve for: (a) hard magnetic materials; (b) soft magnetic materials.

molecules, and the magnetic moments form a magnetic domain that may be thought of as a microscopic magnet. Returning to the hysteresis loop in Figure 1.6, when the material is exposed to an external magnetic field, the magnetic domains line up with the direction of the applied field, thus reinforcing the field inside the material. As the field is further increased, there is less and less opportunity for orientation of the domains and the material becomes saturated; this is part (a) of the loop. If the field is now reduced or removed along path (c), the domains will resist and will retain at least some alignment; this is the residual or remanent magnetism, labelled B_r. The resistance to realignment following saturation is the hysteresis effect.

The coercive force describes the effort involved in rotating the domains to the point where the net field is zero. This is the distinction between soft, easily realigned magnetic materials and hard, difficult to realign, magnetic materials. The materials that retain most of their magnetization and are the most difficult to realign are called permanent magnets. These materials are easily identified by their *B-H* loops, as illustrated in Figure 1.7. Hard magnetic materials have higher remanent magnetism and higher coercive force than their soft counterparts.

It has been observed that when the external field is increased at a uniform rate, the magnetization displays finite step jumps known as the Barkhausen effect, which may be explained by the sudden reorientation of the domains. This observation is often taken to validate the domain theory of ferromagnetism because, in the past, the effect was audible in speakers.

The permeability of a ferromagnetic material is temperature-dependent, as is the saturation flux density. This is illustrated for a Mn-Zn ferrite in Figure 1.8. The saturation flux density falls from approximately 450 mT at 25 °C to about 360 mT at 100 °C.

As with any atomic level activity, the domains are influenced by temperature. At high temperatures (above 760 °C for iron), the thermal motion of the molecules is sufficiently agitated to block the alignment of the domains with the external field. The relative permeability returns to approximately 1; materials with $\mu_r \approx 1$ are called paramagnetic materials, and examples include metals such as aluminium and titanium.

The temperature above which a ferromagnetic material becomes paramagnetic is called the Curie temperature; it is a property of the ferromagnetic material and is normally given by the manufacturer. For ferrites, the Curie temperature may be as low as 200 °C and, for this reason, the control of the temperature in an inductor or transformer core of ferrite material is important.

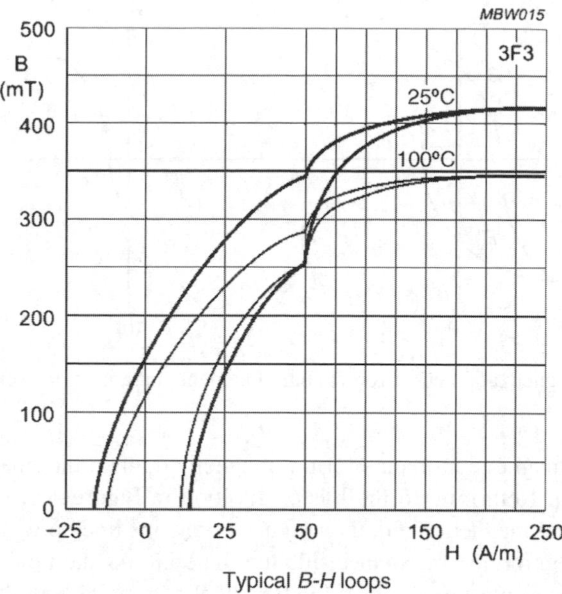

Figure 1.8 Temperature dependence of saturation flux density [Reproduced with permission of Ferroxcube].

In addition to paramagnetic and ferromagnetic materials, there is another broad class called diamagnetic materials. In these materials, the dipoles formed by the orbiting electrons oppose the applied field in accordance with Lenz's law. The effect is very weak; examples include copper, silver, and gold. In silver, the field is reduced by one part in 40 000. It may be argued that the diamagnetic effect exists in all materials, but that the paramagnetic effect and the ferromagnetic effect dominate in some materials.

Under AC operating conditions, the domains are constantly rotating and this requires an input of energy to overcome the molecular resistance or friction created by the changing domains. Over a complete cycle, the net energy expended appears in the form of heat, and this is known at the hysteresis loss. We will take a closer look at hysteresis loss in Section 1.4.

1.4 Losses in Magnetic Components

Losses in inductors and transformers can be classified as core loss and winding or copper loss. In the core there are hysteresis loss and eddy current loss.

1.4.1 Copper Loss

The resistance R of the wire used to build a winding in an inductor or transformer causes heat generation in the form of I^2R loss, where I is the DC or RMS current.

At high frequencies, the copper loss is aggravated by a phenomenon known as 'skin depth'. At high frequencies, the current in a conductor bunches towards the surface of the conductor, due to the AC magnetic field created by the conductor current. This is a direct result of Faraday's law, whereby current will flow inside a conductor to oppose the AC flux, in the form of eddy currents. This increases the effective resistance of the conductor, called R_{ac}, by reducing the net area available for current flow. Skin depth δ can be thought of as the thickness of a hollow conductor which has the same resistance as the solid conductor with skin effect. This topic is covered in Chapter 6. We will see that the skin depth is given by:

$$\delta = \frac{1}{\sqrt{\pi f \mu \sigma}} \tag{1.21}$$

where f is the frequency and σ is the conductivity of the conductor material.

For copper at 50 Hz, $\delta = 9.5$ mm and at 10 kHz, $\delta = 0.56$ mm. Most of the current is contained within one skin depth of the surface, so that the conduction area decreases as the frequency increases. A useful approximation for the AC resistance of a round conductor of radius r_o and DC resistance R_{dc} is:

$$R_{ac} = R_{dc}\left[1 + \frac{(r_o/\delta)^4}{48 + 0.8(r_o/\delta)^4}\right] \tag{1.22}$$

We will take a closer look at this approximation in Chapter 6.

Another effect at high frequencies is the 'proximity effect', where the magnetic field of the current in one conductor interferes with that of another conductor nearby, increasing the resistance further.

The limitations of skin depth and proximity effects can be avoided by using stranded wire, with each strand insulated. Each strand is transposed over the length of the wire, so that it occupies the positions of all the other strands over the cross-section; in this way, all strands are equally exposed to the prevailing magnetic field. Transposition ensures that each strand has equal inductance over the length of the wire. Litz wire is commercially available for this purpose. In cases where only a few turns are required in a coil, thin foil may be used.

1.4.2 Hysteresis Loss

At this point, we can take a closer look at the hysteresis loss that occurs as a result of the application of an AC field to the material. Hysteresis may be measured by uniformly winding insulated wire on a toroidal core, as shown in Figure 1.9. Internal molecular friction resists

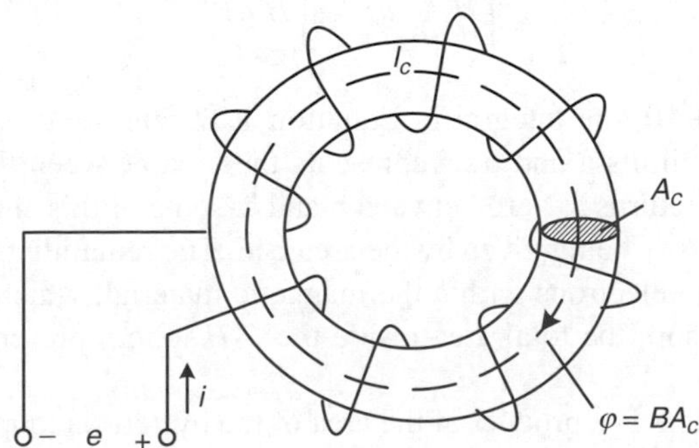

Figure 1.9 Circuit setup to measure hysteresis loss in a ferromagnetic material.

the continuous re-orientation of the microscopic magnets or domains in ferromagnetic materials, and energy is expended in the form of heat as the material undergoes its cyclic magnetization.

The measurement of the *B-H* loop and hysteresis loss will be described in detail in Chapter 8. For present purposes, we can establish that the integral of the applied voltage is related to the flux by Faraday's law and the magnetic field intensity is related to the current by Ampere's law. The applied voltage *e* in Figure 1.9 is:

$$e = Ri + N\frac{d\phi}{dt} \tag{1.23}$$

The instantaneous power supplied to the circuit is:

$$ei = Ri^2 + Ni\frac{d\phi}{dt} \tag{1.24}$$

Taking A as the cross-sectional area, the flux ϕ is BA. Taking l as the mean length of the path around the toroid, from Ampere's law, $H.l = Ni$ and the term representing hysteresis in Equation 1.24 may be rewritten:

$$ei = Ri^2 + \frac{Ni}{\ell}A\ell\frac{dB}{dt} \tag{1.25}$$

or

$$ei = Ri^2 + H\frac{dB}{dt}V \tag{1.26}$$

where $V = Al$ is the volume of the core. The first term on the right hand side of Equation 1.26 represents the copper loss in the windings and the second term represents the hysteresis loss. The hysteresis loss per unit volume is found by integrating the hysteresis term in Equation 1.26, yielding:

$$\int H\frac{dB}{dt}dt = \int H\,dB \tag{1.27}$$

In terms of Figure 1.10, the integral in Equation 1.27 represents a strip of width H and height dB, and for the limits a and b it represents the area between the H curve and the B axis. When the H field returns to zero between b and c, some of this energy is returned to the circuit, the returned energy being given by the area bcd. The remaining unrecoverable energy is stored in the spinning electrons within the magnetic material, which produce the residual flux. Completing the loop, the total area inside the *B-H* loop represents the hysteresis loss over a complete cycle.

The total hysteresis loss is a product of the area of the hysteresis loop, the frequency of the applied signal and the core volume. Manufacturers normally specify loss in the form of watts/m^3 as a function of B for different frequencies. Soft magnetic materials normally have

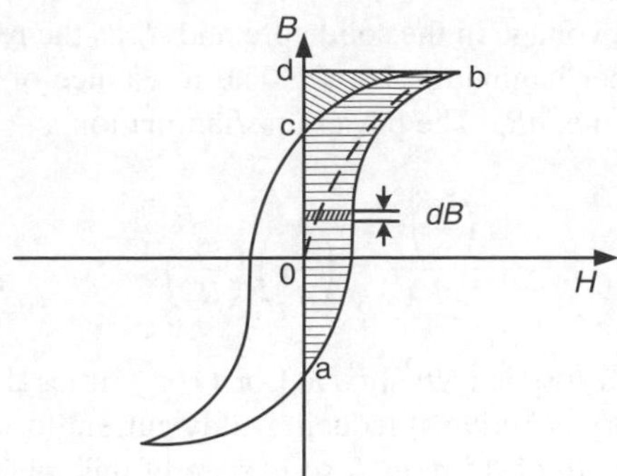

Figure 1.10 Hysteresis loss in a ferromagnetic material.

smaller hysteresis loops than hard magnetic materials, as illustrated in Figure 1.7, and consequently they have lower hysteresis loss at a given frequency.

1.4.3 Eddy Current Loss

Many magnetic cores are made of materials that are, themselves, conductors of electricity; therefore, under AC conditions, currents are induced in the core, giving rise to an I^2R type loss. In particular, silicon steel, used in power transformers, falls into this category. Many ferrites are classified as non-conductors but, at a high enough frequency, they are subjected to eddy currents as a direct consequence of Faraday's law of elcctromagnetic induction.

Cores are laminated to reduce eddy current loss. Essentially, the laminations consist of insulated sheets of magnetic material such as grain-orientated steel. A magnetic field along the lamination induces an emf, which drives a current through a resistance path as shown in Figure 1.11. The resistance is proportional to the length and thickness of the lamination, while the induced voltage is proportional to the cross-sectional area of the lamination.

Consider two equal areas, one solid and the other with n laminations, as shown in Figure 1.11.

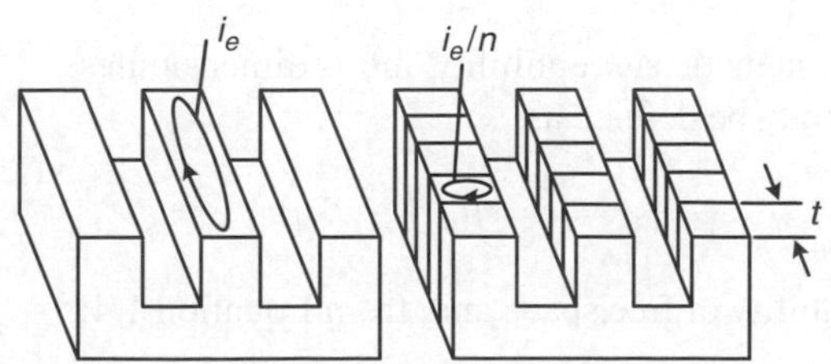

Figure 1.11 Eddy current loss in a ferromagnetic material.

Taking e_c as the induced voltage in the solid core and R_c as the resistance in the solid core, then the voltage induced per lamination is e_c/n. The resistance of one lamination will be n times that of the solid area, i.e. nR_c. The power loss/lamination is:

$$\frac{\left(\frac{e_c}{n}\right)^2}{nR_c} = \left(\frac{1}{n^3}\right)\left(\frac{e_c^2}{R_c}\right) \tag{1.28}$$

For n laminations, the total loss is $(1/n^2)[e_c^2/R_c]$ or $(1/n^2)$ times the loss for the solid core. Clearly, laminations are very effective at reducing eddy current loss. Grain-orientated silicon steel at 50 Hz has a skin depth of 0.5 mm. A solid core of this material would contain all its flux in an outer shell of 0.5 mm thickness, while a core of greater thickness would contain little flux at its centre.

1.4.4 Steinmetz Equation for Core Loss

The celebrated general Steinmetz equation [1] for core loss is commonly used to describe the total core loss under sinusoidal excitation:

$$P_{\text{fe}} = K_c f^{\alpha} B_{\text{max}}^{\beta} \tag{1.29}$$

where: P_{fe} is the time-average core loss per unit volume; B_{max} is the peak value of the flux density with sinusoidal excitation at the frequency f; K_c, α and β are constants that may be found from manufacturers' data (examples are given in Table 1.1).

For power electronics applications, non-sinusoidal excitation is common, and also AC excitation under DC bias conditions. These effects are discussed in Chapter 7.

1.5 Magnetic Permeability

The magnetic flux density is related to the magnetic field intensity by the magnetic permeability in Equation 1.1. We have seen that the relationship is non-linear, as depicted in the *B-H* loop shown in Figure 1.6. At this point, it is worthwhile to revisit permeability and take a closer look.

The magnetization density **M** describes the manifestation of the effects of the magnetic dipoles in the magnetic material:

$$\mathbf{M} = X_m \mathbf{H} \tag{1.30}$$

where X_m is called the magnetic susceptibility and is dimensionless.

The permeability μ may be defined as:

$$\mu = \mu_0(1 + X_m) \tag{1.31}$$

where μ_0 is the permeability of free space and, from Equation 1.4:

$$\mathbf{B} = \mu \mathbf{H}$$

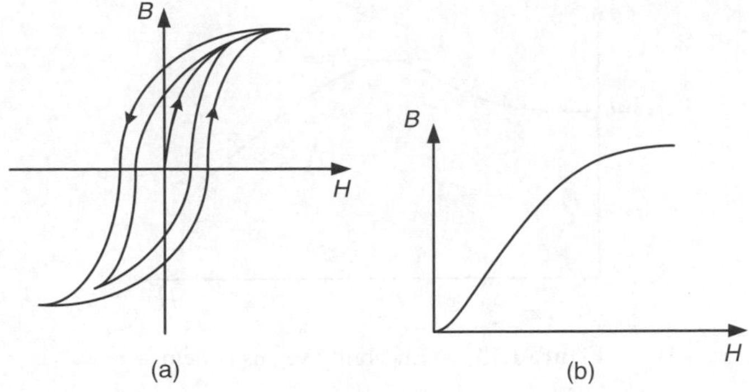

Figure 1.12 Normal magnetization curve of a ferromagnetic material.

The relative permeability μ_r is defined in terms of Equation 1.5:

$$\mu = \mu_r \mu_0$$

and therefore

$$\mu_r = 1 + X_m \tag{1.32}$$

Taking this approach, X_m is slightly greater than 1.0 for paramagnetic materials, slightly less than 1.0 for diamagnetic materials and much greater than 1.0 for ferromagnetic materials operating below their Curie temperature. On a macro scale, we can think of the magnetic field intensity H being produced by electric current as the cause or driving force, and the magnetic field intensity B is the result or effect of H. Permeability quantifies the ease with which H give rise to B.

One of the features of the B-H hysteresis loop shown in Figure 1.6 is that the tips of the loop are a function of the maximum value of H. For any sample of material, we can generate a whole series of loops with different values of H_{max}, as shown in Figure 1.12(a). If we now plot the value of B_{max} and H_{max} corresponding to the tips, we have a plot of the normal magnetization curve as shown in Figure 1.12(b).

The single value of permeability as defined in Equation 1.4 is obtained by taking the ratio of B/H at any point on the magnetization curve. This is sometimes referred to as the **static permeability** or **absolute permeability**. When the material is saturated, this value approaches μ_0, the permeability of free space. Figure 1.13 shows the static permeability for the normal magnetization curve. The value of permeability at very low values of B is called the **initial permeability**, μ_i. The permeability continues to increase from μ_i until it reaches a maximum value μ_{max}, and then decreases in the saturation region with a limiting value of μ_0.

In many applications involving inductors, there may be a DC bias with an AC signal. The AC components of B-H give rise to minor B-H loops that are superimposed on the normal magnetization curve, as shown in Figure 1.14, the DC component of H is different in each loop and the amplitude of the AC component of flux density is the same. The slopes of these minor loops are called **incremental permeability** μ_{inc}.

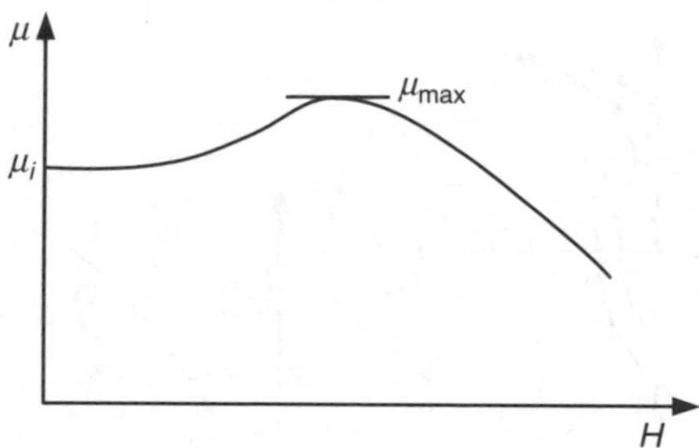

Figure 1.13 Permeability versus *H* field.

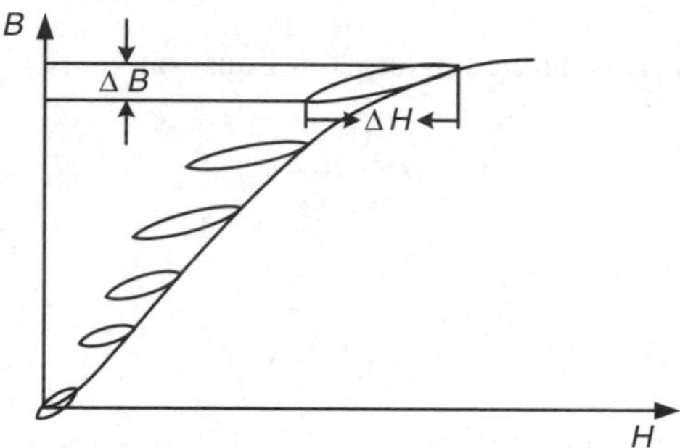

Figure 1.14 Minor *B-H* loops in a ferromagnetic material.

Complex permeability is often used to describe both the ferromagnetic effect and the attendant core loss in inductor design. It is particularly useful to describe high-frequency effects in magnetic cores, and we will return to this topic in Chapter 7.

1.6 Magnetic Materials for Power Electronics

The magnetic materials used in power electronics applications can be classified into soft magnetic materials and hard magnetic materials, the main criteria for classification being the width and slope of the hysteresis loop.

Hard magnetic materials have a wide hysteresis loop, as shown in Figure 1.7(a). The coercive force H_c of hard material is higher than the corresponding value for a soft material illustrated in Figure 1.7(b). Comparing hard and soft materials, a strong field is required to rotate the atomic level domains in a hard material so that when the material is fully magnetized a strong reverse magnetic field is needed to decrease the magnetic flux density in the material to zero. Hard magnetic materials are used in permanent magnets and mainly include an iron alloy of aluminium (Al), nickel (Ni) and cobalt (Co), sometimes called alnico; an alloy of samarium (Sm) and cobalt (SmCo) and an alloy of neodymium (Nd), iron (Fe), and boron

(B) designated NdFeB. Hard magnetic materials are commonly utilized for generating the magnetic field in electrical motors and generators.

Soft magnetic materials, on the other hand, can achieve a high value of flux density in the presence of a relatively low value of magnetic field intensity, as shown in Figure 1.7(b). This means that soft materials can be easily magnetized and demagnetized. The coercive force is low and the *B-H* loop is narrow. Soft magnetic materials include ferrites, silicon steel, nickel iron (Fe-Ni) alloy, iron-cobalt-vanadium (Fe-Co-V) alloy and amorphous alloy.

1.6.1 Soft Magnetic Materials

Soft magnetic materials find many applications in power electronics. They are widely used in high frequency transformers, in isolation transformers for driving and triggering switching components, in filter inductors for rectifiers, power factor correction and EMI control, and in resonance inductors for soft switching, as well as current transformers.

Soft magnetic materials are classified as:

- Ferrites
- Laminated iron alloys
- Powder iron
- Amorphous alloys
- Nanocrystalline materials.

The main features of each material may be summarized as follows:

Ferrites

In power electronic devices, ferrites are the most commonly applied magnetic materials.

Ferrites are deep grey or black ceramic materials and are typically brittle. They are made up of iron oxide (Fe_2O_3), mixed with other metals such as cobalt (Co), copper (Cu), magnesium (Mg), manganese (Mn), nickel (Ni), silicon (Si) and zinc (Zn). The two most common types of ferrite are Mn-Zn and Ni-Zn.

The magnetic and electrical properties vary with each alloy. For example, Ni-Zn ferrite has high electrical resistivity around 10 000 Ωm, which makes it more suitable for high-frequency operation above 1 MHz. On the other hand, the lower resistivity of Mn-Zn ferrite, around 1 Ωm, is offset by higher permeability and saturation flux density, making it a good candidate for applications below 1 MHz.

Ferrites in general have low Curie temperature, and this must be taken into account in the design. Ferrites come in many shapes and are found in inductors, transformers and filters. The saturation flux density of ferrites is considerably lower than that of laminated or powdered iron cores and this restricts their use in high current applications.

Laminated Iron Alloys

Laminated iron is used in magnetic cores for low- to medium-frequency applications. Laminations reduce eddy currents when they are electrically insulated from each other in the core of a transformer or inductor under AC operating conditions. The laminations may be stamped into any shape, with E or C shapes the most common. They are often used in cut cores, and the small gap that inadvertently arises when the cut cores are assembled reduces the likelihood of saturation. Tape-wound toroidal cores are available for higher frequencies up to 20 kHz. The iron alloys may be divided into two broad categories: silicon-iron and nickel-iron alloys.

In silicon-iron alloys, silicon is added to iron in order to reduce the overall conductivity compared with iron and hence to reduce the eddy current loss in the alloy. Additionally, the effects of magnetostriction due to domain wall rotation in AC applications are reduced and this effect is manifested in reduced acoustic noise. However, silicon-iron alloys exhibit reduced saturation flux density, and they are more brittle than iron. The silicon content is normally around 3% due to manufacturing considerations, although it can be as high as 6.5%. Silicon steel has been the workhorse of laminated core materials for power transformers and inductors for over 100 years. The steel is annealed in laminations and grain-orientated to give maximum flux density along the main axis. Silicon steel is also found in generators and motors.

Nickel-iron alloy is usually made up of 80% nickel and 20% iron in laminated and tape-wound cores The alloy is characterized by low coercive force, high saturation flux density, high permeability (up to 100 000) and high Curie temperature. These alloys are mainly found in current transformers, audio transformers and magnetic amplifiers. Increasingly, they are being used in power electronics applications up to 20 kHz. They also exhibit very low levels of magnetostriction, making them suitable for thin film applications. The cores are often encased in non-metallic enclosures to protect the material against winding stresses.

Powder Core

A magnetic powder core is manufactured by having iron or iron alloy powder mixed or glued with an insulation material and compressed into a ring or toroidal form. The combination of the magnetic powder and insulating resin results in a distributed gap, which gives the material its characteristic low value of effective relative permeability. The effective permeability is a function of the size and spacing of the iron particles, their composition and the thickness of the insulation binder. The bonding material has the same effect as that of an air gap distributed along the core. The distributed gap means that high DC current can be tolerated before the iron saturates.

The linear dimension of the iron particles is less than the skin depth at the desired operating frequency, resulting in low eddy current loss. The effective permeability usually ranges from 15 to 550 and the core electrical resistivity is around 1 Ωm. The maximum flux density may be as high as 1.5 T. The resulting inductance values tend to be very stable over a wide temperature range.

Molybdenum permalloy or MPP is one of the most popular materials used in the manufacture of powdered cores, while those made of carbonyl iron are very stable up to 200 MHz. Powder cores are suited to applications where the advantages of an air gap are desired, such as energy storage inductors. They are commonly used in switched mode power supplies, high Q inductors, filters and chokes.

Amorphous Alloys

The chemical composition of amorphous alloys contains two types of elements: ferromagnetic elements (Fe, Ni, Co and their combinations) generate the magnetic properties, while metallic elements Si, B and carbon (C) are introduced to decrease the melting point of the alloy to aid the manufacturing process. The resulting structure is similar to that of glass, and these alloys are often called metallic glass. In general, the resistivity of the amorphous alloy can reach 1.6 $\mu\Omega$m, which is three times of silicon steel, but several orders of magnitude lower than ferrites. Their Curie temperature is around 350 °C, and the saturation magnetic flux density is typically up to 1.6 T, which is much higher than the corresponding values for ferrites. Relative permeability values up to 100 000 are not unusual. Amorphous alloys are also low in coercive force. The core loss is reduced by the lamination effect of the tape-wound (thin ribbon) cores. However, amorphous alloys do not share the temperature stability of nanocrystalline materials. When the temperature goes from 25 °C to 250 °C, the saturation flux density may be reduced by as much as 30%.

Iron-based amorphous alloys have found application in low-frequency transformers and high-power inductors due to their low loss compared with grain-orientated steel, but with comparable saturation flux density levels. They may be found in pulse transformers, current transducers and magnetic amplifiers.

Nickel-iron-based amorphous alloys can achieve very high relative permeability, with saturation flux densities around 1 T. These are used in low- to medium-frequency transformers to replace iron cores. Cobalt-based amorphous alloys tend to be expensive, with very high relative permeability, but the maximum value of the saturation flux density is below 1 T. They tend to be used in specialist applications.

Nanocrystalline Materials

Nanocrystalline materials contain ultra-fine crystals, typically 7–20 nm in size, that are iron (Fe) based. In addition to Fe, there are traces of Si, B, Cu, molybdenum (Mo) and niobium (Nb). Among these, Cu, B and Nb are nearly always present. These materials combine the high saturation magnetic flux density of silicon steels with the low loss of ferrites at high frequencies. The relative permeability is typically 20 000, and the saturation flux density could be as high as 1.5 T. Core loss due to eddy currents is low, because these materials are supplied in nano-ribbon form with a thickness of 15–25 μm and an electrical resistivity of 0.012 $\mu\Omega$m. The thin ribbon is a form of lamination and reduces eddy current loss. The nanocrystalline material is very stable over a wide temperature range, and the Curie temperature at 600 °C is much higher than that for ferrites.

Nanocrystalline tape-wound cores are used in applications up to 150 kHz. Their high relative permeability makes them suitable for applications in current transformers, pulse transformers and common-mode EMI filters. In some cases, nanocrystalline materials are favoured over ferrites in military applications.

1.6.2 The Properties of some Magnetic Materials

Table 1.1 shows the magnetic and operating properties of some magnetic materials.

Table 1.1 Soft magnetic materials

Materials	Ferrites	Nanocrystalline	Amorphous	Si iron	Ni-Fe (permalloy)	Powdered iron
Model	Epcos N87	Viroperm 500 F	Metglas 2605	Unisil 23M3	Magnetics Permalloy 80	Micro-metals 75 μ
Permeability, μ_i	2200	15 000	10 000–150 000	5000–10 000	20 000–50 000	75
B_{peak}, T	0.49	1.2	1.56	2.0	0.82	0.6–1.3
ρ, $\mu\Omega$m	10×10^6	1.15	1.3	0.48	0.57	10^6
Curie temp. T_c, °C	210	600	399	745	460	665
P_{fe} mW/cm^3	288 at 0.2 T 50 kHz	312 at 0.2 T 100 kHz	294 at 0.2 T 25 kHz	5.66 at 1.5 T 50 Hz	12.6 at 0.2 T 5 kHz	1032 at 0.2 T 10 kHz
K_c	16.9	2.3	0.053	3.388	0.448	1798
α	1.25	1.32	1.81	1.70	1.56	1.02
β	2.35	2.1	1.74	1.90	1.89	1.89

1.7 Problems

1.1 Derive the *B*-*H* relationship in the toroidal setup of Figure 1.9.

1.2 Derive the *B*-*H* relationship in a long solenoid with a uniformly wound coil of *N* turns per metre, neglect fringing at the ends of the solenoid.

1.3 Calculate the H field in the dielectric of a coaxial cable with the following dimensions: the radius of the inner conductor is r_i and the inner and outer radii of the outer conductor are r_{oi} and r_{oo} respectively.

1.4 Describe the three types of power loss in a magnetic component.

Reference

1. Steinmetz, C.P. (1984) On the law of hysteresis. *Proceedings of the IEEE* **72** (2), 197–221.

Further Reading

1. Blume, L.F. (1982) *Transformer Engineering*, John Wiley & Sons, New York.
2. Bueno, M.D.A. (2001) *Inductance and Force Calculations in Electrical Circuits*, Nova Science Publishers, Huntington.
3. Del Vecchio, R.M., Poulin, B., Feghali, P.T. *et al.* (2001) *Transformer Design Principles: With Applications to Core-Form Power Transformers*, 1st edn, CRC Press, Boca Raton, FL.
4. Erickson, R.W. (2001) *Fundamentals of Power Electronics*, 2nd edn, Springer, Norwell, MA.
5. Flanagan, W.M. (1992) *Handbook of Transformer Design and Application*, 2nd edn, McGraw-Hill, New York.
6. Georgilakis, P.S. (2009) *Spotlight on Modern Transformer Design (Power Systems)*, 1st edn, Springer, New York.
7. Hui, S.Y.R. and Zhu, J. (1995) Numerical modelling and simulation of hysteresis effects in magnetic cores using transmission-line modelling and the Preisach theory. *IEE Proceedings – Electric Power Applications B*, **142** (1), 57–62.
8. Jiles, D.C. and Atherton, D.L. (1984) Theory of ferromagnetic hysteresis (invited). *Journal of Applied Physics* **55** (6), 2115–2120.
9. Kaye, G.W.C. and Laby, T.H. (2008) *Tables of Physical and Chemical Constants*, vol. Section 2.6.6, National Physical Laboratory, Teddington, Middlesex, UK.
10. Kazimierczuk, M.K. (2009) *High-Frequency Magnetic Components*, John Wiley & Sons, Chichester.
11. Krein, P.T. (1997) *Elements of Power Electronics (Oxford Series in Electrical and Computer Engineering)*, Oxford University Press, Oxford.
12. Kulkarni, S.V. (2004) *Transformer Engineering: Design and Practice*, 1st edn, CRC Press, New York.

Section One
Inductors

2

Inductance[1]

Inductors and transformers are present in almost every power electronics circuit. Broadly speaking, inductors are dynamic energy storage devices and, as such, they are employed to provide stored energy between different operating modes in a circuit. They also act as filters for switched current waveforms. In snubber circuits, they are used to limit the rate of change of current and to provide transient current limiting. Transformers, on the other hand, are energy transfer devices, for example converting power at a high voltage and a low current to power at a lower voltage and a higher current. In a mechanical analogy, an inductor is like a flywheel which stores energy, while a transformer is like a gearbox which trades speed for torque. Besides transforming voltage levels, transformers are also used to provide electrical isolation between two parts of a circuit; to provide impedance matching between circuits for maximum power transfer; and to sense voltage and currents (potential and current transformers).

2.1 Magnetic Circuits

We have seen in Chapter 1 that Ampere's law describes the relationship between the electric current and magnetic field in a magnetic circuit. Figure 2.1 shows a coil with N turns wound on a closed magnetic core of mean length l_c and cross-sectional area A_c, from Equation 2.1.

$$\oint_C = \mathbf{H} \cdot d\mathbf{l} = Ni \tag{2.1}$$

Recall that the direction of $\mathbf{H}$ around the loop is related to i by the right hand screw rule.

At this point we need not concern ourselves with the shape of the core or the shape of the cross-section. For simplicity, we assume that the cross-section is constant over the length of

[1] Part of this chapter is reproduced with permission from [1] Hurley, W.G. and Duffy, M.C. (1995) Calculation of self and mutual impedances in planar magnetic structures. *IEEE Transactions on Magnetics* **31** (4), 2416–2422.

Transformers and Inductors for Power Electronics: Theory, Design and Applications, First Edition.
W. G. Hurley and W. H. Wölfle.

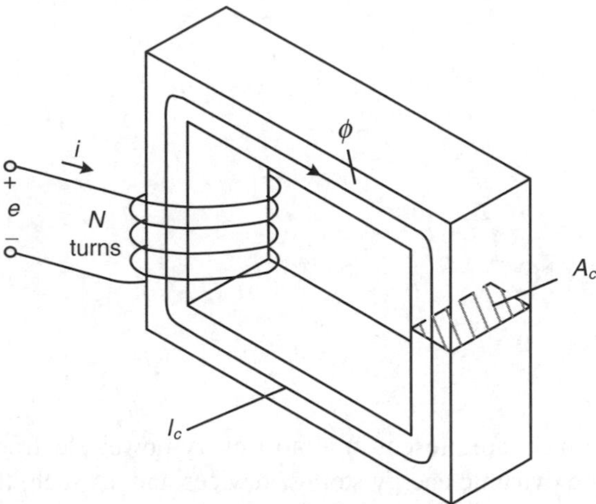

Figure 2.1 Magnetic circuit.

the core. We also assume that the flux density is uniform over the cross-sectional area and that the value of the magnetic field intensity H_c is constant around the loop C. This assumption is reasonable as long as there is no flux 'leaking' from the core; we can take leakage effects into account later. Thus, applying Equation 2.1 yields the expression for H_c in terms of the product Ni:

$$H_c l_c = Ni \tag{2.2}$$

The magnetic field intensity H_c produces a magnetic flux density B_c:

$$B_c = \mu_r \mu_0 H_c \tag{2.3}$$

where μ_r is the relative permeability and μ_0 is the permeability of free space ($\mu_0 = 4\,\pi \times 10^{-7}$ H/m), as before.

The ampere turns product Ni drives the magnetic field around the core in a manner analogous to the emf in an electrical circuit driving current through the conductors. Ni is called the magnetomotive force or mmf, and it is normally denoted by the symbol Φ.

The flux in the core in Figure 2.1 is:

$$\phi = B_c A_c = \mu_r \mu_0 H_c A_c \tag{2.4}$$

H_c may found from Equation 2.4:

$$H_c = \frac{\phi}{\mu_r \mu_0 A_c} \tag{2.5}$$

We can rewrite Ampere's law in Equation 2.2 with this value of H_c:

$$Ni = H_c l_c = \phi \frac{l_c}{\mu_r \mu_0 A_c} = \phi \mathcal{R}_c \tag{2.6}$$

$\mathcal{R}$ is called the reluctance of the magnetic circuit. This form of the magnetic circuit equation is analogous to the Ohm's law for electrical circuits. The mmf Φ is related to the reluctance $\mathcal{R}$ by:

$$\mathcal{F} = \phi \mathcal{R}_c \tag{2.7}$$

and the reluctance is defined as:

$$\mathcal{R}_c = \frac{l_c}{\mu_r \mu_0 A_c} \tag{2.8}$$

Reluctance is a measure of the effect of the magnetic circuit in impeding the flow of the magnetic field; its units are At/Wb. The inverse of reluctance is permeance.

There is an analogy here between reluctance in a magnetic circuit and resistance in an electrical circuit. Just like resistance, the reluctance of the magnetic circuit is proportional to the length of the circuit and inversely proportional to the cross-sectional area, as shown by Equation 2.8. However, it is important to note that while resistance represents power dissipation in an electric circuit, reluctance is closer to electrical impedance, in the sense that it does not involve power loss. We shall see later that it is inherently connected to inductance and energy storage.

The shape of a wire in an electrical circuit does not affect Ohm's law; likewise, the shape of the magnetic circuit does not affect the magnetic circuit law in Equation 2.7. The analogy with the electrical circuit can be used to analyze more complex magnetic circuits. In the same way that individual parts of a circuit are represented by electrical resistance, a magnetic circuit can be made up of different reluctances. The total mmf is then:

$$\mathcal{F} = Ni = \phi_1 \mathcal{R}_1 + \phi_2 \mathcal{R}_2 + \phi_3 \mathcal{R}_3 + \ldots \tag{2.9}$$

In many cases, the flux is common to all parts of the magnetic circuit in the same way that current could be common to a set of series resistances. Equation 2.9 is then:

$$Ni = \phi(\mathcal{R}_1 + \mathcal{R}_2 + \mathcal{R}_3 + \ldots) \tag{2.10}$$

We can extend the analogy to both series and parallel reluctances. One drawback with reluctance is that it is a function of the relative permeability of the material, and we saw in Chapter 1 that this is a non-linear function of the applied magnetic field intensity in ferromagnetic materials. In the magnetic core of Figure 2.1, the flux is limited by the core material, and a small current with sufficient turns can easily drive the magnetic core material into saturation. However, if a gap is introduced into the core, as shown in Figure 2.2, a larger current will be required to achieve saturation of the magnetic material because of the increased reluctance

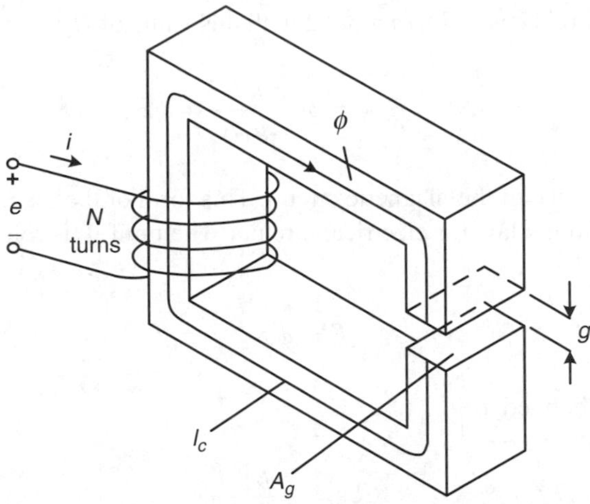

Figure 2.2 Magnetic circuit with an air gap.

introduced by the air gap. The analysis to follow also shows that the air gap gives rise to more energy storage. This arrangement is the basis for most inductors encountered in power electronics.

In the structure of Figure 2.2, there are two reluctances in series, so:

$$\mathcal{F} = \phi(\mathcal{R}_c + \mathcal{R}_g) \tag{2.11}$$

The individual reluctances are given by their respective physical properties of length and cross-section; subscript c refers to the core and subscript g refers to the air gap.

For the core:

$$\mathcal{R}_c = \frac{l_c}{\mu_r \mu_0 A_c} \tag{2.12}$$

and for the air gap $\mu_r = 1$:

$$\mathcal{R}_g = \frac{g}{\mu_0 A_g} \tag{2.13}$$

Where g is the length of the air gap.

The equivalent reluctance of the gapped core is:

$$\mathcal{R}_{\text{eq}} = \mathcal{R}_c + \mathcal{R}_g = \frac{l_c}{\mu_r \mu_0 A_c} + \frac{g}{\mu_0 A_g} \tag{2.14}$$

At this point, we will set the cross-sectional area of the core equal to the cross-sectional area of the gap. This is a reasonable assumption when g is small compared to the dimensions of the cross-section, and we will return to it later when we discuss fringing effects. Equation 2.14 becomes:

$$\mathcal{R}_{\text{eq}} = \frac{l_c}{\mu_0 A_c}\left[\frac{1}{\mu_r} + \frac{1}{l_c/g}\right] = \frac{l_c}{\mu_{\text{eff}}\mu_0 A_c} \tag{2.15}$$

This means that the reluctance of the gapped core is equivalent to the reluctance of a core of length l_c and relative permeability μ_{eff}, with:

$$\mu_{\text{eff}} = \frac{1}{\dfrac{1}{\mu_r} + \dfrac{1}{l_c/g}} \tag{2.16}$$

If $\mu_r \gg 1$, then the effective relative permeability is:

$$\mu_{\text{eff}} \approx \frac{l_c}{g} \tag{2.17}$$

One way to interpret this result is to say that if the air gap (length g) is distributed over the whole core of length l_c, the effective permeability of the core is reduced from μ_r to μ_{eff}.

Example 2.1

The magnetic circuit of Figure 2.2 has the following dimensions: $A_c = A_g = 12\,\text{mm} \times 15\,\text{mm}$, $g = 0.5\,\text{mm}$, $l_c = 10.3\,\text{cm}$, $N = 5$ turns. The core material is ferrite, with relative permeability $\mu_r = 2000$.

Find:

(a) the reluctance of the core;
(b) the reluctance of the air gap;
(c) the flux for a magnetic flux density of $B_c = 0.2\,\text{T}$;
(d) the magnetic field intensity inside the core and inside the air gap.

The area of the gap is $12 \times 15 \times 10^{-2} = 1.80\,\text{cm}^2$.
The reluctances are given by Equation 2.12 and Equation 2.13 for the core and air gap respectively:

(a) $$\mathcal{R}_c = \frac{l_c}{\mu_r \mu_0 A_c} = \frac{10.3 \times 10^{-2}}{(2000)(4\pi \times 10^{-7})(1.8 \times 10^{-4})}$$
$$= 0.228 \times 10^6\ \text{At/Wb}$$

(b) $$\mathcal{R}_g = \frac{g}{\mu_0 A_g} = \frac{0.5 \times 10^{-3}}{(4\pi \times 10^{-7})(1.8 \times 10^{-4})}$$
$$= 2.21 \times 10^6\ \text{At/Wb}$$

The flux is given by Equation 2.4:

(c) $\phi = B_c A_c = (0.2)(1.8 \times 10^{-4}) = 3.6 \times 10^{-5}$ Wb

(d) $$H_c = \frac{B_c}{\mu_r \mu_0} = \frac{0.2}{(2000)(4\pi \times 10^{-7})} = 79.6\ \text{A/m}$$

$$H_g = \frac{B_c}{\mu_0} = \frac{0.2}{4\pi \times 10^{-7}} = 159.2 \times 10^3\ \text{A/m}$$

As expected, the H field is much smaller inside the core than in the gap.

The discussion so far has concerned itself with series reluctance. Many practical designs have parallel reluctance paths and these can be handled by noting the analogy with the electrical circuit.

2.2 Self and Mutual Inductance

The quasi-static magnetic field system depicted in Figure 2.3 consists of a coil with N turns in a magnetic field.

If the magnetic flux is allowed to vary with time, then a voltage is produced in accordance with Faraday's law:

$$e = -N\frac{d\phi}{dt} \tag{2.18}$$

In other words, an emf is induced in the circuit as long as the flux is changing. Furthermore, the magnitude of the induced emf is equal to the time rate of change of flux linkage in the circuit. Lenz's law tells us that the induced emf due to the changing flux is always in such a direction that the current it produces tends to oppose the changing flux that is causing it.

Figure 2.4 shows a current carrying coil near a solenoid. The solenoid has N turns. The current in the coil creates flux as a direct consequence of Ampere's law. The total flux

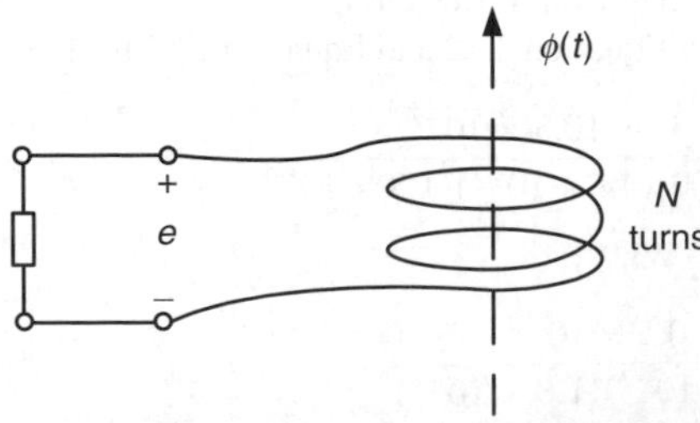

Figure 2.3 Coil of N turns in a changing magnetic field.

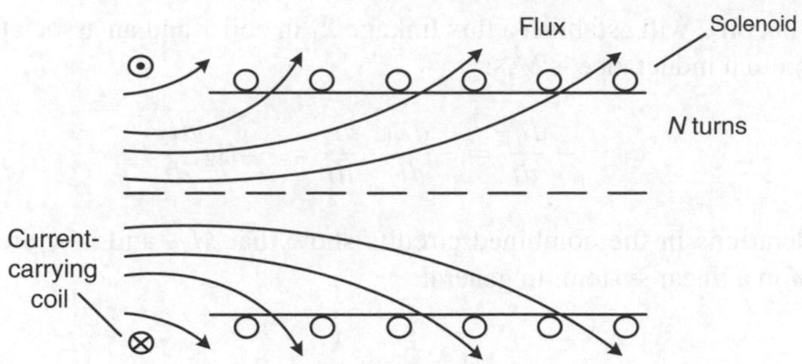

Figure 2.4 Current carrying coil and solenoid.

linking the solenoid is the sum of the fluxes linking each turn of the solenoid, in this case the flux linking each turn is not the same, but the total flux linkage for the N turn solenoid is

$$\lambda = \phi_1 + \phi_2 + \ldots + \phi_n + \ldots + \phi_N \tag{2.19}$$

where ϕ_n is the flux linking the nth turn. If ϕ is the average flux linking each turn, then the total flux linkage is:

$$\lambda = N\phi \tag{2.20}$$

Let us now take two coils: coil 1 has a current i_1, and the flux linkage in coil 2 is λ_2, as shown in Figure 2.5.

The flux linkage of coil 2 due to the current i_1 in coil 1 gives rise to an induced emf in coil 2, given by:

$$e_2 = -\frac{d\lambda_2}{dt} = -\frac{d\lambda_2}{di_1} \cdot \frac{di_1}{dt} = -M_{21}\frac{di_1}{dt} \tag{2.21}$$

M_{21} is called the coefficient of mutual inductance of coil 2 w.r.t. coil 1. If the flux linkage is proportional to the current, which is the case unless there are saturation effects at play, then $\frac{d\lambda_2}{di_1}$ is constant and M_{21} is constant.

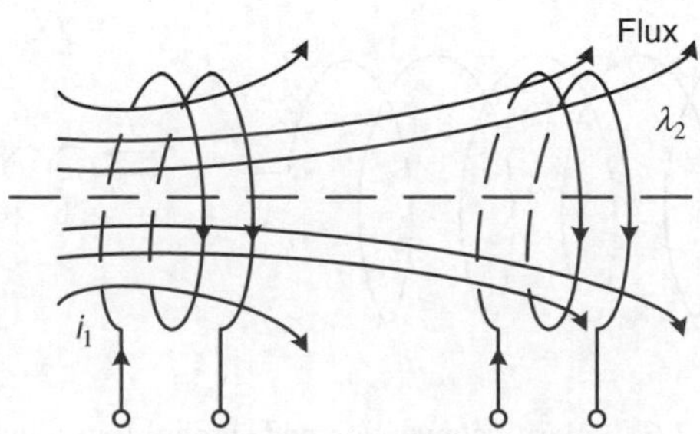

Figure 2.5 Coils with mutual flux.

A current i_2 in coil 2 will establish a flux linkage λ_1 in coil 1 and an associated emf e_1. The coefficient of mutual inductance is M_{12}:

$$e_1 = -\frac{d\lambda_1}{dt} = -\frac{d\lambda_1}{di_2} \cdot \frac{di_2}{dt} = -M_{12}\frac{di_2}{dt} \tag{2.22}$$

Energy considerations in the combined circuits show that M_{12} and M_{21} are equal so that $M_{12} = M_{21} = M$ in a linear system. In general:

$$M = \frac{\lambda_2}{i_1} = \frac{\lambda_1}{i_2} \tag{2.23}$$

Two coils have a mutual inductance of one Henry (H) if a current of one ampere (A) flowing in either coil sets up a flux linkage of one Weber (Wb) in the other coil. Alternatively, two coils have a mutual inductance of one Henry if a uniformly changing current at the rate of one ampere per second (A/s) in one coil induces an emf of one volt (V) in the other. The mutual inductance is dependent on the shapes of the coils involved, their proximity to each other and the number of turns in each coil.

Self inductance is a special case of mutual inductance. A current carrying coil has a flux linkage due to its own current i, apart from any outside sources of flux.

The self inductance L is defined as:

$$L = \frac{\lambda}{i} \tag{2.24}$$

A coil carrying 1 A that sets up a flux linkage of 1 Wb has a self inductance of 1 H. If the current changes at a steady rate of 1 A/s and induces an emf of 1 V, then the self inductance is 1 H.

The circuit symbol for inductance is shown in Figure 2.6. The question of the polarity of the voltage across the inductor can be resolved by adopting the convention shown in Figure 2.6 for the polarity of e:

$$e = \frac{d\lambda}{dt} = L\frac{di}{dt} \tag{2.25}$$

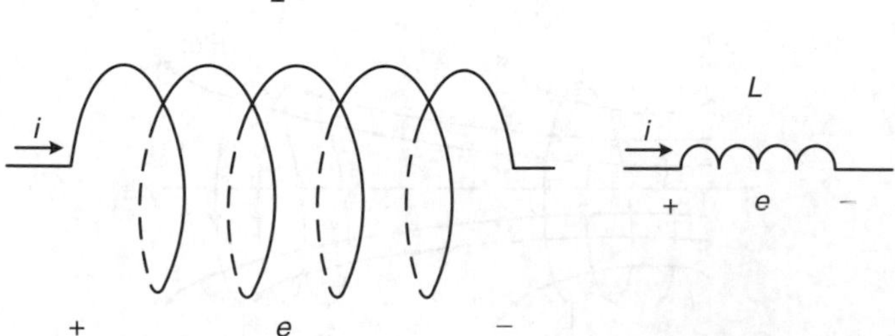

Figure 2.6 Polarity convention and symbol for inductance.

If the inductor represented by the inductance in Figure 2.6 were connected to an external circuit, with i increasing with time, then e would generate an induced current out of the positive terminal through the external circuit and into the negative terminal, correctly opposing i as expected from Lenz's law.

Let us take a closer look at Equation 2.24, using Equations 2.20, 2.4 and 2.2:

$$L = \frac{\lambda}{i} = \frac{N\phi}{i} = \frac{N^2BA}{Ni} = \frac{N^2BA}{Hl} = \frac{N^2\mu_r\mu_0A}{l} = \frac{N^2}{\mathcal{R}} \tag{2.26}$$

This result shows that the self inductance may be computed from the total number of turns and the reluctance of the magnetic circuit.

This approach may be applied to the inductance of a gapped core:

$$L = \frac{N^2}{\mathcal{R}_c + \mathcal{R}_g} = \frac{N^2}{\mathcal{R}_g}\frac{1}{1 + \dfrac{\mu_{\text{eff}}}{\mu_r}} \tag{2.27}$$

Thus, the value of the inductance can be made independent of the magnetic properties of the core material if $\mu_r \gg \mu_{\text{eff}}$, as long as the core material has not saturated. In saturation, there is no guarantee that the inequality $\mu_r \gg \mu_{\text{eff}}$ holds.

Example 2.2

For the magnetic circuit in Example 2.1, find: (a) the self inductance and (b) the induced emf for a sinusoidal flux density of magnitude 0.2 T at 50 kHz.

(a) The self inductance is found from Equation 2.27

The reluctance of the core and the reluctance of the air gap were found in Example 2.1. The reluctance of the series combination of the core and air gap is:

$$\mathcal{R} = \mathcal{R}_c + \mathcal{R}_g = (0.228 \times 10^6) + (2.21 \times 10^6) = 2.438 \times 10^6 \text{ At/Wb}$$

And the self inductance for 5 turns is:

$$L = \frac{5^2}{2.438 \times 10^6} \times 10^3 = 0.01 \text{ mH}$$

In manufacturers' data books, the quantity $1/\mathcal{R}$ (inverse of reluctance) is sometimes called the A_L value or specific inductance; this is the inductance for a single turn, and the inductance for N turns is $L = A_LN^2$.

(b) The core is operating at 50 kHz and the induced emf is:

$$
\begin{aligned}
e &= N\frac{d\phi}{dt} = NA_c\frac{dB}{dt} = NA_c\omega B_{\max}\cos(\omega t)\\
&= (5)(1.8\times10^{-4})((2\pi)(50\times10^{3}))(0.2)\cos(2\pi 50\times10^{3}t)\\
&= (56.55)\cos(314.159\times10^{3}t)\ \text{V}
\end{aligned}
$$

2.3 Energy Stored in the Magnetic Field of an Inductor

One way to interpret Equation 2.25 is that a voltage is required to change the current when inductance is present – or, in other words, the current in the inductor has inertia. A consequence of this state of affairs is that electrical energy is stored in the magnetic field of the inductor.

Consider the inductor represented in Figure 2.7, with a current i and voltage e. The power or rate of energy flow into the winding is:

$$p = ie = i\frac{d\lambda}{dt} \tag{2.28}$$

The change in magnetic field energy between t_1 and t_2 is:

$$W_m = \int_{t_1}^{t_2} p dt = \int_{\lambda_1}^{\lambda_2} i\, d\lambda \tag{2.29}$$

Noting that $\lambda = Li$

$$W_m = \int_0^i Li\, di = \frac{1}{2}Li^2 \tag{2.30}$$

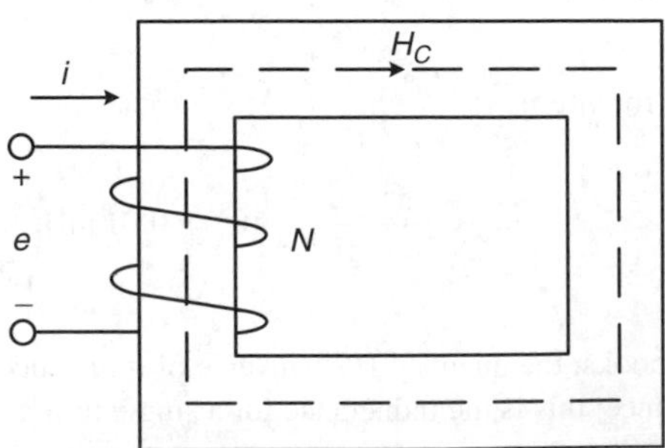

Figure 2.7 A closed magnetic circuit.

Furthermore, $\lambda = N\phi$ and $id\lambda = Nid\phi$, the mmf $Ni = H_c l_c$ (Ampere's law), $\phi = B_c A_c$. Substituting these relationships into Equation 2.30 allows us to express the stored energy in terms of B and H:

$$W_m = A_c l_c \int_{B_1}^{B_2} H_c \, dB_c \tag{2.31}$$

$A_c l_c = V_c$ is the volume of the core, so that the stored energy per unit volume is the integral of the magnetic field intensity between the limits of the flux density. This is similar to the expression we had in Chapter 1 for hysteresis loss.

For an inductor with flux B, the integration of Equation 2.31 yields:

$$W_m - \frac{B^2 V_c}{2\mu_r \mu_0} \tag{2.32}$$

where V_c is the volume of the core.

The result in Equation 2.32 may be applied to a gapped core:

$$W_m = \frac{B^2 V_c}{2\mu_r \mu_0} + \frac{B^2 V_g}{2\mu_0} \tag{2.33}$$

where $V_g = A_c g$ is the volume of the gap.

Equation 2.33 may be rewritten:

$$W_m = \frac{B^2 V_g}{2\mu_0}\left[1 + \frac{\mu_{\text{eff}}}{\mu_r}\right] \tag{2.34}$$

with $\mu_{\text{eff}} = l_c/g$ as before.

The result in Equation 2.34 shows that, in a gapped inductor, most of the energy is stored in the gap itself. This arises because the H field is diminished inside the core (see Example 2.1), and it raises the obvious question: why do we need a core if there is so little energy stored in it?

2.3.1 *Why Use a Core?*

Consider a core made of ferrite material which saturates at about 0.3 T.

The *B-H* magnetization curve of the ferrite has the non-linear characteristic shown in Figure 2.8, with saturation occurring at about $H = 2$ A/m. The *B-H* characteristic of the air gap is linear, with $\mu_r = 1$. If the air gap were spread out over the whole core, then for the same reluctance:

$$\mathcal{R} = \frac{g}{\mu_0 A_g} = \frac{l_c}{\mu_{\text{eff}} \mu_0 A_g} \tag{2.35}$$

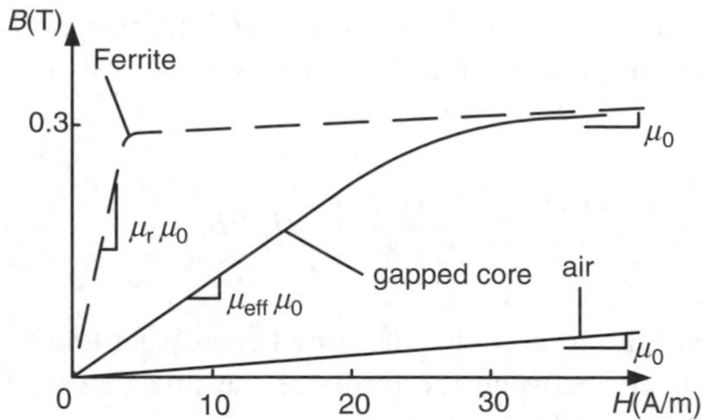

Figure 2.8 *B-H* magnetization curve of a ferrite material.

For a typical gapped core with $l_c = 100g$, the effective relative permeability of the core is 100. The gap dominates the core characteristics if $\mu_r \gg \mu_{\text{eff}}$, as shown by Equation 2.27.

The *B-H* characteristic of the combination is shown in Figure 2.8, which is dominated by the air gap but eventually reaches saturation. It is clear from the combination that saturation occurs at a much higher value of *H*, in this case about 30 A/m. This means that much higher currents can be tolerated before the onset of saturation at the expense of a lower value of inductance. From Equation 2.17, the slope of the *B-H* curve in the combination is $\mu_{\text{eff}}\mu_0$. Recall that Equation 2.26 shows that higher reluctance leads to lower overall inductance. Remembering that $\mu_{\text{eff}} = l_c/g$, it is worth noting that increasing *g* reduces the slope of the *B-H* characteristic and allows more stored energy even though the inductance is reduced (see Equation 2.27).

The energy stored in the magnetic field is the area to the left of the *B-H* curve from the origin up to the operating point. Evidently, the area in question is increased by the presence of the air gap and more energy can be stored in the gapped core.

The properties of most core materials vary with temperature, flux level, sample and manufacture. The gap reduces the dependence of the overall inductance on these parameters to make the value of inductance more predictable and stable. The issues associated with saturation, hysteresis and remanance are mitigated by the introduction of an air gap as long as $\mu_{\text{eff}} \gg \mu_r$.

The advantages of the air gap can thus be summarized:

- Saturation occurs at higher values of current.
- More energy can be stored, which is the basic function of an inductor.
- The inductor is less susceptible to variations in the magnetic properties of the core.

In the case of a transformer, it is energy transfer and not energy storage that counts, so that although a gap is not normally used, a small gap may store some energy to feed parasitic capacitances.

With problems of saturation, hysteresis and eddy current losses, and with the expense and weight of magnetic materials, one might ask why not build an air-core inductor with sufficient turns? First, the overall size of an air-core inductor is much larger. Also, the lower reluctance of a ferromagnetic core ensures that most of the flux is inside the core and in an air-core coil there is more leakage flux, which may give rise to electromagnetic interference (EMI). For a given value of inductance and a given overall volume, a lot fewer turns are required when a core is used and, therefore, there is less winding resistance.

Example 2.3

For the magnetic circuit in Example 2.1, find the energy stored (a) in the core; and (b) in the air gap for B = 0.2 T.

(a) The energy stored in the core is:

$$W_c = \frac{B^2 V_c}{2\mu_r \mu_0} = \frac{(0.2)^2(1.8 \times 10^{-4})(10.3 \times 10^{-2})}{(2)(2000)(4\pi \times 10^{-7})} \times 10^3 = 0.148\ \text{mJ}$$

(b) The energy stored in the gap is:

$$W_g = \frac{B^2 V_g}{2\mu_0} = \frac{(0.2)^2(1.8 \times 10^{-4})(0.5 \times 10^{-3})}{(2)(4\pi \times 10^{-7})} \times 10^3 = 1.43\ \text{mJ}$$

As expected, most of the energy in this inductor is stored in the air gap.

The total energy stored is 1.578 mJ.

Example 2.4

In this example, we take a look at a more practical structure that is often encountered in power electronics. The inductor shown in Figure 2.9 consists of two E-shaped cores with a gap in the centre leg.

Calculate the equivalent reluctance of the structure.

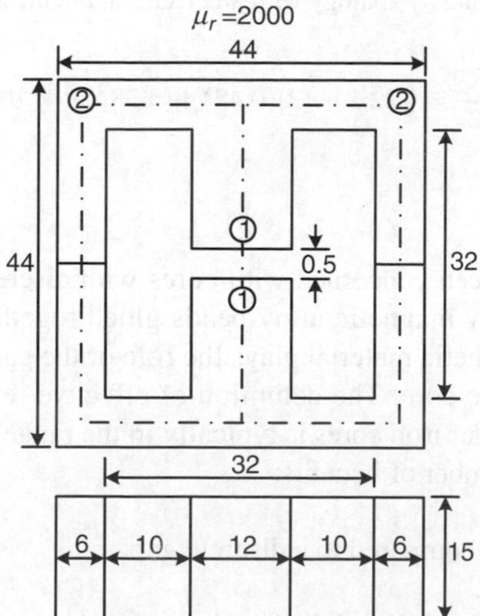

Figure 2.9 Core dimensions given in mm.

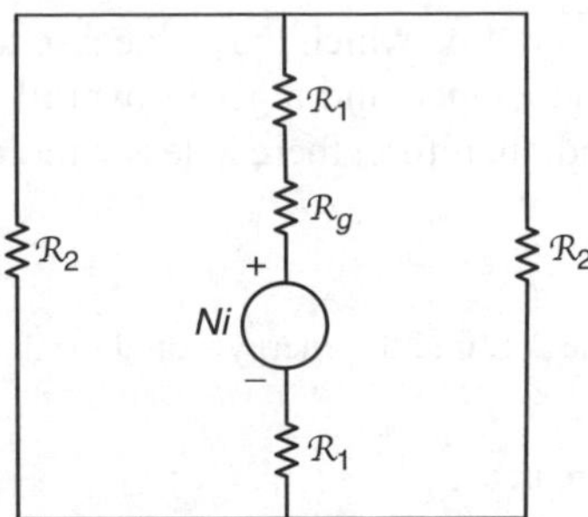

Figure 2.10 Equivalent magnetic circuit.

The equivalent magnetic circuit is shown in Figure 2.10. $\mathcal{R}_1$ represents the reluctance of the centre leg on either side of the air gap. $\mathcal{R}_2$ represents the reluctance of the outer leg and $\mathcal{R}_g$ represents the reluctance of the air gap.

$$\mathcal{R}_1 = \frac{l_1}{\mu_r \mu_0 A_c} = \frac{19 \times 10^{-3}}{(2000)(4\pi \times 10^{-7})(12)(15 \times 10^{-6})} = 0.42 \times 10^5 \text{ At/Wb}$$

$$\mathcal{R}_2 = \frac{l_2}{\mu_r \mu_0 A_c} = \frac{76.5 \times 10^{-3}}{(2000)(4\pi \times 10^{-7})(6)(15 \times 10^{-6})} = 3.38 \times 10^5 \text{ At/Wb}$$

$$\mathcal{R}_g = \frac{g}{\mu_0 A_c} = \frac{0.5 \times 10^{-3}}{(4\pi \times 10^{-7})(12)(15 \times 10^{-6})} = 22.1 \times 10^5 \text{ At/Wb}$$

The equivalent reluctance is found by analogy with an electrical circuit with resistances.

$$\mathcal{R}_{eq} = (\mathcal{R}_g + 2\mathcal{R}_1)||\frac{\mathcal{R}_2}{2} = [(22.1 + (2)(0.42))||(3.38/2)] \times 10^5 = 1.57 \times 10^5 \text{ At/Wb}$$

2.3.2 *Distributed Gap*

The discussion so far has been concerned with cores with discrete gaps. Powder iron cores consist of high-permeability magnetic alloy beads glued together with non-ferromagnetic material. The non-ferromagnetic material plays the role of the gap except that, in this case, it is distributed over the whole core. The definition of effective relative permeability remains valid and the value for powder iron cores is typically in the range of 15–550. Distributing the gap in this manner has a number of benefits:

- There is less flux fringing compared to a discrete gap.
- EMI effects are mitigated.
- Winding loss due to fringing effects is reduced.
- Core assembly is mechanically more stable.

However, ferrites with discrete gaps tend to be less expensive and there is a greater variety of shapes and sizes.

There is another important distinction. In a gapped ferrite, when the onset of saturation is reached ($\mu_{\text{eff}} \approx \mu_r$), the permeability of the ferrite rolls off quite sharply, resulting in a dramatic lowering of inductance. With the distributed gap, this effect is less severe, because the intrinsic permeability of the beads is much higher than that of ferrite to begin with. The ferrite is also more dependent on temperature than the beads, and saturation in the ferrite occurs at a lower value of H at higher temperatures.

2.4 Self and Mutual Inductance of Circular Coils

2.4.1 Circular Filaments

The fundamental building block for inductance calculations in coils is the mutual inductance between two filaments, as shown in Figure 2.11 [1].

A filamentary energizing turn of radius a at $z = 0$ carries a sinusoidal current represented by $i_\phi(t) = I_\phi e^{j\omega t}$. A second filamentary turn, of radius r, is located an axial distance z from the energizing turn.

For a magnetoquasistatic system, the following forms of Maxwell's equations hold in a linear homogeneous isotropic medium:

$$\nabla \times \mathbf{H} = \mathbf{J}_f \tag{2.36}$$

$$\nabla \times \mathbf{E} = -\frac{\partial \mathbf{B}}{\partial t} \tag{2.37}$$

On the basis of cylindrical symmetry, the following identities apply to the electric field intensity E and the magnetic field intensity H:

$$E_r = 0;\ E_z = 0;\ \frac{\partial E_\phi}{\partial \phi} = 0 \tag{2.38}$$

$$H_\phi = 0;\ \frac{\partial H_r}{\partial \phi} = 0;\ \frac{\partial H_z}{\partial \phi} = 0 \tag{2.39}$$

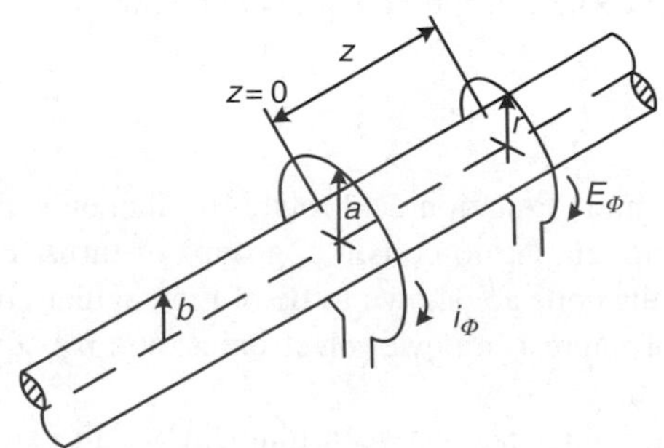

Figure 2.11 Circular filaments.

Maxwell's equations reduce to:

$$\frac{\partial H_r}{\partial z} - \frac{\partial H_z}{\partial r} = I_\phi \delta(r-a)\delta(z) \tag{2.40}$$

$$\frac{\partial E_\phi}{\partial z} = j\omega\mu_0 H_r \tag{2.41}$$

$$\frac{1}{r}\frac{\partial(rE_\phi)}{\partial r} = -j\omega\mu_0 H_z \tag{2.42}$$

Eliminating H gives the following result for E_ϕ:

$$\frac{\partial^2 E_\phi}{\partial z^2} + \frac{\partial^2 E_\phi}{\partial r^2} + \frac{1}{r}\frac{\partial E_\phi}{\partial r} - \frac{E_\phi}{r^2} = j\omega\mu_0 I_\phi \delta(r-a)\delta(z) \tag{2.43}$$

where $\delta(r\text{-}a)$ and $\delta(z)$ are Dirac unit impulse functions which locate the energizing current at the co-ordinates (a, 0). The mutual inductance between the two filaments is found from the solution to the Bessel equation in Equation 2.43 and is well known; it was first proposed by Maxwell [2]:

$$M = \mu_0 \pi a r \int_0^\infty J_1(kr)J_1(ka)e^{-k|z|}\,dk \tag{2.44}$$

where J_1 is a Bessel function of the first kind.

The solution of Equation 2.43 may also be written in the form of elliptic integrals given by Gray [3]:

$$M = \mu_0\sqrt{ar}\frac{2}{f}\left[\left(1-\frac{f^2}{2}\right)K(f) - E(f)\right] \tag{2.45}$$

$$f = \sqrt{\frac{4ar}{z^2 + (a+r)^2}} \tag{2.46}$$

where $K(f)$ and $E(f)$ are complete elliptic integrals of the first and second kind, respectively.

2.4.2 Circular Coils

The mutual inductance formula Equation 2.44 relates to filaments. As designers, however, in practice, we need formulae relating to coils, i.e. groups of turns. Figure 2.12 indicates two such coils. In this figure, the coils are shown to be of rectangular cross-section (the shape of cross-sections normally of interest in power electronics, and it is understood that coil 1 has N_1 turns and coil 2 has N_2 turns.

The formula for mutual inductance between filaments in air can be formally modified to apply to coils of rectangular cross-section. This is achieved by integrating the filament

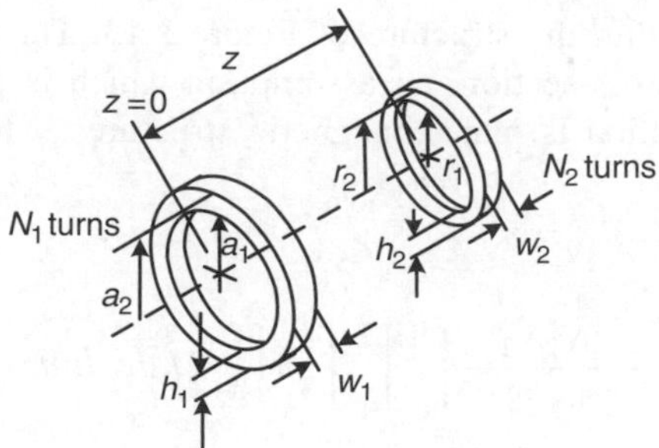

Figure 2.12 Circular coils.

formula over the coil cross-section, assuming uniform current density across the cross-section. We shall see later, in Chapter 9, that this is not necessarily true when the width to-height ratio of the coil cross-section is very large and must be taken into account.

Figure 2.13 shows the arrangement and dimensions of two illustrative coils, which are assumed to be circular and concentric.

The impedance formula for this arrangement is derived from the filamentary formula by an averaging process. Consider a representative point (τ_1,a) within the cross-section of coil 1 and a representative point ($z + \tau_2$, r) within the cross-section of coil 2, as indicated in Figure 2.13. If these two points are taken to define two filaments, Equation 2.44 gives the following formula for the mutual inductance between them:

$$M = \mu_0 \pi a r \int_0^\infty J_1(kr) J_1(ka) e^{-k|z+\tau_2-\tau_1|}\, dk \tag{2.47}$$

The basic mutual inductance formula for coils, rather than filamentary turns, may then be obtained from Equation 2.46 by taking this mutual inductance as the average effect of all

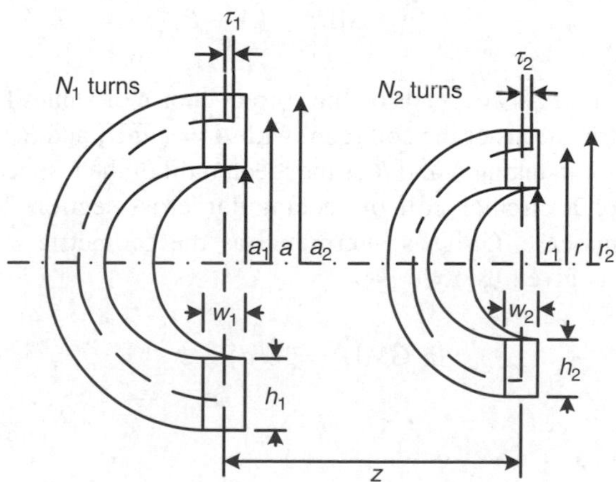

Figure 2.13 Filaments in circular coils.

such filamentary turns within the structure of Figure 2.13. This assumes uniform current density over either coil cross-section, an assumption which is justified in most practical cases. However, an exception is planar magnetic structures, which will be dealt with in Chapter 9.

The proposed formula is:

$$M' = \frac{N_1 N_2}{h_1 w_1 h_2 w_2} \int_{a_1}^{a_2} \int_{r_1}^{r_2} \int_{\frac{-w_1}{2}}^{\frac{w_1}{2}} \int_{\frac{-w_2}{2}}^{\frac{w_2}{2}} M \, da \, dr \, d\tau_1 \, d\tau_2 \tag{2.48}$$

The integration of Equation 2.48 has been carried out in the past, but the exact integral solution converges slowly and an alternative solution, such as the elliptic integral formula, does not exist. Approximations to the integral formula for coils in air have been developed by Gray [3] and Lyle [4]. Dwight [5] and Grover [6] have published extensive tables to calculate the inductance between coil sections.

Maxwell [2] proposed that the elliptic integral formula for filaments could be used if the coil separation is replaced by the Geometric Mean Distance (GMD) between the coil sections. Basically, if conductor 1 is divided into n identical parallel filaments, each carrying an equal share of the current, and if conductor 2 is similarly divided into m identical parallel filaments, then the GMD between the two conductors is the mnth root of mn terms. These are the products of the distances from all the n filaments in conductor 1 to all the m filaments in conductor 2.

Using GMD to replace coil separation z in Equation 2.45 gives:

$$M = \mu_0 \sqrt{AR} \frac{2}{f} \left[\left(1 - \frac{f^2}{2} \right) K(f) - E(f) \right] \tag{2.49}$$

$$f = \sqrt{\frac{4AR}{\mathrm{GMD}^2 + (A+R)^2}} \tag{2.50}$$

The accuracy of Equation 2.49 depends on the correct choice of values for A, R and GMD. A and R are the geometric means of the coil radii, with $A = \sqrt{a_1 a_2}$ and $R = \sqrt{r_1 r_2}$. In practice, there is negligible error in taking A and R as the mean radii of the respective coils.

Self inductance of a circular coil of rectangular cross-section is a special case of Equation 2.49. In this case, GMD is interpreted as the geometric mean distance of the coil from itself and is given by Lyle [4]:

$$\ln(\mathrm{GMD}) = \ln(d) - \theta \tag{2.51}$$

where:

$$d = \sqrt{h^2 + w^2} \tag{2.52}$$

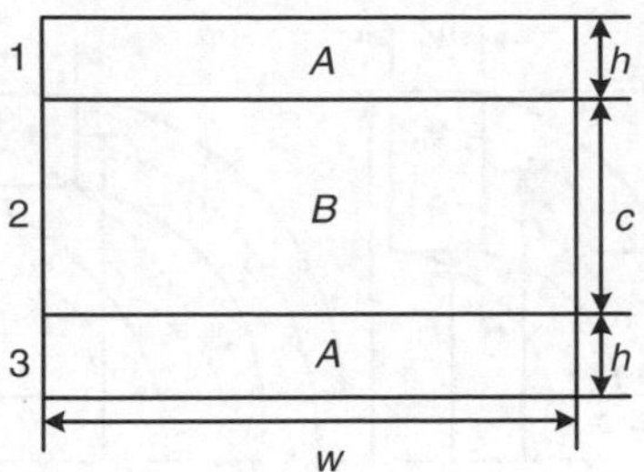

Figure 2.14 GMD between coils of equal area.

and:

$$\theta = \frac{u + v + 25}{12} - \frac{2}{3}(x + y) \tag{2.53}$$

with:

$$u = \frac{w^2}{h^2} \ln\left(\frac{d^2}{w^2}\right) \tag{2.54}$$

and:

$$v = \frac{h^2}{w^2} \ln\left(\frac{d^2}{h^2}\right) \tag{2.55}$$

and:

$$x = \frac{w}{h} \tan^{-1}\left(\frac{h}{w}\right) \tag{2.56}$$

and:

$$y = \frac{h}{w} \tan^{-1}\left(\frac{w}{h}\right) \tag{2.57}$$

A simple approximation for the GMD of a coil from itself is GMD $\approx 0.2235\ (h + w)$.

For coils of equal section, as shown in Figure 2.14, the GMD between the coils may be obtained using the formula for the GMD of a coil from itself.

We want to calculate the GMD between areas 1 and 3, each having an area $A = w \times h$. Area 2, between Sections 2.1 and 2.3, has an area $B = w \times c$. R_s is the GMD of the total area $(1 + 2 + 3)$ from itself. R_1 is the GMD of area 1 from itself (the same as R_2) and R_{12} is the

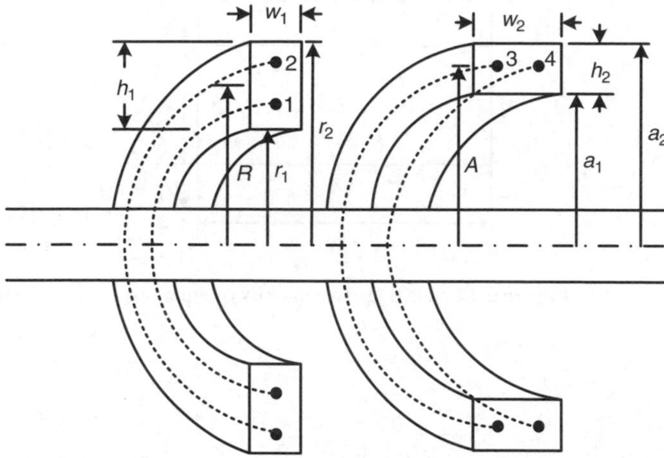

Figure 2.15 Lyle's method of equivalent filaments.

GMD of the combined area 1 and 2 from itself. It may be shown that R_{13}, the GMD between Sections 2.1 and 2.3, is given by Grover [6]:

$$\ln(R_{13}) = \frac{(2w+c)}{2w^2}\ln(R_s) - \ln(R_1) - \frac{c^2}{2w^2}\ln(R_2) - \frac{2c}{w}\ln(R_{12}) \tag{2.58}$$

where:

$$\begin{aligned} R_1 &= 0.2235(w+h) \\ R_2 &= 0.2235(w+c) \\ R_{12} &= 0.2235(w+h+c) \\ R_s &= 0.2235(w+2h+c) \end{aligned} \tag{2.59}$$

The elliptic integral formula can also be used for coils of unequal sections with Lyle's method [4], where each coil is replaced by two equivalent filaments, each carrying half the ampere-turns of the actual coil, as shown in Figure 2.15.

For $w < h$, the equivalent filaments are co-planar, with radii $r + \alpha$ and $r - \alpha$:

$$r = R\left(1 + \frac{w_1^2}{24R^2}\right) \tag{2.60}$$

$$\alpha = \sqrt{\frac{h_1^2 - w_1^2}{12}} \tag{2.61}$$

For $w > h$ the equivalent filaments have the same radius at a distance β on either side of the mid-plane of the coil.

$$r = A\left(1 + \frac{h_2^2}{24A^2}\right) \tag{2.62}$$

$$\beta = \sqrt{\frac{w_2^2 - h_2^2}{12}} \tag{2.63}$$

The mutual inductance is then given by:

$$M = N_1 N_2 \frac{M_{13} + M_{14} + M_{23} + M_{24}}{4} \tag{2.64}$$

where the M_{ij} terms are given by the Elliptic Integral Formula (2.45).

Example 2.5

For the sections shown in Figure 2.16, calculate the self inductance of Section 2.0 and the mutual inductance between Sections 2.1 and 2.2 and between Sections 2.1 and 2.4, with the dimensions shown.

The required dimensions for the mutual inductance calculations are:

Coils 1 and 2: $r_1 = a_1 = 1.15\,\text{mm}, r_2 = a_2 = 1.75\,\text{mm}, h_1 = h_2 = 15\,\text{mm}$
$z = 0.$
Coils 1 and 4: $r_1 = 2.00\,\text{mm}, r_2 = 2.60\,\text{mm}, h_1 = 15\,\text{mm},$
$a_1 = 2.00\,\text{mm}, a_2 = 2.60\,\text{mm}, h_2 = 15\,\text{mm}$
$z = 55\,\text{mm}.$

Formulae 2.49 and 2.58 may be used to find L_1 and M_{12}. Formula 2.64 may be used to find M_{14}.

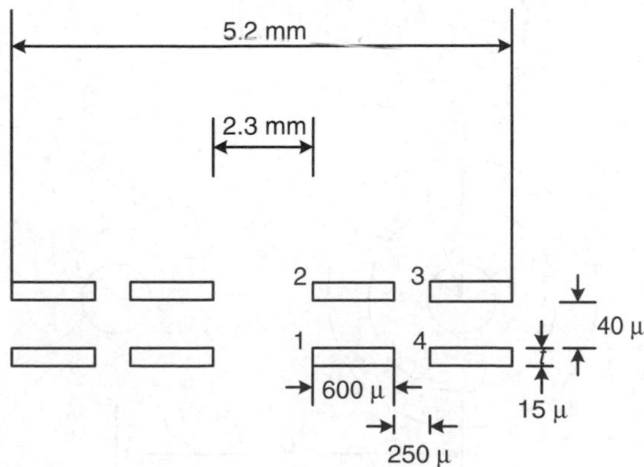

Figure 2.16 Planar coils.

The MATLAB program is listed at the end of this chapter, yielding the follow results:

$$L_1 = 4.316\,\text{nH}$$
$$M_{12} = 4.009\,\text{nH}$$
$$M_{14} = 2.195\,\text{nH}$$

Example 2.6

Calculate the inductance of a two-wire line consisting of a go-and-return circuit, as shown in Figure 2.17. The straight conductor is round, with a radius r_o, and the separation is D. Assume the length of the conductor is infinite in both directions.

In Example 1.1, we found expressions for the flux density inside and outside a straight round conductor. Consider an annular ring of inside radius r_1 and outside radius r_2 concentric with conductor 1 as shown.

The flux linking the annular element between r_1 and r_2 of length one metre is found from $B(r)$ in Equation 1.12:

$$d\phi = \frac{\mu_0 I}{2\pi r}\,dr$$

Integrating from r_1 to r_2 yields the flux linkage (in effect, the number of turns $N = 1$) due to the current I as:

$$\lambda_{12} = \int_{r_1}^{r_2} \frac{\mu_0 I}{2\pi r}\,dr = \frac{\mu_0 I}{2\pi}\ln\left(\frac{r_2}{r_1}\right)$$

The corresponding inductance per metre is then:

$$L_{12} = \frac{\mu_0}{2\pi}\ln\left(\frac{r_2}{r_1}\right)$$

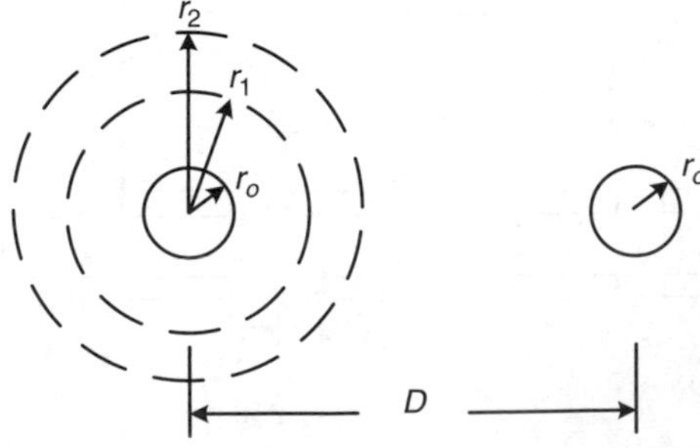

Figure 2.17 Two-wire transmission line.

This only accounts for the flux linkage outside the conductor. We need to add the contribution from the internal flux, based on the flux density derived in Example 1.1. Taking an annular section inside the conductor, the flux density due to the current inside the radius r is, by Equation 1.23:

$$B(r) = \frac{\mu_0 r}{2\pi r_o^2} I$$

The flux ϕ is:

$$d\phi = \frac{\mu_0 r I}{2\pi r_o^2} dr$$

In calculating the flux linkage, the number of turns is interpreted as the proportion of the total current that is linked ($N = 1$ represents the total current), i.e. $(\pi r^2 / \pi r_o^2)$. This yields the flux linkage as:

$$d\lambda = \frac{\mu_0 r^3 I}{2\pi r_o^4} dr$$

Performing the integration from $r = 0$ to $r = r_o$ results in the internal inductance per metre in terms of the exponential function:

$$L_{\text{int}} = \frac{\mu_0}{8\pi} = \frac{\mu_0}{2\pi} \ln\left(\varepsilon^{1/4}\right)$$

Combining the internal and external contributions gives a compact expression for the inductance at a radius r_2 outside a conductor of radius r_o:

$$L_{12} = \frac{\mu_0}{2\pi} \ln\left(\frac{r_2}{r_o \varepsilon^{-1/4}}\right) = \frac{\mu_0}{2\pi} \ln\left(\frac{r_2}{r'_o}\right)$$

Taking the return conductor at $r_2 = D$, the inductance resulting from the flux linking conductor 2 due to the current is 1 is given by this expression, with $r_2 = D$. The inductance resulting from the flux linking conductor 1 due to the current in conductor 2 is given by the same expression. The two components of flux linkage add together and the total inductance per unit length becomes:

$$L_{12} = \frac{\mu_0}{\pi} \ln\left(\frac{D}{r'_o}\right)$$

r'_o is called the geometric mean radius (GMR) or self Geometric Mean Distance (GMD) of the round conductor cross-section from itself.

We have already used GMD to calculate self and mutual inductance of circular coils. Strictly speaking, the concept only applies to straight conductors; it works for circular coils with rectangular cross-sections when the radius of curvature is much greater than the dimensions of the coil cross-section.

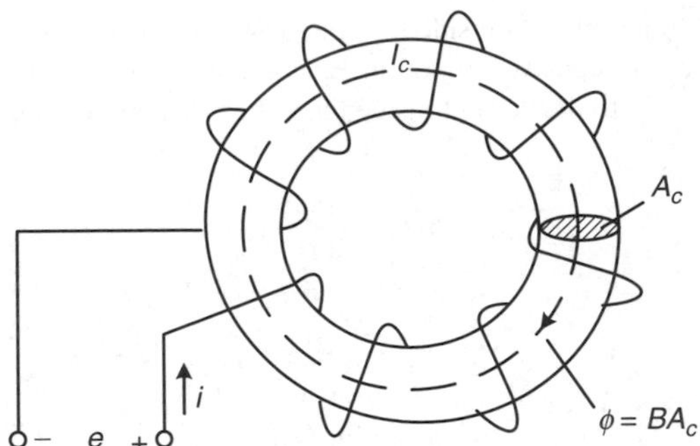

Figure 2.18 Circuit setup to measure hysteresis loss in a ferromagnetic material.

Example 2.7

Calculate the inductance of the coil on a toroidal core shown in Figure 2.18 of mean length l_c and with a cross-sectional area A_c. The relative permeability of the core material is μ_r.

The toroid is shown with N turns tightly wound over the whole core to ensure that there is no leakage of flux from the core.

We begin by applying Ampere's law along the contour of the mean path of the core:

$$H_c = \frac{NI}{l_c}$$

The corresponding flux linkage is:

$$\lambda = \frac{\mu_r \mu_0 N^2 A_c I}{l_c}$$

and the self inductance of the toroid is:

$$L = \frac{\mu_r \mu_0 N^2 A_c}{l_c}$$

2.5 Fringing Effects around the Air Gap

The expression for self inductance of a gapped core in Equation 2.27 assumes that there are no fringing effects around the air gap. In practice, the flux lines do not cross the gap in straight lines, but rather bulge out around the gap as shown in Figure 2.19. Fringing increases loss in neighbouring conductors, and the overall inductance is increased because the reluctance of the air gap is reduced.

This fringing of the flux field is a function of the geometry of the gap and can be evaluated by resorting to magnetic field plotting using finite element analysis tools.

In practice, a useful rule of thumb is to add the gap length to each dimension in the cross-section. For a rectangular cross-section with dimensions a by b, the cross-section would

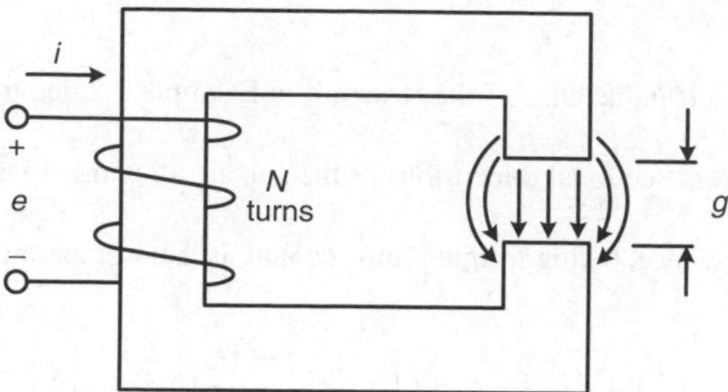

Figure 2.19 Fringing effect in an air gap.

become $(a+g)$ by $(b+g)$, where g is the length of the air gap and does not change. The overall inductance is then increased because the reluctance of the air gap is reduced:

$$L' = \frac{N^2}{\mathcal{R}_c + \mathcal{R}'_g} \approx \frac{N^2}{\mathcal{R}'_g} = L\frac{A'_g}{A_g} = L\frac{(a+g)(b+g)}{ab} \tag{2.65}$$

We have inherently assumed that $g \ll a$ or b in a gapped core, so we can neglect the g^2/ab term and the expression is reduced to:

$$L' \approx L\left(1 + \frac{a+b}{ab}g\right) \tag{2.66}$$

Finally, in a square cross-section, the self inductance is increased by fringing by the factor $(1+2g/a)$, so if g is 5% of a, then the overall inductance is estimated to increase by 10%. In the case of a round core, the diameter should be increased by g.

Referring to Figure 2.20, the reluctance term due to the fringing ($\mathcal{R}_{\text{fringing}}$) is parallel to the reluctance of the gap without fringing, and the parallel combination reduces the overall reluctance of the gap.

Fringing flux can introduce eddy currents into neighbouring conductors, so winding placement must be considered to reduce associated loss.

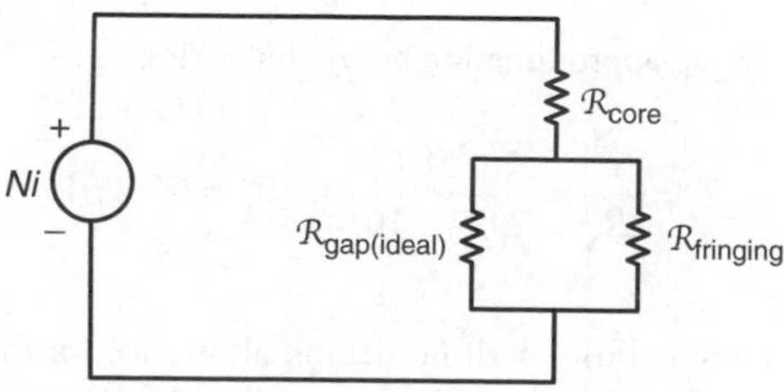

Figure 2.20 Equivalent magnetic circuit for fringing effects in an air gap.

Example 2.8

Estimate the increase in self inductance of the structure in Example 2.2 due to fringing effects around the air gap.

In Example 2.2, the cross-sectional dimensions of the gap are 12 mm × 15 mm and the gap length is 0.5 mm.

The new value of inductance, taking fringing into account, is from Equation 2.66:

$$L' \approx L\left(1 + \frac{a+b}{ab} g\right) = L\left(1 + \frac{12+15}{(12)(15)}(0.5)\right) = 1.075L$$

This represents an increase of 7.5% in the self inductance.

Let us look at this in more detail by considering four options:

1. Base inductance on the equivalent reluctance without fringing at the gap:

$$L = \frac{5^2}{2.438 \times 10^6} \times 10^6 = 10.3\ \mu\text{H}$$

2. Base inductance on the equivalent reluctance with fringing at the air gap. Taking the fringing into account, the reluctance of the air gap is:

$$\begin{aligned} \mathcal{R}'_g &= \frac{0.5 \times 10^{-3}}{(4\pi \times 10^{-7})(12.5)(15.5 \times 10^{-6})} \\ &= 2.05 \times 10^6\ \text{At/Wb} \end{aligned}$$

$$\mathcal{R}_{\text{eq}} = (0.228 + 2.05) \times 10^6 = 2.278 \times 10^6\ \text{At/Wb}$$

$$L = \frac{N^2}{\mathcal{R}'_{\text{eq}}} = \frac{5^2}{2.278 \times 10^6} \times 10^6 = 11.0\ \mu\text{H}$$

3. Base inductance on the air gap approximation without fringing:

$$L = \frac{N^2}{\mathcal{R}_g} = \frac{5^2}{2.21 \times 10^6} \times 10^6 = 11.3\ \mu\text{H}$$

4. Base inductance on the air gap approximation but include fringing:

$$L = \frac{N^2}{\mathcal{R}'_g} = \frac{5^2}{2.05 \times 10^6} \times 10^6 = 12.2\ \mu\text{H}$$

This example shows that there is little merit in making allowance for the fringing effect around the air gap while at the same time neglecting the reluctance of the core. Or, if we look at it in another way, the effect of the core on the inductance may be just as important at the fringing effect.

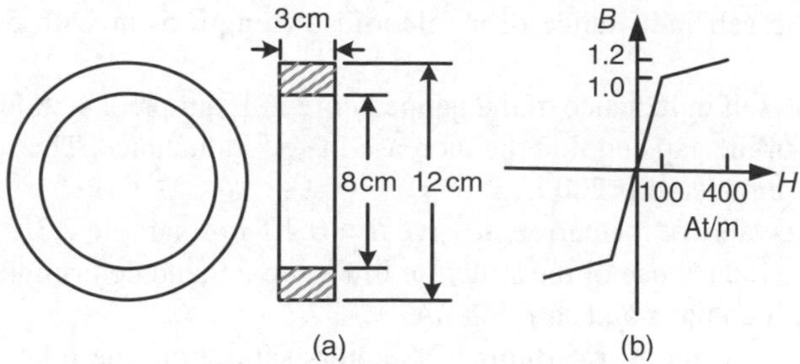

Figure 2.21 Problem 2.2 toroidal core.

2.6 Problems

2.1 List five factors that affect the inductance of a coil.

2.2 A toroidal core has the dimensions shown in Figure 2.21. It is wound with a coil having 100 turns. The *B-H* characteristic of the core may be represented by the linearized magnetization curve shown (*B* is in T, *H* in At/m).

(a) Determine the inductance of the coil if the flux density in any part of the core is below 1.0T.

(b) Determine the maximum value of current for the condition of part (a).

(c) Calculate the stored energy for the conditions in part (b).

2.3 The centre limb of the magnetic structure in Figure 2.22 is wound with 400 turns and has a cross-sectional area of 750 mm^2. Each of the outer limbs has a cross-sectional area of 450 mm^2. The air gap has a length of 1 mm. The mean lengths of the various magnetic paths are shown in Figure 2.22. The relative permeability of the magnetic material is 1500.

Calculate:

(a) The reluctances of the centre limb, the air gap and the two outer limbs;

(b) The current required to establish a flux of 1.4 mWb in the centre limb, assume there is no magnetic leakage or fringing;

(c) The energy stored in the air gap in (b).

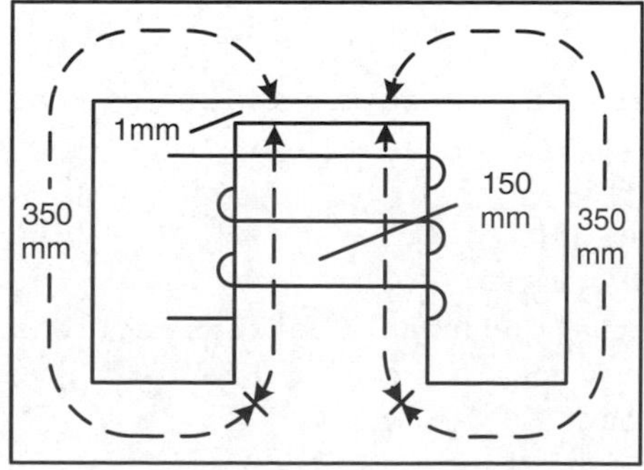

Figure 2.22 Problem 2.3 gapped core.

2.4 Calculate the self inductance of a solenoid of length 5 cm with 500 turns. Neglect fringing.
2.5 Calculate the self inductance of the gapped core in Example 2.4. Add 10% to the linear dimensions of the gap and find the increase in self inductance. The relative permeability of the core material is 2000.
2.6 Calculate the value of the current to give $B = 0.2\,\mathrm{T}$ in Example 2.3.
2.7 Calculate the inductance of the structure in Figure 2.9 and determine the stored energy for a coil with 25 turns and carrying 2A.
2.8 When the core material in Example 2.8 enters saturation, the relative permeability is reduced, which will counter the effect of increase inductance due to fringing. Calculate the value of the relative permeability to offset the increase the inductance.
2.9 Use Lyle's method to calculate M12 in Example 2.5.
2.10 Use the GMD method to calculate M14 in Example 2.5.

MATLAB Program for Example 2.5

```
%This MATLAB program is used to calculate the self and mutual
inductances
%in Example 2.5
%The parameters are shown in Figure 2.16
a1_in = 1150e-6;
a1_out = 1750e-6;
a2_in = 1150e-6;
a2_out = 1750e-6;
a4_in = 2000e-6;
a4_out = 2600e-6;
h1 = 15e-6;
h2 = 15e-6;
h4 = 15e-6;
w = 600e-6;
z12 = 55e-6;
c12 = z12-(h1+h2)/2;
z14 = 0;

%This section the self inductance of sections 1 is calculated R1=sqrt
(a1_in*a1_out);
GMD_1 = 0.2235*(w+h1);
L1 = inductance(R1,R1,GMD_1)

%This section the mutual inductance between sections 1 and 2 is
calculated
R2 = sqrt(a2_in*a2_out);
GMD_b = 0.2235*(w+c12);
GMD_a = 0.2235*(w+h1+c12);
```

```
GMD_s = 0.2235*(w+h1+h2+c12);
GMD_12 = (2*w+c12)^2/(2*w^2)*log(GMD_s)-log(GMD_1)-
((c12/w)^2)/2*log(GMD_b)-((2*c12)/w)*log(GMD_a);
GMD_12 = exp(GMD_12);
M_12 = inductance(R1,R2,GMD_12)

%This section the mutual inductance between sections 1 and 4 is
calculated
%The dimension calculation results with Lyle's method applied r1_lyle =
R1*(1+h1^2/(24*R1^2));
alpha1_lyle = sqrt((w^2-h1^2)/12);
R4 = sqrt(a4_in*a4_out);
r4_lyle = R4*(1+h4^2/(24*R4^2));
alpha4_lyle = sqrt((w^2-h4^2)/12);
r1_filament = r1_lyle-alpha1_lyle;
r2_filament = r1_lyle+alpha1_lyle;
r3_filament = r4_lyle-alpha4_lyle;
r4_filament = r4_lyle+alpha4_lyle;
M13 = inductance(r1_filament,r3_filament,z14);
M14 = inductance(r1_filament,r4_filament,z14);
M23 = inductance(r2_filament,r3_filament,z14);
M24 = inductance(r2_filament,r4_filament,z14);
M_14 = (M13+M14+M23+M24)/4

%File to define the function inductance

function M = inductance(a,r,GMD)
% This function is used to calculate the self and mutual inductance
applying the GMD method

uo = 4*pi*1e-7;
f = sqrt((4*a*r)/(GMD^2+(a+r)^2));
%MATLAB definition of ellipke integal uses square function
[K,E] = ellipke(f^2);
M = uo*(2*sqrt(a*r)/f)*((1-(f^2/2))*K-E);
end
```

References

1. Hurley, W.G. and Duffy, M.C. (1995) Calculation of self and mutual impedances in planar magnetic structures. *IEEE Transactions on Magnetics* **31** (4), 2416–2422.
2. Maxwell, J.C. (1881) *A Treatise on Electricity and Magnetism*, Clarendon Press, Oxford.
3. Gray, A. (1893) *Absolute Measurements in Electricity and Magnetism*, MacMillan, London.
4. Lyle, T.R. (1914) *Philosophical Transactions of the Royal Society of London. Series A, Containing Papers of a Mathematical or Physical Character*, The Royal Society, London.

5. Dwight, H.B. (1919) Some new formulas for reactance coils. *Transactions of the American Institute of Electrical Engineers* **XXXVIII** (2), 1675–1696.
6. Grover, F.W. (2004) *Inductance Calculations: Working Formulas and Tables*, Dover Publications Inc., New York.

Further Reading

1. Blume, L.F. (1982) *Transformer Engineering*, John Wiley & Sons, New York.
2. Bueno, M.D.A. (2001) *Inductance and Force Calculations in Electrical Circuits*, Nova Science Publishers, Huntington.
3. Del Vecchio, R.M., Poulin, B., Feghali, P.T. *et al.* (2001) *Transformer Design Principles: With Applications to Core-Form Power Transformers*, 1st edn, CRC Press, Boca Raton, FL.
4. Dowell, P.L. (1966) Effects of eddy currents in transformer windings. *Proceedings of the Institute of Electrical and Electronic Engineers* **113** (8), 1387–1394.
5. Erickson, R.W. (2001) *Fundamentals of Power Electronics*, 2nd edn, Springer, Norwell, MA.
6. Flanagan, W.M. (1992) *Handbook of Transformer Design and Application*, 2nd edn, McGraw-Hill, New York.
7. Georgilakis, P.S. (2009) *Spotlight on Modern Transformer Design (Power Systems)*, 1st edn, Springer, New York.
8. Goldberg, A.F., Kassakian, J.G., and Schlecht, M.F. (1989) Issues related to 1–10-MHz transformer design. *IEEE Transactions on Power Electronics* **4** (1), 113–123.
9. Kassakian, J.G., Schlecht, M.F., and Verghese, G.C. (1991) *Principles of Power Electronics (Addison-Wesley Series in Electrical Engineering)*, Prentice Hall, Reading, MA.
10. Kazimierczuk, M.K. (2009) *High-Frequency Magnetic Components*, John Wiley & Sons, Chichester.
11. Krein, P.T. (1997) *Elements of Power Electronics (Oxford Series in Electrical and Computer Engineering)*, Oxford University Press, Oxford.
12. Kulkarni, S.V. (2004) *Transformer Engineering: Design and Practice*, 1st edn, CRC Press, New York.
13. Kusko, A. (1969) *Computer-aided Design of Magnetic Circuits*, The MIT Press, Cambridge, MA.
14. B.H.E. Limited (2004) *Transformers: Design, Manufacturing, and Materials (Professional Engineering)*, 1st edn, McGraw-Hill, New York.
15. McLachlan, N.W. (1955) *Bessel Functions for Engineers*, 2nd edn, Clarendon Press, Oxford.
16. E.S. MIT (1943) *Magnetic Circuits and Transformers (MIT Electrical Engineering and Computer Science)*, The MIT Press, Cambridge, MA.
17. Niemela, V.A., Owen, H.A. Jr, and Wilson, T.G. (1990) Cross-coupled-secondaries model for multiwinding transformers with parameter values calculated from short-circuit impedances. *Proceedings of the IEEE Power Electronics Specialists Conference, PESC*, pp. 822–830.
18. Niemela, V.A., Skutt, G.R., Urling, A.M. *et al.* (1989) Calculating the short-circuit impedances of a multiwinding transformer from its geometry. *Proceedings of the IEEE Power Electronics Specialists Conference, PESC*, pp. 607–617.
19. Paul, C.R. (2010) *Inductance: Loop and Partial*, John Wiley & Sons, Hoboken, NY.
20. Pressman, A.I., Bellings, K., and Morey, T. (2009) *Switching Power Supply Design*, 3rd edn, McGraw-Hill, New York.
21. Ramo, S., Whinnery, J.R., and Van Duzer, T. (1965) *Fields and Waves in Communications Electronics*, John Wiley & Sons, New York.
22. Smith, B. (2009) *Capacitors, Inductors and Transformers in Electronic Circuits (Analog Electronics Series)*, Wexford College Press, Wexford.
23. Snelling, E.C. (1988) *Soft Ferrites: Properties and Applications*, 2nd edn, Butterworths, London.
24. Van den Bossche, A. (2005) *Inductors and Transformers for Power Electronics*, 1st edn, CRC Press, New York.
25. Van den Bossche, A., Valchev, V., and Filchev, T. (2002) Improved approximation for fringing permeances in gapped inductors. *Proceedings of the IEEE Industry Applications Conference, IAS*, pp. 932–938.
26. Wilcox, D.J., Conlon, M., and Hurley, W.G. (1988) Calculation of self and mutual impedances for coils on ferromagnetic cores. *IEE Proceedings A, Physical Science, Measurement and Instrumentation, Management and Education* **135** (7), 470–476.
27. Wilcox, D.J., Hurley, W.G., and Conlon, M. (1989) Calculation of self and mutual impedances between sections of transformer windings. *Generation, Transmission and Distribution, IEE Proceedings C* **136** (5), 308–314.

3

Inductor Design

In Chapter 2, we described the fundamentals of inductance, and this will form the basis of the design of practical inductors. Essentially, inductors can have a discrete gap, typically in a laminated or ferrite core, or a distributed gap in an iron powder core. The design outlined in this chapter will apply to both types of cores. The approach is to adopt the concept of effective permeability, as enunciated in Chapter 2. The physical core selection is based on the energy stored in the inductor, the maximum flux density in the core and the temperature rise in the inductor. The core selection is followed by the winding design including turns and wire size.

3.1 The Design Equations

3.1.1 Inductance

The inductance of a coil with N turns wound on a core of length l_c and cross section A_c with effective relative permeability μ_{eff} to account for the gap (discrete or distributed) is given by Equation 2.26:

$$L = \frac{\mu_{\text{eff}} \mu_0 N^2 A_c}{l_c} \tag{3.1}$$

3.1.2 Maximum Flux Density

Applying Ampere's law to the closed loop around a core of length l_c with N turns establishes the relationship between the magnetic field intensity and the current in the core:

$$H_{\max} = \frac{N\hat{I}}{l_c} \tag{3.2}$$

Transformers and Inductors for Power Electronics: Theory, Design and Applications, First Edition.
W. G. Hurley and W. H. Wölfle.

Here, $\hat{I}$ is the peak value of current. For simplicity, we will relate it to the rms value of the current waveform:

$$I_{\text{rms}} = K_i \hat{I} \tag{3.3}$$

where K_i is the current waveform factor.

H_{max} is related to the maximum flux density, so:

$$B_{\text{max}} = \mu_{\text{eff}} \mu_0 H_{\text{max}} = \frac{\mu_{\text{eff}} \mu_0 N \hat{I}}{l_c} \tag{3.4}$$

The effective relative permeability in Equation 3.4 includes the effect of an air-gap as described in Section 2.1.

By rearranging Equation 3.4 we can obtain a relationship for $\hat{I}$ in terms of the maximum flux density in the core:

$$\hat{I} = \frac{B_{\text{max}} l_c}{\mu_{\text{eff}} \mu_0 N} \tag{3.5}$$

B_{max} is limited by the saturation flux density, B_{sat} for the core material. Typical values for different materials are listed in Table 1.1.

3.1.3 *Winding Loss*

The resistive or I^2R loss P_{cu} in a winding is

$$P_{\text{cu}} = \rho_w \frac{l_w}{A_w} I_{\text{rms}}^2 = \rho_w \frac{N\,\text{MLT}\left(K_i \hat{I}\right)^2}{A_w} = \rho_w \frac{N^2\,\text{MLT}\left(K_i \hat{I}\right)^2}{N A_w} \tag{3.6}$$

The resistivity of the conductor is ρ_w and the length of the conductor in the winding is l_w, which is the product of the number of turns N and the mean length of a turn (MLT). The rms value of the coil current is related to the peak current by K_i as before.

Extracting $\hat{I}$ from Equation 3.6:

$$\hat{I} = \frac{1}{N K_i} \sqrt{\frac{P_{\text{cu}} N A_w}{\rho_w \text{MLT}}} \tag{3.7}$$

The maximum dissipation P_D in the core is related to the temperature rise and the heat transfer from the surface of the inductor to the surroundings. This relationship is normally described by the thermal resistance:

$$\Delta T = R_\theta P_D \tag{3.8}$$

This relationship will be examined in more detail in Section 3.2, but for now P_D represents the maximum allowable dissipation in the inductor. In choosing the best core for the application, the choice is ultimately limited by P_D, which is the limiting value in Equation 3.7. Equally, B_{sat} is the limiting value of flux density in Equation 3.5.

3.1.4 Optimum Effective Permeability

The maximum energy stored in an inductor is $\frac{1}{2}L\hat{I}^2$. Combining Equations 3.1 and 3.5 gives an expression for the stored energy in terms of B_{max}:

$$\frac{1}{2}L\hat{I}^2 = \frac{1}{2}\frac{A_c l_c}{\mu_{eff}\mu_0}B_{max}^2 \tag{3.9}$$

Combining Equations 3.1 and 3.7 yields another expression for the stored energy in terms of copper loss:

$$\frac{1}{2}L\hat{I}^2 = \frac{1}{2}\frac{\mu_{eff}\mu_0 A_c N A_w}{\rho_w \text{ MLT } K_i^2 l_c}P_{cu} \tag{3.10}$$

Both Equations 3.9 and 3.10 express the stored energy in terms of the effective permeability of the core, the maximum flux density in the core and the winding dissipation. All of the other parameters are derived from the physical dimensions of the core. Figure 3.1 shows the stored energy as a function of both B_{max} and P_{cu}.

The design space of the inductor is bounded by $L\hat{I}^2$, P_{cumax} and B_{sat}. The curves for $L\hat{I}^2$, as a function of P_{cu} and B_{max} respectively, are drawn for three values of μ_{eff}.

The solid lines in Figure 3.1 (point 1) show the stored energy for the optimum value of the effective permeability, which is the maximum stored energy for that particular core. The core is operating at its maximum permissible flux density and its maximum permissible dissipation. The application may require less stored energy. Point 2 represents a design at which the

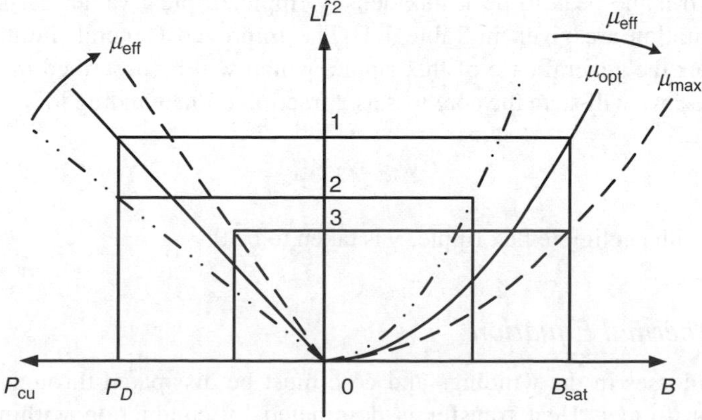

Figure 3.1 Stored energy as a function of flux density and dissipation.

current is set at the maximum dissipation. The core is operating at a lower value of flux density at point 2, and μ_{eff} is smaller than the optimum value. On the other hand, at point 3, the core is operating at the maximum allowed value of flux density but the dissipation is less than the maximum permissible value. This means that μ_{eff} is greater than the optimum value.

The design space in Figure 3.1 shows that increasing μ_{eff} above the optimum value while maintaining maximum dissipation will cause the core to go into saturation. Reducing μ_{eff} below the optimum value while maintaining the maximum flux density will cause the core to overheat.

Equating the expressions for $\hat{I}$ in Equations 3.5 and 3.7, using the maximum values of B_{sat} and P_{cumax} for $B_{\max}$ and P_{cu} respectively, shows that there is an optimum value of μ_{eff}, to be designated μ_{opt}:

$$\mu_{\text{opt}} = \frac{B_{\text{sat}} l_c K_i}{\mu_0 \sqrt{\dfrac{P_{\text{cu max}} N A_w}{\rho_w \text{MLT}}}} \tag{3.11}$$

The expression for μ_{opt} in Equation 3.11 shows that for a given core with a known window winding area and mean turn length, the value of the optimum value of the effective permeability is determined by the maximum flux density B_{sat} in the core and the maximum dissipation P_D.

3.1.5 Core Loss

In inductor design, the core or iron loss is often negligible compared to the winding loss. This is a common situation where there is a small ripple in the presence of a DC current. The iron loss occurs as a result of the ripple current and, as a first estimation, may be treated with the Steinmetz equation. The time-average core loss per unit volume is usually stated as:

$$P_{\text{fe}} = K_c f^{\alpha} \left(\frac{\Delta B}{2}\right)^{\beta} \tag{3.12}$$

In this case, ΔB is the peak to peak flux density ripple, Typical values for k_c, α and β for sinusoidal excitation are given in Table 1.1. The Improved General Steinmetz Equation (iGSE) [1] treats the general case of flux ripple, which will be described in Chapter 7. For present purposes, we will state the core loss as a fraction of the winding loss:

$$P_{\text{fe}} = \gamma P_{\text{cu}} \tag{3.13}$$

In an inductor with negligible flux ripple, γ is taken to be 0.

3.1.6 The Thermal Equation

The combined losses in the windings and core must be dissipated through the surface of the wound transformer. Heat transfer is dominated by conduction within the core and winding, and from the transformer surface by convection. Newton's equation of convection relates heat flow to temperature rise (ΔT), surface area (A_t) and the coefficient of heat transfer h_c, by:

$$Q = h_c A_t \Delta T \tag{3.14}$$

where Q represents the total power loss, i.e. the combined winding loss and the core loss.

In terms of an equivalent circuit analogy with Ohm's law, the heat flow is linearly proportional to the temperature difference and inversely proportional to the thermal resistance, so:

$$\Delta T = R_\theta Q = \frac{1}{h_c A_t} Q \tag{3.15}$$

In the electrical analogy, Q represents current (and not power), and ΔT represents the potential difference. Manufacturers will sometimes provide the thermal resistance for a core, but normally the value is based on empirical data. For example, the thermal resistance may be related to the volume of the core V_c:

$$R_\theta = \frac{0.06}{\sqrt{V_c}} \tag{3.16}$$

In this empirical equation, R_θ is in °C/W for V_c in m^3.

There are many empirical formulas for estimating the h_c value in natural convection for different configurations. For a vertical object of height H, one such formula is [2]:

$$h_c = 1.42 \left[\frac{\Delta T}{H} \right]^{0.25} \tag{3.17}$$

For an ETD55 core, $H = 0.045$ m and $h_c = 8.2$ W/m^2°C for a 50 °C temperature rise. Evidently, the position of the transformer relative to other components will have a profound effect on the value of h_c. In fact, the value of h_c is probably the most uncertain parameter in the entire design. However, the typical value of $h_c = 10$ W/m^2°C is often used for cores encountered in switching power supplies [3]. In the case of forced convection with fan-assisted airflow, the value of h_c will be much higher – in the range of 10–30 W/m^2 °C.

3.1.7 *Current Density in the Windings*

The window utilization factor, k_u, is defined as the ratio of the total conduction area W_c, for all conductors in all windings to the total window winding area W_a of the core:

$$k_u = \frac{W_c}{W_a}. \tag{3.18}$$

The window utilization factor may vary from 0.2–0.8. In a core with a bobbin, it is possible to wind the coil very tightly, and k_u may be as high as 0.8. On the other hand, in the case of a toroidal core, automatic winding may necessitate a large free window winding area for the winding arm, and k_u may be as low as 0.2. Insulation spacing to achieve creepage and clearance distances required by the relevant safety standards are included in the specification of the window utilization factor.

The total conduction area may be expressed in terms of the individual conductor area and the number of turns:

$$W_c = NA_w = k_u W_a \tag{3.19}$$

Substituting Equation 3.19 into Equation 3.11 yields an expression for the optimum value of the effective relative permeability μ_{opt}, i.e. point 1 in Figure 3.1:

$$\mu_{\text{opt}} = \frac{B_{\text{sat}} l_c K_i}{\mu_0 \sqrt{\dfrac{P_{\text{cu max}} k_u W_a}{\rho_w \text{MLT}}}} \tag{3.20}$$

In this form, the optimum value of the effective relative permeability is a function of the saturation flux density and the maximum dissipation.

The current density in the winding is:

$$J_o = \frac{I_{\text{rms}}}{A_w} \tag{3.21}$$

Noting that the volume of the windings (fully wound $k_u = 1$) is $V_w = \text{MLT} \times W_a$, then the copper loss given by Equation 3.6, using Equation 3.19 with Equations 3.21 and 3.3, may be expressed as:

$$P_{\text{cu}} = \rho_w \frac{N^2\, \text{MLT}(J_o A_w)^2}{NA_w} = \rho_w V_w k_u J_o^2 \tag{3.22}$$

Combining Equations 3.13 and 3.22 in Equation 3.14:

$$Q = P_{\text{cu}} + P_{\text{fe}} = (1+\gamma)\left[\rho_w V_w k_u J_0^2\right] = h_c A_t \Delta T \tag{3.23}$$

Extracting the current density:

$$J_o = \sqrt{\frac{1}{1+\gamma} \frac{h_c A_t \Delta T}{\rho_w V_w k_u}} \tag{3.24}$$

3.1.8 Dimensional Analysis

The physical quantities V_w, V_c and A_t may be related to the product of the core window winding area and the cross-sectional area (core window area product). It follows by dimensional analysis that:

$$V_w = k_w A_p^{\frac{3}{4}} \tag{3.25}$$

$$V_c = k_c A_p^{\frac{3}{4}} \tag{3.26}$$

$$A_t = k_a A_p^{\frac{1}{2}} \tag{3.27}$$

The coefficients k_w, k_c and k_a are dimensionless. The exponents of A_p are chosen so that the dimensions are consistent; for example V_w has the dimensions of m^3 when A_p has the dimensions of m^4, and so on. The values of k_a, k_c and k_w vary for different types of cores [4]. However, the combinations required for the inductor design are approximately constant. Further accuracy in establishing these general constants is somewhat offset by the lack of accuracy around the heat transfer coefficient in Equation 3.14.

Based on extensive studies of several core types and sizes, it was found that typically $k_a = 40$, $k_c = 5.6$ and $k_w = 10$. An exception is the pot core, for which k_w is typically 6. Of course, it is perfectly reasonable and straightforward to carry out a detailed study of a particular core type, or for the cores of a particular manufacturer to establish more accurate values of these coefficients.

The dimensional analysis resulting in Equations 3.25 and 3.27, substituted into Equation 3.24, gives an expression for the current density in terms of the temperature rise in the windings and the core-window winding area product:

$$J_o = K_t \sqrt{\frac{\Delta T}{k_u(1+\gamma)}} \frac{1}{\sqrt[8]{A_p}} \tag{3.28}$$

where

$$K_t = \sqrt{\frac{h_c k_a}{\rho_w k_w}}. \tag{3.29}$$

Inserting typical values: $\rho_w = 1.72 \times 10^{-8}$ Ω-m, $h_c = 10$ W/m^2°C, $k_a = 40$, $k_w = 10$ gives $K_t = 48.2 \times 10^3$, A_p is in m^4 and the current density is in A/m^2.

3.2 The Design Methodology

From Section 2.3, Equation 2.33, the energy stored in an inductor with a gap of length g is:

$$W_m = \frac{B^2 V_c}{2\mu_r \mu_0} + \frac{B^2 V_g}{2\mu_0}$$

The volume of the core is $V_c = A_c l_c$ and the volume of the gap is $V_g = A_g g$. Adopting the notation from Section 3.1, and assuming that the cross-sectional area of the gap is equal to the cross-sectional area of the core, then:

$$\frac{1}{2}L\hat{I}^2 = \frac{B_{\max}^2 A_c}{2\mu_0}\left[\frac{l_c}{\mu_r} + g\right] \tag{3.30}$$

Ampere's law for the gapped core is:

$$Ni = H_c l_c + H_g g = \frac{B}{\mu_0}\left[\frac{l_c}{\mu_r} + g\right] \tag{3.31}$$

For peak current $\hat{I}$ and for maximum flux density $B_{\max}$, Equation 3.31 becomes, after rearranging:

$$\frac{l_c}{\mu_r} + g = \frac{N\hat{I}}{\dfrac{B_{\max}}{\mu_0}} \tag{3.32}$$

Substituting the result in Equation 3.32 into Equation 3.30 yields:

$$\frac{1}{2}L\hat{I}^2 = \frac{1}{2}B_{\max}A_c N\hat{I} \tag{3.33}$$

Invoking the definition for current density in Equation 3.21 with Equation 3.3, together with Equation 3.19:

$$L\hat{I}^2 = \frac{B_{\max}A_c k_u J_o W_a}{K_i} \tag{3.34}$$

Taking J_o from Equation 3.28 yields an expression for the core window winding area × cross sectional area product A_p:

$$A_p = \left[\frac{\sqrt{1+\gamma}K_i L\hat{I}^2}{B_{\max}K_t\sqrt{k_u \Delta T}}\right]^{\frac{8}{7}} \tag{3.35}$$

The A_p value determines the selection of the core. The parameters of the selected core determine μ_{opt} given by Equation 3.11, and the design may proceed from there.

The overall design methodology is shown in flow-chart form in Figure 3.2. The core manufacturer normally supplies the core data: cross-section, A_c, core length l_c, window winding area, W_a, the mean length of a turn, MLT and the core volume V_c.

The selected core from standard designs may not correspond exactly to the value of A_p given by Equation 3.35, and therefore the actual current density based on the selected core may be calculated using Equation 3.28.

There are two basic designs: a core with a discrete gap or a core with a distributed gap.

The effective permeability for the core with the discrete gap determines the gap length; the manufacturer normally supplies cores with pre-selected gaps. In the case of cores with

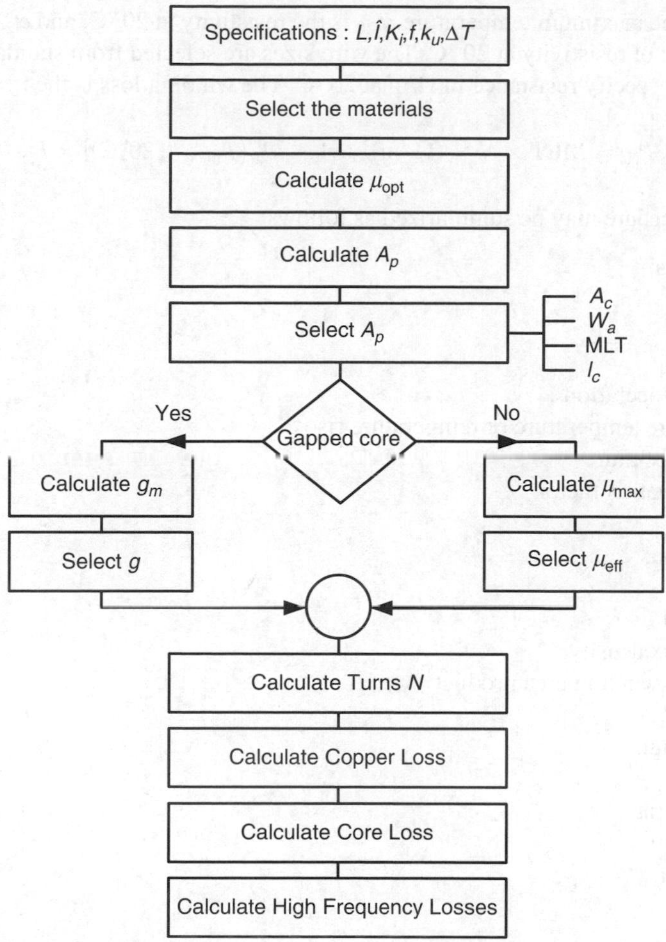

Figure 3.2 Flow chart of design process.

distributed gaps, the manufacturer normally supplies a set of cores with pre-selected values of effective permeability. The correct choice for gap length or effective permeability will be illustrated in the designs to follow.

The next step is to calculate the number of turns N in the winding. The manufacturer will normally supply the A_L value of the core for the corresponding gap length, i.e. the inductance per turn:

$$N = \sqrt{\frac{L}{A_L}} \tag{3.36}$$

Alternatively, it may be supplied in the form of inductance per 1000 turns.

The following step is to select the conductor, based on the current density.

The resistivity of the conductor at the maximum operating temperature is given by:

$$\rho_w = \rho_{20}[1 + \alpha_{20}(T_{\max} - 20°\text{C})], \tag{3.37}$$

where T_{max} is the maximum temperature, ρ_{20} is the resistivity at 20 °C, and α_{20} is the temperature coefficient of resistivity at 20 °C. The wire sizes are selected from standard wire tables, which normally specify resistance in Ω/m at 20 °C. The winding loss is then:

$$P_{cu} = \text{MLT} \times N \times (\Omega/\text{m}) \times [1 + \alpha_{20}(T_{max} - 20°\text{C})] \times I^2. \qquad (3.38)$$

The design procedure may be summarized as follows:

Specifications:

- Inductance
- DC current
- Frequency of operation
- Maximum core temperature or temperature rise
- Ambient temperature
- Window utilization factor

Core selection:

- Core material
- Maximum flux density
- Core window winding area product

Winding design:

- Number of turns
- Current density
- Wire selection

Losses:

- Copper loss
- Winding loss if applicable

In Chapter 6, we will refine the design to take high-frequency skin and proximity effects into account.

3.3 Design Examples

3.3.1 Example 3.1: Buck Converter with a Gapped Core

Design an inductor for a buck converter with the specifications listed in Table 3.1.

Circuit Parameters

The circuit for a buck converter is shown in Figure 3.3 and the associated current and voltage waveforms are shown in Figure 3.4.

Table 3.1 Specifications

Input voltage	12 V
Output voltage	6 V
Inductance	34 μH
DC Current	20 A
Frequency, f	80 kHz
Temperature rise, ΔT	15 °C
Ambient temperature, T_a	70 °C
Window utilization factor	0.8

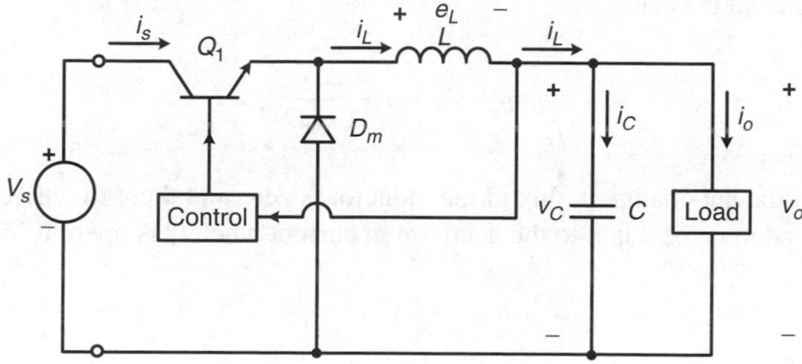

Figure 3.3 Buck converter circuit diagram.

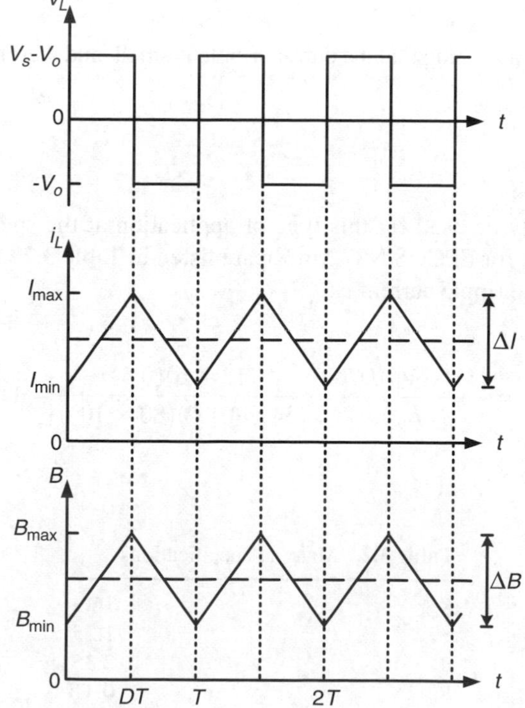

Figure 3.4 Buck converter voltage and current waveforms.

The basic operation of the buck converter is treated in power electronics text, and the main features are summarized here for clarity. In the time period between 0 and DT, the switch Q is closed and the diode D is open and the voltage $V_i - V_o$ is applied across the inductor. Between DT and T the applied voltage across the inductor is – V_o, resulting in the current waveform in Figure 3.4 with the current ripple ΔI.

When Q is closed and D is open:

$$v_L = V_s - V_o = L\frac{\Delta I_{L+}}{DT}$$

When Q is open and D is closed:

$$v_L = -V_o = L\frac{\Delta I_{L-}}{(1-D)T}$$

In steady state, the net change in flux in the inductor is zero and there the increase in current while Q is closed must be equal to the decrease in current when Q is open. It follows that:

$$|\Delta I_{L+}| = |\Delta I_{L-}| = \Delta I$$

and:

$$V_o = DV_s$$

In this example, we are assuming that the current ripple is small, and therefore γ in Equation 3.13 is negligible.

Core Selection

Ferrite would normally be used for this type of application at the specified frequency. The material specifications for EPCOS N87 Mn-Zn are listed in Table 3.2.

The amplitude of the ripple current is:

$$\Delta I_L = \frac{(V_s - V_o)DT}{L} = \frac{(12-6)(0.5)}{(34\times 10^{-6})(80\times 10^{3})} = 1.1\text{ A}$$

Table 3.2 Material specifications

K_c	16.9
α	1.25
β	2.35
B_{sat}	0.4T

Table 3.3 Core and winding specifications

A_c	2.09 cm^2
l_c	11.4 cm
W_a	2.69 cm^2
A_p	5.62 cm^4
V_c	23.8 cm^3
k_f	1.0
k_u	0.8
K_i	1.0
MLT	8.6 cm
ρ_{20}	1.72 $\mu\Omega$-cm
α_{20}	0.00393

The peak current is:

$$\hat{I} = I_{\text{dc}} + \frac{\Delta I}{2} = 20.0 + \frac{1.1}{2} = 20.55\text{ A}$$

$$L\hat{I}^2 = (34 \times 10^{-6})(20.55)^2 = 0.0144\text{ J}$$

For Mn-Zn ferrite, select $B_{\text{max}} = 0.25$T.

This type of core will allow a very tight winding, so therefore the window utilization factor is set to $k_u = 0.8$.

A_p from Equation 3.35 with $\gamma = 0$ is:

$$A_p = \left[\frac{\sqrt{1+\gamma}K_i L\hat{I}^2}{B_{\text{max}}K_t\sqrt{k_u \Delta T}}\right]^{8/7} = \left[\frac{0.0144}{(0.25)(48.2 \times 10^3)\sqrt{(0.8)(15)}}\right]^{8/7} \times 10^8 = 4.12\text{ cm}^4$$

The ETD49 core is suitable. The core specifications are given in Table 3.3.

The thermal resistance of this core according to the manufacturer is 11 °C/W. Therefore, the maximum dissipation given by Equation 3.8 is:

$$P_D = \frac{\Delta T}{\text{R}_\theta} = \frac{15}{11} = 1.36\text{ W}$$

The optimum value of the effective permeability for this core is given by Equation 3.20:

$$\mu_{\text{opt}} = \frac{B_{\text{max}} l_c K_i}{\mu_0 \sqrt{\dfrac{P_{\text{cu max}} k_u W_a}{\rho_w \text{MLT}}}} = \frac{(0.25)(11.4 \times 10^{-2})(1.0)}{(4\pi \times 10^{-7})\sqrt{\dfrac{(1.36)(0.8)(2.69 \times 10^{-4})}{(1.72 \times 10^{-8})(8.6 \times 10^{-2})}}} = 51$$

Winding Design

Gap g

Referring to Figure 3.1, increasing the value of effective permeability above the optimum value means that we can operate the core at its maximum flux density and below the maximum dissipation. This makes sense, since we have not included any core loss at this point.

The maximum gap length is found from Section 2.1:

$$g_{\max} = \frac{l_c}{\mu_{\min}} = \frac{11.4 \times 10^{-2}}{51} \times 10^3 = 2.24\ \text{mm}$$

The manufacturer provides a standard core set with $g = 2$ mm, a corresponding A_L value of 188 nH and an effective permeability of 81.

Turns

For the core, the manufacturer supplies the A_L value (i.e. the inductance per turn):

$$N = \sqrt{\frac{L}{A_L}} = \sqrt{\frac{(34 \times 10^{-6})}{(188 \times 10^{-9})}} = 13.5\ \text{turns}$$

Select 13 turns.

Wire Size

The current density is:

$$J_o = K_t \frac{\sqrt{\Delta T}}{\sqrt{k_u(1+\gamma)}\sqrt[8]{A_p}} = 48.2 \times 10^3 \frac{\sqrt{(15)}}{\sqrt{(0.8)}\sqrt[8]{(5.62 \times 10^{-8})}} \times 10^{-4} = 168\ \text{A/cm}^2$$

The cross-sectional area of the conductor is (for the purposes of this calculation, we are neglecting the current ripple):

$$A_w = I_{\text{rms}}/J = 20/168 = 0.119\ \text{cm}^2$$

An 8 mm × 2 mm wire meets this specification, with a DC resistance of 10.75×10^{-6} Ω/cm at 70 °C.

Copper Loss

Use Equation 3.37 to correct the winding resistance for temperature:

$$T_{\max} = 70 + 15 = 85\ ^\circ\text{C}.$$

$$R_{\text{dc}} = (13)(8.6)(10.75 \times 10^{-6})[1 + (0.00393)(85 - 20)] \times 10^3 = 1.51\ \text{m}\Omega$$

The copper loss is:

$$P_{cu} = R_{dc} I_{rms}^2 = (1.51 \times 10^{-3})(20.0)^2 = 0.604 \text{ W}$$

Core Loss

The flux density ripple ΔB may be calculated by using Faraday's law, noting that the voltage across the inductor in the time period 0 to DT is $(V_i - V_o)$:

$$\Delta B = \frac{(V_i - V_o)DT}{NA_c} = \frac{(12-6)(0.5)}{(13)(2.09 \times 10^{-4})(80 \times 10^3)} = 0.014 \text{ T}$$

Using the General Steinmetz Equation (GSE) (Equation 1.29), the peak value of the flux density B_{max} is $\Delta B/2$ and the core loss is:

$$P_{fe} = V_c K_c f^{\alpha} B_{max}{}^{\beta} = (23.8 \times 10^{-6})(16.9)(80\,000)^{1.25}(0.014/2)^{2.35} = 0.005 \text{ W}$$

As expected, the core loss is much smaller than the copper loss and well within the margins allowed by our core selection.

Total Losses :	Copper loss	0.604 W
	Core loss	0.005 W
	Total losses	0.609 W

3.3.2 *Example 3.2: Forward Converter with a Toroidal Core*

Specifications

Design an inductor for a forward converter with the specifications listed in Table 3.4.

Circuit Parameters

The circuit for a forward converter is shown in Figure 3.5 and the associated current and voltage waveforms are shown in Figure 3.6.

Table 3.4 Specifications

Input voltage	12 V
Output voltage	9 V
Inductance	1.9 mH
DC Current	2.1 A
Frequency, f	60 kHz
Temperature rise, ΔT	20 °C
Ambient temperature, T_a	60 °C
Window utilization factor	0.2

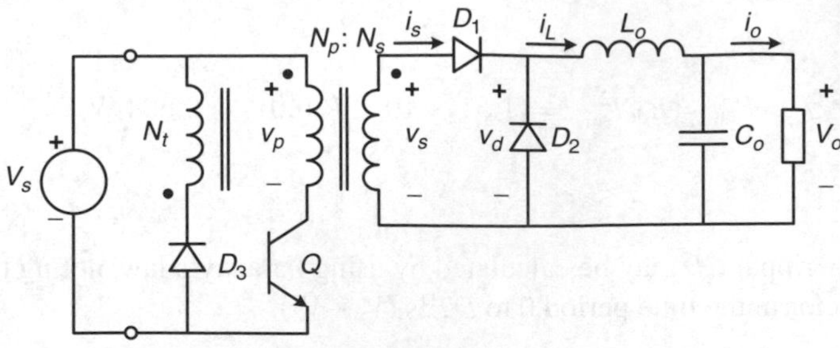

Figure 3.5 Forward converter circuit diagram.

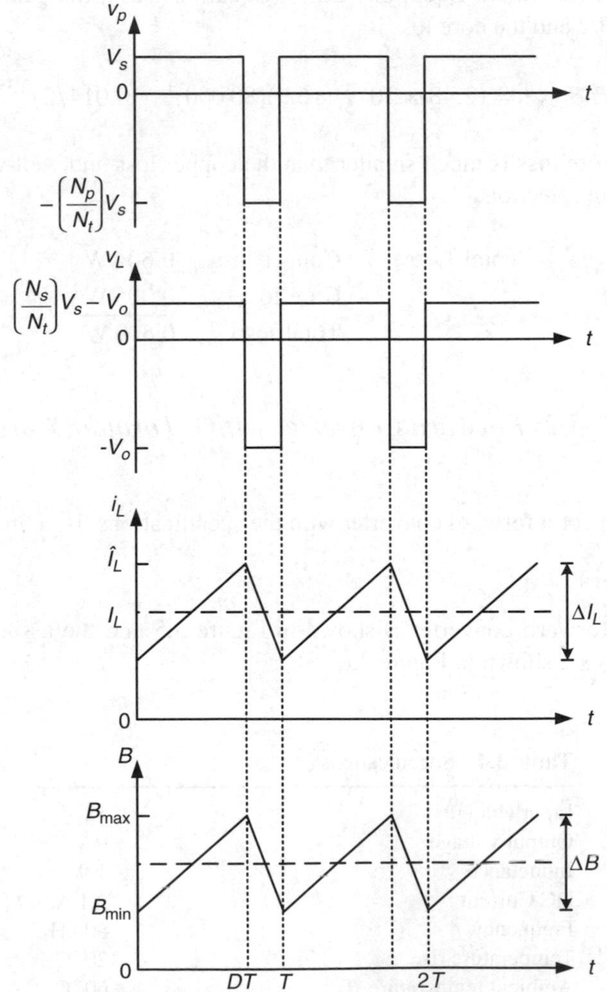

Figure 3.6 Forward converter current and flux waveforms.

Table 3.5 Material specifications

K_c	231.8
α	1.41
β	2.56
B_{sat}	0.5 T

The operation of the forward converter is very similar to the buck converter, with electrical isolation provided by the transformer. We will examine its operation in detail in Chapter 5 (transformer design). For present purposes, we are interested in the output inductor and we will assume the transformer turns ratio is 1: 1. From the analysis of the buck converter in Example 3.1:

$$V_o = DV_i,$$

The duty cycle in this case is $D = 9/12 = 0.75$.

We assume the flux ripple in the output inductor is negligible, therefore $\gamma = 0$.

Core Selection

In this case, we select an MPP powder iron core. The material specifications for the MPP core are listed in Table 3.5.

The amplitude of the ripple current is:

$$\Delta I_L = \frac{(V_s - V_o)DT}{L} = \frac{(12-9)(0.75)}{(1.6\times 10^{-3})(60\times 10^3)} = 0.0234\ \text{A}$$

$$I_{\text{peak}} = I_{\text{dc}} + \frac{\Delta I}{2} = 1.9 + \frac{0.0234}{2} = 1.912\ \text{A}$$

$$L\hat{I}^2 = (1.6\times 10^{-3})(1.912)^2 = 0.0058\ \text{J}$$

In this example, $B_{\text{max}} = 0.35$ T.

A_p, from Equation 3.35, with $\gamma = 0$, is:

$$A_p = \left[\frac{\sqrt{1+\gamma}K_i L\hat{I}^2}{B_{\text{max}}K_t\sqrt{k_u\Delta T}}\right]^{8/7} = \left[\frac{0.0058}{(0.35)(48.2\times 10^3)\sqrt{(0.2)(20)}}\right]^{8/7}\times 10^8 = 1.875\ \text{cm}^4$$

The Magnetics MPP toroidal core with dimensions given in Table 3.6 is suitable. The core specifications are given in Table 3.6, $A_p = 2.58\ \text{cm}^4$.

In this case, the manufacturer did not provide the thermal resistance. We can estimate it from Equation 3.16:

$$R_\theta = \frac{0.06}{\sqrt{V_c}} = \frac{0.06}{\sqrt{6.09\times 10^{-6}}} = 24.3\ ^\circ\text{C/W}$$

Table 3.6 Core and winding specifications

A_c	0.678 cm^2
W_a	3.8 cm^2
A_p	2.58 cm^4
l_c	8.98 cm
V_c	6.09 cm^3
k_f	1.0
k_u	0.2
K_i	1.0
MLT	5.27 cm
ρ_{20}	1.72 $\mu\Omega$-cm
α_{20}	0.00393

and the maximum dissipation is:

$$P_D = \frac{\Delta T}{R_\theta} = \frac{20}{24.3} = 0.823\ \text{W}$$

The optimum value of the effective permeability for this core is given by Equation 3.20:

$$\mu_{\text{opt}} = \frac{B_{\max} l_c k_i}{\mu_0 \sqrt{\dfrac{P_{\text{cu max}} k_u W_a}{\rho_w \text{MLT}}}} = \frac{(0.35)(8.98 \times 10^{-2})(1.0)}{(4\pi \times 10^{-7})\sqrt{\dfrac{(0.823)(0.2)(3.8 \times 10^{-4})}{(1.72 \times 10^{-8})(5.27 \times 10^{-2})}}} = 95$$

In this case, there is no provision for core loss.

Taking Equation 3.9, with the core specifications in Table 3.6, yields the maximum value of the effective relative permeability. This is based on point 3 of Figure 3.1:

$$\mu_{\text{eff}} = \frac{B_{\max}^2 A_c l_c}{\mu_0 L \hat{I}^2} = \frac{(0.35)^2 (0.678 \times 10^{-4})(8.98 \times 10^{-2})}{(4\pi \times 10^{-7})(0.0058)} = 101$$

The manufacturer provides a core with effective permeability of 125.

Winding Design

The MPP core with effective permeability of 125 has an inductance of 117 mH per 1000 turns.

Turns

$$N = 1000\sqrt{\frac{L}{L_{1000}}} = 1\,000\sqrt{\frac{1.6}{117}} = 117\ \text{Turns}$$

The magnetic field intensity at the maximum value of the DC bias is:

$$H = \frac{NI}{l_c} = \frac{(117)(1.9)}{(8.98 \times 10^{-2})} = 2\,490\ \text{A/m} = 31.3\ \text{oersted}.$$

At this value of H, the inductance has fallen to 80%.

Wire Size

The current density is:

$$J_o = K_t \frac{\sqrt{\Delta T}}{\sqrt{k_u(1+\gamma)}\sqrt[8]{A_p}} = (48.2 \times 10^3)\frac{\sqrt{(20)}}{(\sqrt{0.2})(\sqrt[8]{2.58 \times 10^{-8}})} \times 10^{-4} = 428.1\ \text{A/cm}^2$$

The cross-sectional area of the conductor is:

$$A_w = I_{dc}/J_o = 1.9/428.1 = 0.0044\ \text{cm}^2$$

This corresponds to a 0.79 mm diameter. A 1 mm diameter copper wire with a DC resistance of 218×10^{-6} Ω/cm at 20 °C (Table A.1) will suffice.

Copper Loss

Use Equation 3.37 to correct the winding resistance for temperature:

$$T_{max} = 60 + 20 = 80\ ^\circ\text{C}.$$

$$R_{dc} = (117)(5.27)(218 \times 10^{-6})[1 + (0.00393)(80 - 20)] \times 10^3 = 166\ \text{m}\Omega$$

The copper loss is:

$$P_{cu} = R_{dc}I_{rms}^2 = (0.166)(1.9)^2 = 0.600\ \text{W}$$

Core loss

The flux density ripple ΔB may be calculated by using Faraday's law, noting that the voltage across the inductor in the time period 0 to DT is $(V_i - V_o)$:

$$\Delta B = \frac{(V_i - V_o)DT}{NA_c} = \frac{(12 - 9)(0.75)}{(117)(0.678 \times 10^{-4})(60 \times 10^3)} = 0.005\ \text{T}$$

Using the General Steinmetz Equation (GSE) (Equation 1.29), the peak value of the flux density B_{max} is $\Delta B/2$ and the core loss is:

$$P_{fe} = V_c K_c f^\alpha B_{max}^\beta = (6.09 \times 10^{-6})(231.8)(60\,000)^{1.41}(0.005/2)^{2.56} = 0.002\ \text{W}$$

As expected, the core loss is much smaller than the copper loss and well within the margins allowed by our core selection.

Total Losses :	Copper loss	0.600 W
	Core loss	0.002 W
	Total losses	0.602 W

3.4 Multiple Windings

In many applications, such as a flyback inductor, there are two windings. The total current is divided in the ratio m:$(1-m)$ and the areas are distributed in the ratio n:$(1-n)$. In other words, mI flows in an area nW_c and current $(1-m)I$ flows in an area $(1-n)W_c$. Taking the individual I^2R loss in each winding, using Equation 3.6, the total copper loss becomes:

$$P_{cu} = \rho_w \frac{l_w}{W_c}\left[\frac{m^2}{n} + \frac{(1-m)^2}{1-n}\right] I_{rms}^2 \tag{3.39}$$

Taking P_o as the loss for the current I in area W_c, Equation 3.39 may be normalized as:

$$P_{cu} = P_o\left[\frac{m^2}{n} + \frac{(1-m)^2}{1-n}\right] \tag{3.40}$$

The optimum value of P_{cu} in relation to m is obtained by taking the partial derivative with respect to m and setting it to zero:

$$\frac{\partial P_{cu}}{\partial m} = P_o\left[\frac{2m}{n} - \frac{2(1-m)}{1-n}\right] = 0 \tag{3.41}$$

The minimum loss occurs when $m = n$.

The current density in each of the windings is:

$$\frac{mI}{nW_c} = \frac{(1-m)I}{(1-n)W_c} = \frac{I}{W_c} = J \tag{3.42}$$

The result in Equation 3.42 shows that the optimum distribution of current in the available area is to have the same current density in each winding. Of course, this result may be generalized by further subdivision of any of the sections above so that, for any number of windings, the optimum distribution of current is set to the same current density in each winding.

Consider winding i in an inductor with multiple windings and with current density J_o

$$J_o = \frac{I_{rms_i}}{A_{w_i}} = \frac{N_i I_{rms_i}}{N_i A_{w_i}} = \frac{N_i I_{rms_i}}{W_{c_i}} \tag{3.43}$$

For two windings with the same current density, we can deduce from Equation 3.43:

$$\frac{W_{ci}}{W_{c1}+W_{c2}}=\frac{W_{c1}}{W_c}=\frac{N_1 I_{\text{rms1}}}{N_1 I_{\text{rms1}}+N_2 I_{\text{rms2}}}=\frac{1}{1+\dfrac{I_{\text{rms2}}}{aI_{\text{rms1}}}}. \tag{3.44}$$

Where a is the ratio N_1/N_2, and W_c is the total winding area. This means that the area assigned to each winding is directly related to the rms current in that winding.

Note that from Equation 3.19, it follows that the window utilization factor for each individual winding may be related to the overall window utilization factor:

$$\frac{W_{c1}}{W_c}=\frac{W_{c1}}{W_a}\frac{W_a}{W_c}=\frac{k_{u1}}{k_u}=\frac{1}{1+\dfrac{I_{\text{rms2}}}{aI_{\text{rms1}}}} \tag{3.45}$$

and:

$$k_u = k_{u1}+k_{u2} \tag{3.46}$$

3.4.1 *Example 3.3: Flyback Converter*

The flyback converter is the isolated version of the buck-boost converter. The circuit diagram is shown in Figure 3.7 and the main voltage and current waveforms are shown in Figure 3.8. The inductor has N_p turns in the primary winding and N_s turns in the secondary winding. We assume continuous conduction.

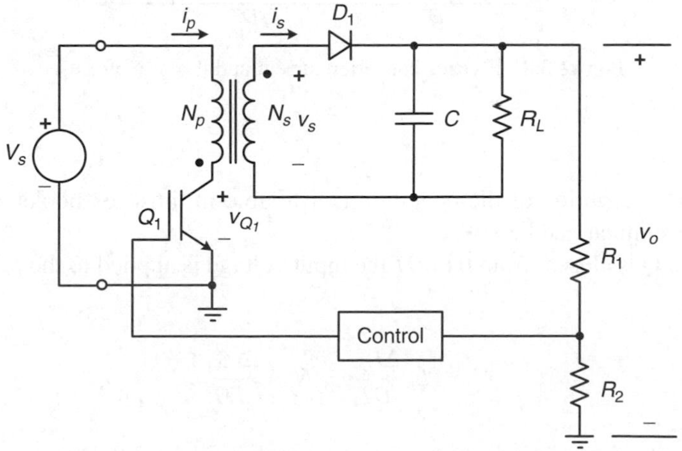

Figure 3.7 Flyback converter circuit diagram.

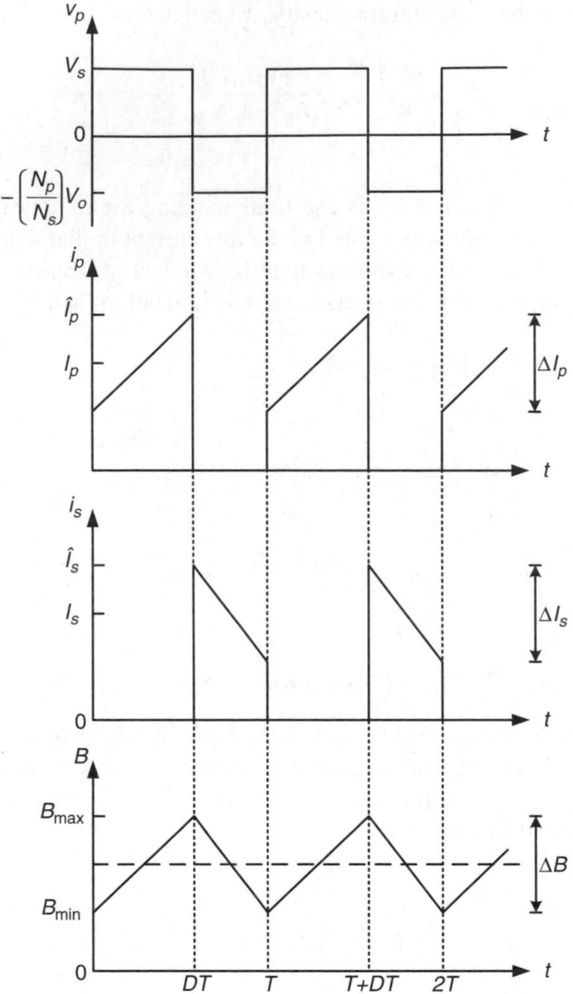

Figure 3.8 Flyback converter current and flux waveforms.

The detailed description of the operation is available in most textbooks on power electronics and are summarized here.

When switch Q is closed, from 0 to DT the input voltage is applied to the primary winding and:

$$V_i = L_p \frac{\Delta I_{L+}}{DT} = N_p A_c \frac{\Delta B_+}{DT},$$

The current in the primary winding increases as shown in Figure 3.9. Diode D is off and there is no current in the second winding.

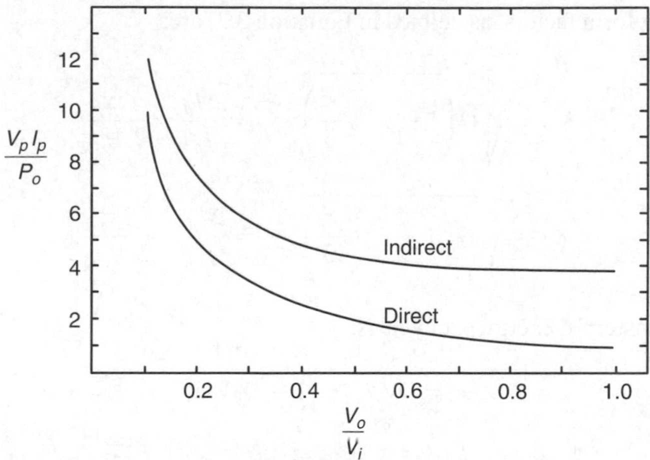

Figure 3.9 Switch stress factors for direct and indirect converters [5]. Kassakian, John G.; Schlecht, Martin F.; Verghese, George C., Principles of Power Electronics, 1st Edition, © 1991. Reprinted by permission of Pearson Education, Inc., Upper Saddle River, NJ.

At time $t = DT$, Q is opened and D is closed so that:

$$v_2 = -V_o = L_s \frac{\Delta I_{L-}}{(1-D)T} = N_s A_c \frac{\Delta B_-}{(1-D)T},$$

In steady state, the net change in flux in the core is zero and, therefore, the increase in flux while Q is closed must be equal to the decrease in flux when Q is open. It follows that:

$$|\Delta B_+| = |\Delta B_-|,$$

and

$$V_o = \frac{N_s}{N_p} \frac{D}{1-D} V_i,$$

The characteristic trapezoidal current waveforms for the primary winding and the secondary winding are shown in Figure 3.8.

The rms values of these waveforms are related to the ripple as:

$$I_{p\,\text{rms}} = I_p \sqrt{D\left(1 + \frac{x_p^2}{12}\right)};\ x_p = \frac{\Delta I_p}{I_p}$$

$$I_{s\,\text{rms}} = I_p \sqrt{D\left(1 + \frac{x_s^2}{12}\right)};\ x_s = \frac{\Delta I_s}{I_s}$$

The current waveform factors, as defined in Equation 3.3, are:

$$K_{ip} = \sqrt{D\left(1 - y_p + \frac{y_p^2}{3}\right)};\ y_p = \frac{\Delta I_p}{\hat{I}_p} = \frac{2x_p}{2 + x_p}$$

$$K_{is} = \sqrt{D\left(1 - y_s + \frac{y_s^2}{3}\right)};\ y_s = \frac{\Delta I_p}{\hat{I}_s} = \frac{2x_s}{2 + x_s}$$

Note that the transferred energy per cycle is:

$$P = DV_iI_p = (1 - D)V_oI_s$$

so that:

$$I_p = \frac{P}{DV_i}$$

$$I_s = \frac{P}{(1 - D)V_o} = \frac{N_p}{N_s}I_p$$

$$\frac{\Delta I_s}{\Delta I_p} = \frac{N_p}{N_s}$$

P is the transferred energy per cycle in the flyback converter.

The value of inductance that ensures continuous conduction is:

$$L_g = \frac{V_i^2D^2}{P}\frac{T}{2} = \frac{V_o}{I_o}\left(\frac{N_p}{N_s}\right)^2\frac{T}{2}(1 - D)^2$$

This proof of this formula is available in texts on power electronics.

Specifications

The design specifications for inductor are listed in Table 3.7.

Table 3.7 Specifications

Input voltage	325.3 V
Output voltage	24 V
Output current	10 A
Inductance	700 μH
Frequency, f	70 kHz
Temperature rise, ΔT	30 °C
Ambient temperature, T_a	60 °C
Window utilization factor	0.235

Circuit Parameters

For continuous conduction, the minimum value of inductance is:

$$L_g = \frac{V_i^2 D^2}{P} \frac{T}{2} == \frac{(325.3)^2 (0.314)^2}{(240)(2)(70 \times 10^3)} \times 10^6 = 310\ \mu\text{H}$$

The input DC voltage is based on a rectified 240 V mains, i.e. $\sqrt{2} \times 240$ V.

The turns ratio is chosen to ensure that the switch stress is minimized, and the ratio V_o/V_i should be approximately 0.5, as shown in Figure 3.9 [5], so that $N_p/N_s \approx 325.3/48 = 6.7$. The value chosen was 6.2, so therefore D is:

$$D = \frac{1}{1 + \frac{V_i}{aV_o}} = \frac{1}{1 + \frac{(\sqrt{2})(230)}{(6.2)(24)}} = 0.314$$

The output power is $P = 24\text{ V} \times 10\text{ A} = 240$ W.

The primary current is:

$$I_p = \frac{P}{DV_i} = \frac{240}{(0.314)(\sqrt{2})(230)} = 2.351\text{ A}$$

The ripple of the primary winding current is found from:

$$V_i = L_p \frac{\Delta I_p}{DT}$$

$$\Delta I_p = \frac{V_i DT}{L_p} = \frac{(\sqrt{2})(230)(0.314)}{(700 \times 10^{-6})(70 \times 10^3)} = 2.084\text{ A}$$

The peak value of the primary current is:

$$\hat{I}_p = I_p + \frac{\Delta I_p}{2} = 2.351 + \frac{2.084}{2} = 3.393\text{ A}$$

The current waveform factor is:

$$y_p = \frac{\Delta I_p}{\hat{I}_p} = \frac{2.084}{3.393} = 0.614$$

$$K_{ip} = \sqrt{D\left(1 - y_p + \frac{y_p^2}{3}\right)} = \sqrt{(0.314)\left(1 - 0.614 + \frac{(0.614)^2}{3}\right)} = 0.4$$

The rms value of the primary current is:

$$I_{p\,\mathrm{rms}} = K_{ip}\,\hat{I}_p = (0.4)(3.393) = 1.357\ \mathrm{A}$$

The secondary current is:

$$I_s = \frac{P}{(1-D)V_o} = \frac{240}{(1-0.314)(24)} = 14.577\ \mathrm{A}$$

The ripple of the secondary winding current is:

$$\Delta I_s = \frac{N_p}{N_s}\Delta I_p = (6.2)(2.084) = 12.92\ \mathrm{A}$$

The peak value of the secondary current is:

$$\hat{I}_s = I_s + \frac{\Delta I_s}{2} = 14.577 + \frac{12.92}{2} = 21.037\ \mathrm{A}$$

The current waveform factor is:

$$y_s = \frac{\Delta I_s}{\hat{I}_s} = \frac{12.92}{21.037} = 0.614$$

$$K_{is} = \sqrt{(1-D)\left(1 - y_s + \frac{y_s^2}{3}\right)} = \sqrt{(1-0.314)\left(1 - 0.614 + \frac{(0.614)^2}{3}\right)} = 0.592$$

The rms value of the secondary current is:

$$I_{\mathrm{srms}} = K_{is}\hat{I}_s = (0.592)(21.037) = 12.454\ \mathrm{A}$$

We are now in a position to calculate k_{up} using Equation 3.46. Select the total window utilization factor as 0.235, due to the voltage insulation requirement:

$$k_{up} = k_u \frac{1}{1 + \dfrac{I_{s\,\mathrm{rms}}}{aI_{p\,\mathrm{rms}}}} = (0.235)\frac{1}{1 + \dfrac{12.454}{(6.2)(1.359)}} = (0.235)(0.4035) = 0.0948$$

Core Selection

Ferrite would normally be used for this type of application at the specified frequency. The material specifications for EPCOS N87 Mn-Zn are listed in Table 3.8.

Table 3.8 Material specifications

K_c	16.9
α	1.25
β	2.35
B_{sat}	0.4 T

The maximum stored energy in the primary is:

$$L\hat{I}_p^2 = (700 \times 10^{-6})(3.393)^2 = 0.0081\,\text{J}$$

For Mn-Zn N87 ferrite, select $B_{max} = 0.2$ T.

At this point, we do not know the core loss but, since we are dealing with a large current ripple, we can expect large core loss, so a good design rule is to set the iron loss at two times the copper loss, that is $\gamma = 2$.

A_p, from Equation 3.35, is corrected for the two windings (see Problem 1):

$$A_p = \left[\frac{\sqrt{1+\gamma}K_{ip}L_p\hat{I}_p^2}{B_{max}K_t(k_{up}/\sqrt{k_u})\sqrt{\Delta T}}\right]^{8/7}$$
$$= \left[\frac{\sqrt{(1+2)}(0.4)(0.0081)}{(0.2)(48.2\times 10^3)(0.0948/\sqrt{0.235})(\sqrt{30})}\right]^{8/7} \times 10^8 = 6.89\,\text{cm}^4$$

The E55/28/21 core with $A_p = 9.72\,\text{cm}^4$ is suitable. The core specifications are given in Table 3.9.

The thermal resistance of this core, according to the manufacturer, is 10 °C/W; therefore, the maximum dissipation, given by Equation 3.8, is:

$$P_D = \frac{\Delta T}{R_\theta} = \frac{30}{10} = 3.0\,\text{W}$$

Since $\gamma = 2$, the total copper loss in 1.0 W and half of this is assigned to the primary winding giving $P_{cu_p max} = 0.5$ W.

Table 3.9 Core and winding specifications

A_c	3.51 cm^2
l_c	12.4 cm
W_a	2.77 cm^2
A_p	9.72 cm^4
V_c	43.5 cm^3
k_f	1.0
k_u	0.235
MLT	11.3 cm
ρ_{20}	1.72 $\mu\Omega$-cm
α_{20}	0.00393

The optimum value of the effective permeability for this core is given by Equation 3.20, corrected for two windings (see Problem 2):

$$\mu_{\text{opt}} = \frac{B_{\max} l_c K_{ip}}{\mu_0 \sqrt{\dfrac{P_{cu_p\max} k_{up} W_a}{\rho_w \text{MLT}}}} = \frac{(0.2)(12.4 \times 10^{-2})(0.4)}{(4\pi \times 10^{-7})\sqrt{\dfrac{(0.5)(0.155)(2.77 \times 10^{-4})}{(1.72 \times 10^{-8})(11.3 \times 10^{-2})}}} = 75$$

Winding Design

Gap g

Referring to Figure 3.1, increasing the value of effective permeability above the optimum value means that we can operate the core at its maximum flux density and below the maximum dissipation. This makes sense, since we have not included any core loss at this point.

The maximum gap length is found from Section 2.1:

$$g_{\max} = \frac{l_c}{\mu_{\min}} = \frac{(12.4 \times 10^{-2})}{75} 10^3 = 1.65\,\text{mm}$$

The manufacturer provides a standard core set, with $g = 1$ mm, a corresponding A_L value of 496 nH and an effective permeability of 138.

Turns

For the core, the manufacturer supplies the A_L value (i.e. the inductance per turn):

$$N = \sqrt{\frac{L}{A_L}} = \sqrt{\frac{(700 \times 10^{-6})}{(496 \times 10^{-9})}} = 37.6\, Turns$$

Select 38 primary turns.

The number of turns in the secondary winding is 38/6.2 = 6 turns

Wire Size

The current density is:

$$J_o = K_t \frac{\sqrt{\Delta T}}{\sqrt{k_u(1+\gamma)}\sqrt[8]{A_p}} = 48.2 \times 10^3 \frac{\sqrt{30}}{\sqrt{(0.235)(1+2)}\sqrt[8]{(9.72 \times 10^{-8})}} \times 10^{-4}$$
$$= 236.6\ \text{A/cm}^2$$

The cross-sectional area of the conductor in the primary winding is:

$$A_{wp} = I_{p\,\text{rms}}/J_o = 1.357/236.6 = 0.00574\,\text{cm}^2$$

This corresponds to four 0.428 mm diameter copper wires in parallel. A 0.5 mm diameter wire has a DC resistance of 871×10^{-6} Ω/cm at 20 °C (Table A.1).

The operating frequency is 70 kHz, corresponding to 0.25 mm skin depth. The radius of the wire is less than 1.5 skin depths, as required (see Figure 6.3).

The cross-sectional area of the conductor in the secondary winding is:

$$A_{ws} = I_{s\,\mathrm{rms}}/J_o = 12.454/236.6 = 0.0526\ \mathrm{cm}^2$$

This corresponds to a copper foil 25.4 × 0.2 mm, with a DC resistance of 33.86×10^{-6} Ω/cm at 20 °C (Table A.1).

Copper Loss

Primary winding:

Use Equation 3.37 to correct the winding resistance for temperature:

$$T_{\mathrm{max}} = 60 + 30 = 90\ {}^\circ\mathrm{C}.$$

$$R_{\mathrm{dc}} = (38)(11.3)((871/4) \times 10^{-6})[1 + (0.00393)(90 - 20)] \times 10^3 = 119.2\ \mathrm{m}\Omega$$

The copper loss in the primary winding is:

$$P_{\mathrm{cu}} = R_{\mathrm{dc}} I_{\mathrm{rms}}^2 = (119.2 \times 10^{-3})(1.357)^2 = 0.220\ \mathrm{W}$$

Secondary winding:

Use Equation 3.38 to correct the winding resistance for temperature:

$$T_{\mathrm{max}} = 60 + 30 = 90\ {}^\circ\mathrm{C}.$$

$$R_{\mathrm{dc}} = (6)(11.3)(33.86 \times 10^{-6})[1 + (0.00393)(90 - 20)] \times 10^3 = 2.927\ \mathrm{m}\Omega$$

The copper loss in the secondary winding is:

$$P_{\mathrm{cu}} = R_{\mathrm{dc}} I_{\mathrm{rms}}^2 = (2.927 \times 10^{-3})(12.454)^2 = 0.454\ \mathrm{W}$$

The total copper loss is 0.674 W.

Core Loss

The flux density ripple ΔB may be calculated by using Faraday's law, noting that the voltage across the inductor in the time period 0 to DT is V_i:

$$\Delta B = \frac{V_i DT}{N_p A_c} = \frac{325.3 \times 0.314}{38 \times 3.51 \times 10^{-4} \times 70 \times 10^3} = 0.109\ \mathrm{T}$$

Using the General Steinmetz Equation (GSE), (Equation 1.29), the peak value of the flux density $B_{\max}$ is $\Delta B/2$ and the core loss is:

$$P_{\text{fe}} = V_c K_c f^{\alpha} B_{\max}^{\beta} = (43.5 \times 10^{-6})(16.9)(70000)^{1.25}(0.109/2)^{2.35} = 0.898\ \text{W}$$

Total Losses :	Copper loss	0.674 W
	Core loss	0.898 W
	Total losses	1.572 W

These losses are less than P_D max of 3 W.

3.5 Problems

3.1 Derive the formula for A_p for a the two winding inductor.

3.2 Derive the formula for μ_{opt} for a two winding inductor.

3.3 Calculate the H field in the dielectric of a coaxial cable with the following dimensions: the radius of the inner conductor is r_i, and the inner and outer radii of the outer conductor are r_{oi} and r_{oo} respectively.

3.4 Describe the three types of power loss in a magnetic component.

3.5 A buck converter with 25A 100 ms pulse of output current operates at 80 kHz with a DC input voltage of 180 V and an output voltage of 24 V DC. Calculate the inductance required for a current ripple of 5%. Select a toroidal powder iron core to meet these requirements. The maximum allowable temperature rise in the inductor is 35 °C. The transient thermal impedance of the inductor is 0.5 °C/W for the 100 ms pulse.

3.6 A zero current switching quasi-resonant converter (ZCS-QRC) operates at 85 kHz with a nominal load current of 0.8A, and the DC input voltage is 48 V. The resonant inductor is 86 μH and the resonant capacitor is 33 nF. Select a ferrite core to meet these requirements, with the inductor isolation material not exceeding a temperature rise of 110 °C with an ambient temperature of 65 °C.

3.7 Design a common-mode choke for an induction hob which delivers 3000 W at 230 V. The purpose of the common-mode choke is to attenuate the common-mode current which generates EMI noise. The attenuation should be effective at 20 kHz, with common-mode impedance of 1000 Ω. The peak value of the common-mode current is 100 mA. The ambient temperature is 40 °C and the maximum allowable temperature rise is 70 °C. Use Mn-Zn ferrite.

3.8 A 30 W flyback converter operates at 70 kHz with a DC input of 325 V, rectified mains and a DC output voltage of 5 V. Calculate the minimum value of the primary coil inductance to ensure continuous conduction. For a primary coil inductance of 620 μH, select a ferrite ETD core (N87 material in Table 1.1) and calculate the copper loss and core loss. The maximum temperature rise allowed in the winding is 30 °C and the ambient temperature is 50 °C.

MATLAB Program for Example 3.1

```
%example 3.1: Buck Converter with a gapped core

alpha = 1.25
alpha20 = 0.00393
beta = 2.35
deltaT = 15
gamma = 0
muo = 4*pi*10^-7
row = 1.72e-8;

Ac = 2.09e-4
AL = 188e-9
Ap2 = 5.62*10^-8
Bmax = 0.25
Bsat = 0.4
D = 0.5
DT = D/(80e3)
f = 80e3
Idc = 20
Kc = 16.9
Ki = 1.0
Kt = 48.2e3
ku = 0.8
L = 34e-6
lc = 11.4e-2
MLT = 8.6e-2
N1 = 13
Pcumax = 1.36
Rteta = 11
Tmax = 70+15
Vc = 23.8*10^-6
Vi = 12
Vo = 6
Vs = 12
Wa = 2.69e-4
wire_Rdc=10.75*10^-6;

deltaIL = (Vs-Vo)*DT/L
Ipeak = Idc+deltaIL/2
LIpeak2 = L*Ipeak^2
Ap = [(sqrt(1+gamma)*Ki*L*Ipeak^2)/(Bmax*Kt*sqrt(ku*deltaT))]^(8/7)
*10^8
```

```
PD = deltaT/Rteta
muopt = (Bmax*lc*Ki)/(muo*sqrt((Pcumax*ku*Wa)/(row*MLT)))
gmax = lc/muopt
N = sqrt(L/AL)
Jo = Kt*(sqrt(deltaT))/(sqrt(ku*(1+gamma))*Ap2^(1/8))*10^-4
Aw = Idc/Jo
Rdc = N1*MLT*wire_Rdc*[1+alpha20*(Tmax-20)]*10^5
Pcu = (Rdc*Idc^2)*10^-3
deltaB = ((Vi-Vo)*DT)/(N1*Ac)
Bm = deltaB/2
Pfe = Vc*Kc*f^alpha*Bm^beta
Ptot = Pfe+Pcu]]>
```

MATLAB Program for Example 3.2

```
%example 3.2 Forward Converter Toroidal Core

alpha = 1.41
alpha20 = 0.00393
beta = 2.56
deltaT = 20
gamma = 0
muo = 4*pi*10^-7
row = 1.72e-8;

Ac = 0.678e-4
Ap2 = 2.58e-8
Bmax = 0.35
Vo = 9
Vi = 12
D = Vo/Vi
f = 60e3
DT = D/f
Idc = 1.9
Kc = 231.8
Ki = 1
Kt = 48.2*10^3
ku = 0.2
L = 1.6e-3
L1000 = 117e-3
lc = 8.98e-2
MLT = 5.27e-2
N1 = 117
Pcumax = 0.823
```

```
Tmax = 60+20
Vc = 6.09e-6
Vs = 12
Wa = 3.8e-4
wire_Rdc = 218e-6;

deltaIL = (Vs-Vo)*DT/L
Ipeak = Idc+deltaIL/2
LIpeak2 = L*Ipeak^2
Ap = [(sqrt(1+gamma)*Ki*L*Ipeak^2)/(Bmax*Kt*sqrt(ku*deltaT))]^(8/7)
*10^8
Rteta = 0.06/sqrt(Vc)
PD = deltaT/Rteta
muopt = (Bmax*lc*Ki)/(muo*sqrt((Pcumax*ku*Wa)/(row*MLT)))
mueff = (Bmax^2*Ac*lc)/(muo*L*Ipeak^2)
N = 1000*sqrt(L/L1000)
H = N*Ipeak/lc
Jo = Kt*(sqrt(deltaT))/(sqrt(ku*(1+gamma))*Ap2^(1/8))*10^-4
Aw = Idc/Jo
Rdc = N1*MLT*wire_Rdc*[1+alpha20*(Tmax-20)]*10^5
Pcu = (Rdc*Idc^2)*10^-3
deltaB = ((Vi-Vo)*DT)/(N*Ac)
Bm = deltaB/2
Pfe = Vc*Kc*f^alpha*Bm^beta
Ptot = Pfe+Pcu]]>
```

MATLAB Program for Example 3.3

```
%example 3.3 Flyback Converter

alpha = 1.25
alpha20 = 0.00393
beta = 2.35
deltaT = 30
gamma = 2
mumin = 67
muo = 4*pi*10^-7
row = 1.72e-8;

a = 6.2
Ac = 3.51e-4
AL = 496e-9
Ap2 = 9.72e-8
Bm = 0.109/2
```

```
Bmax = 0.2
Bsat = 0.4
D = 0.314
f = 70e3
Kc = 16.9
kc = 16.9
Kt = 48.2e3
ku = 0.235
kup2 = 0.155
lc = 12.4e-2
Lp = 700e-6
MLT = 11.3e-2
N1 = 38
N2 = 6
Np = 38
NpoverNs = 6.2
P = 24*10
Pcumax = 0.5
Rteta = 10
T = 1/70e3
Vc = 43.5e-6
Vi = sqrt(2)*230
Vo = 24
Wa = 2.77e-4
Tmax = 60+30
wire_Rdc = (871/4)*10^-6
wire_Rdc2 = 33.86e-6;

Lg = ((Vi^2*D^2)/P)*(T/2)
D = 1/(1+Vi/(a*Vo))
Ip = P/(D*Vi)
deltaIp = (Vi*D*T)/Lp
Ippeak = Ip+deltaIp/2
yp = deltaIp/Ippeak
Kip = sqrt(D*(1-yp+yp^2/3))
Iprms = Kip*Ippeak
Is = P/((1-D)*Vo)
deltaIs = NpoverNs*deltaIp
Ispeak = Is+deltaIs/2
ys = deltaIs/Ispeak
Kis = sqrt((1-D)*(1-ys+ys^2/3))
Isrms = Kis*Ispeak
kup = ku*1/(1+Isrms/(a*Iprms))
LIppeak2 = Lp*Ippeak^2
Ap = [(sqrt(1+gamma)*Kip*Lp*Ippeak^2)/(Bmax*Kt*(kup/sqrt(ku))*sqrt
(deltaT))]^(8/7)
```

```
PD = deltaT/Rteta
muopt = (Bmax*lc*Kip)/(muo*sqrt((Pcumax*kup2*Wa)/(row*MLT)))
gmax = lc/mumin
N = sqrt(Lp/AL)
Jo = Kt*sqrt(deltaT)/(sqrt(ku*(1+gamma))*Ap2^(1/8))
Awp = Iprms/Jo
Aws = Isrms/Jo
Rdc1 = N1*MLT*wire_Rdc*[1+alpha20*(Tmax-20)]*10^5
Pcu1 = (Rdc1*Iprms^2)*10^-3
Rdc2 = N2*MLT*wire_Rdc2*[1+alpha20*(Tmax-20)]*10^5
Pcu2 = (Rdc2*Isrms^2)*10^-3
deltaB = (Vi*D*T)/(Np*Ac)
Pfe = Vc*kc*f^alpha*Bm^beta
Ptot = Pcu1+Pcu2+Pfe]]>
```

References

1. Venkatachalam, K., Sullivan, C.R., Abdallah, T., and Tacca, H. (2002) Accurate prediction of ferrite core loss with nonsinusoidal waveforms using only Steinmetz parameters. *Proceedings of IEEE Workshop on Computers in Power Electronics, COMPEL*, pp. 36–41.
2. McAdams, W.H. (1954) *Heat Transmission*, 3rd edn, McGraw-Hill, New York.
3. Judd, F. and Kressler, D. (1977) Design optimization of small low-frequency power transformers. *IEEE Transactions on Magnetics*, **13** (4), 1058–1069.
4. McLyman, C.W.T. (2004) *Transformer and Inductor Design Handbook*, 3rd edn, Marcel Dekker Inc., New York.
5. Kassakian, J.G., Schlecht, M.F., and Verghese, G.C. (1991) *Principles of Power Electronics (Addison-Wesley Series in Electrical Engineering)*, Prentice Hall, Reading, MA.

Further Reading

1. Bartoli, M., Reatti, A., and Kazimierczuk, M.K. (1994) High-frequency models of ferrite core inductors. *Proceedings of the IEEE Industrial Electronics, Control and Instrumentation, IECON*, pp. 1670–1675.
2. Bartoli, M., Reatti, A., and Kazimierczuk, M.K. (1994) Modelling iron-powder inductors at high frequencies. *Proceedings of the IEEE Industry Applications Conference, IAS*, pp. 1225–1232.
3. Bennett, E. and Larson, S.C. (1940) Effective resistance to alternating currents of multilayer windings. *Transactions of the American Institute of Electrical Engineers*, **59** (12), 1010–1017.
4. Blume, L.F. (1982) *Transformer Engineering*, John Wiley & Sons, New York.
5. Bueno, M.D.A. (2001) *Inductance and Force Calculations in Electrical Circuits*, Nova Science Publishers, Huntington.
6. Carsten, B. (1986) High frequency conductor losses in switchmode magnetics. *Proceedings of the High Frequency Power Converter Conference*, pp. 155–176.
7. Cheng, K.W.E. and Evans, P.D. (1994) Calculation of winding losses in high-frequency toroidal inductors using single strand conductors. *IEE Proceedings on Electric Power Applications, B*, **141** (2), 52–62.
8. Cheng, K.W.E. and Evans, P.D. (1995) Calculation of winding losses in high frequency toroidal inductors using multistrand conductors. *IEE Proceedings on Electric Power Applications, B*, **142** (5), 313–322.
9. Del Vecchio, R.M., Poulin, B., Feghali, P.T. *et al.* (2001) *Transformer Design Principles: With Applications to Core-Form Power Transformers*, 1st edn, CRC Press, Boca Raton, FL.
10. Dowell, P.L. (1966) Effects of eddy currents in transformer windings. *Proceedings of the Institution of Electrical Engineers*, **113** (8), 1387–1394.
11. Erickson, R.W. (2001) *Fundamentals of Power Electronics*, 2nd edn, Springer, Norwell, MA.

12. Evans, P.D. and Chew, W.M. (1991) Reduction of proximity losses in coupled inductors. *IEE Proceedings on Electric Power Applications, B*, **138** (2), 51–58.
13. Ferreira, J.A. (2010) *Electromagnetic Modelling of Power Electronic Converters (Power Electronics and Power Systems)*, 1st edn, Springer, Norwell, MA.
14. Fitzgerald, A.E., Kingsley, C., and Umans, S.D. (2002) *Electric Machinery*, 6th edn, McGraw-Hill, New York.
15. Flanagan, W.M. (1992) *Handbook of Transformer Design and Application*, 2nd edn, McGraw-Hill, New York.
16. Georgilakis, P.S. (2009) *Spotlight on Modern Transformer Design (Power Systems)*, 1st edn, Springer, New York.
17. Hanselman, D.C. and Peake, W.H. (1995) Eddy-current effects in slot-bound conductors. *IEE Proceedings on Electric Power Applications, B*, **142** (2), 131–136.
18. Hoke, A.F. and Sullivan, C.R. (2002) An improved two-dimensional numerical modeling method for E-core transformers. *Proceedings of the IEEE Applied Power Electronics Conference and Exposition, APEC*, pp. 151–157.
19. Hurley, W.G., Wolfle, W.H., and Breslin, J.G. (1998) Optimized transformer design: inclusive of high-frequency effects. *IEEE Transactions on Power Electronics*, **13** (4), 651–659.
20. Jieli, L., Sullivan, C.R., and Schultz, A. (2002) Coupled-inductor design optimization for fast-response low-voltage. *Proceedings of the IEEE Applied Power Electronics Conference and Exposition, APEC*, pp. 817–823.
21. Kassakian, J.G. and Schlecht, M.F. (1988) High-frequency high-density converters for distributed power supply systems. *Proceedings of the IEEE*, **76** (4), 362–376.
22. Kazimierczuk, M.K. (2009) *High-Frequency Magnetic Components*, John Wiley & Sons, Chichester.
23. Krein, P.T. (1997) *Elements of Power Electronics (Oxford Series in Electrical and Computer Engineering)*, Oxford University Press, Oxford.
24. Kulkarni, S.V. (2004) *Transformer Engineering: Design and Practice*, 1st edn, CRC Press, New York.
25. B.H.E. Limited (2004) *Transformers: Design, Manufacturing, and Materials (Professional Engineering)*, 1st edn, McGraw & -Hill, New York.
26. McLyman, C.W.T. (1997) *Magnetic Core Selection for Transformers and Inductors*, 2nd edn, Marcel Dekker Inc., New York.
27. McLyman, C.W.T. (2002) *High Reliability Magnetic Devices*, 1st edn, Marcel Dekker Inc., New York.
28. E.S. MIT (1943) *Magnetic Circuits and Transformers (MIT Electrical Engineering and Computer Science)*, The MIT Press, Cambridge, MA.
29. Muldoon, W.J. (1978) Analytical design optimization of electronic power transformers. *Proceedings of Power Electronics Specialists Conference, PESC*, pp. 216–225.
30. Pentz, D.C. and Hofsajer, I.W. (2008) Improved AC-resistance of multiple foil windings by varying foil thickness of successive layers. *The International Journal for Computation and Mathematics in Electrical and Electronic Engineering*, **27** (1), 181–195.
31. Perry, M.P. (1979) Multiple layer series connected winding design for minimum losses. *IEEE Transactions on Power Apparatus and Systems*, **PAS-98** (1) 116–123.
32. Petkov, R. (1996) Optimum design of a high-power, high-frequency transformer. *IEEE Transactions on Power Electronics*, **11** (1), 33–42.
33. Pollock, J.D., Lundquist, W., and Sullivan, C.R. (2011) Predicting inductance roll-off with dc excitations. *Proceedings of the IEEE Energy Conversion Congress and Exposition, ECCE*, pp. 2139–2145.
34. Pressman, A.I., Bellings, K., and Morey, T. (2009) *Switching Power Supply Design*, 3rd edn, McGraw-Hill, New York.
35. Ramo, S., Whinnery, J.R., and Van Duzer, T. (1984) *Fields and Waves in Communication Electronics*, 2nd edn, John Wiley & Sons, New York.
36. Smith, B. (2009) *Capacitors, Inductors and Transformers in Electronic Circuits (Analog Electronics Series)*, Wexford College Press, Wexford.
37. Snelling, E.C. (1988) *Soft Ferrites: Properties and Applications*, 2nd edn, Butterworths, London.
38. Sullivan, C.R. (1999) Optimal choice for number of strands in a litz-wire transformer winding. *IEEE Transactions on Power Electronics*, **14** (2), 283–291.
39. Sullivan, C.R. and Sanders, S.R. (1996) Design of microfabricated transformers and inductors for high-frequency power conversion. *IEEE Transactions on Power Electronics*, **11** (2), 228–238.
40. Urling, A.M., Niemela, V.A., Skutt, G.R., and Wilson, T.G. (1989) Characterizing high-frequency effects in transformer windings-a guide to several significant articles. *Proceedings of the IEEE Applied Power Electronics Conference and Exposition, APEC*, pp. 373–385.

41. Van den Bossche, A. (2005) *Inductors and Transformers for Power Electronics*, 1st edn, CRC Press, New York.
42. Vandelac, J.P. and Ziogas, P.D. (1988) A novel approach for minimizing high-frequency transformer copper losses. *IEEE Transactions on Power Electronics*, **3** (3), 266–277.
43. Venkatraman, P.S. (1984) Winding eddy current losses in switch mode power transformers due to rectangular wave currents. *Proceedings of the 11th National Solid-State Power Conversion Conference, Powercon 11*, pp. A1.1–A1.11.
44. Williams, R., Grant, D.A., and Gowar, J. (1993) Multielement transformers for switched-mode power supplies: toroidal designs. *IEE Proceedings on Electric Power Applications, B*, **140** (2), 152–160.
45. Ziwei, O., C. Thomsen, O., and Andersen, M. (2009) The analysis and comparison of leakage inductance in different winding arrangements for planar transformer. *Proceedings of the IEEE Power Electronics and Drive Systems, PEDS*, pp. 1143–1148.

Section Two

Transformers

4

Transformers[1]

The transformer is used in three broad areas of application:

- Mains power transmission, which involves raising or lowering voltage in an AC circuit with a corresponding decrease or increase in current.
- Signals transmission with impedance matching for maximum power transfer.
- In power electronics, for energy conversion and control.

In power generation and transmission, the generator may operate in the 10–20 kV range, whereas high voltage transmission is normally above 200 kV; distribution will be at, say, 10 kV, being further stepped down to 110 V or 230 V for residential supplies.

Transformers are also ideally suited to impedance matching for transfer maximum power in, say, an audio system, where the speaker load resistance might be 8 Ω. This would be matched to the output impedance of an amplifier measured in kΩ. An important application for power electronics is electrically isolating one circuit from another to satisfy safety regulatory requirements.

A fundamental principle of transformer operation is that the size is inversely proportional to the operating frequency (up to a point, as we shall see later in Chapter 5), and this has opened up the role of the transformer from its more traditional role at power frequencies. When the transformer is used in power electronics applications that incorporate electrical isolation, the voltage that appears across a switch can be adjusted by the transformer to reduce the stresses on the switch.

A transformer consists of two or more mutually coupled windings. An alternating voltage source is connected to one of the windings – usually referred to as the primary winding – and

[1] Parts of this chapter are reproduced with permission from [1] Hurley, W.G., Wilcox, D.J., and McNamara, P.S. (1991) Calculation of short circuit impedance and leakage impedance in transformer windings. *Proceedings of the IEEE Power Electronics Specialists Conference, PESC*, pp. 651–658; [2] Hurley, W.G. and Wilcox, D.J. (1994) Calculation of leakage inductance in transformer windings. *IEEE Transactions on Power Electronics* **9** (1), 121–126; [3] Hurley, W.G., Wolfle, W.H., and Breslin, J.G. (1998) Optimized transformer design: inclusive of high-frequency effects. *IEEE Transactions on Power Electronics* **13** (4), 651–659.

Transformers and Inductors for Power Electronics: Theory, Design and Applications, First Edition.
W. G. Hurley and W. H. Wölfle.

this produces a changing magnetic flux field in accordance with the laws outlined in Chapter 2. The resultant flux will depend on the number of turns in the primary winding. Normally, the windings are wound on a core of magnetic material without an air gap to obtain high flux levels, so therefore the flux will depend on the reluctance of the core, including the physical dimensions of the core length and cross-sectional area in addition to the number of turns.

The manufacturing process may produce a very small air gap, which has the advantage of being able to control the magnitude of the inrush current. The magnetic flux is coupled to the other winding – called the secondary winding – and a voltage is induced in accordance with laws of electromagnetic induction. An inductor stores energy, whereas, in a transformer, the energy is transferred from the primary to the secondary load.

Normally, iron laminations are used in the construction of large mains transformers to reduce eddy current loss in the core. Compressed ferromagnetic alloys (ferrites) are used in power electronic circuit applications for high frequency operation.

4.1 Ideal Transformer

A basic two-winding transformer is shown in Figure 4.1, where the windings are wound on a magnetic core.

Sinusoidal excitation is applied to the input winding and the second winding is on open-circuit. These windings are usually referred to as the primary and secondary windings respectively. Drawing on the development of the concept of inductance in Chapter 2, the primary winding has an inductance L_m called the magnetizing inductance. This is given by:

$$L_m = \frac{N_1^2}{\mathcal{R}_c} \tag{4.1}$$

and the reluctance of the core is

$$\mathcal{R}_c = \frac{l_c}{\mu_r \mu_0 A_c} \tag{4.2}$$

where l_c is the mean length of the magnetic path around the closed core and A_c is the cross-sectional area of the core.

Evidently, as the relative permeability of the core increases the reluctance becomes smaller, this in turn means that the mmf ($N_1 I_1$) required to establish the flux in the core also

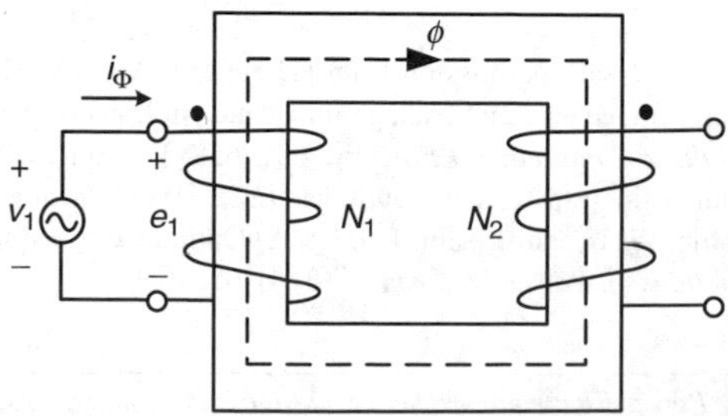

Figure 4.1 Two-winding transformer: no load conditions.

becomes smaller. For this reason, it is normally assumed that the magnetizing current to establish the flux in the core is infinitesimally small.

4.1.1 No Load Conditions

The secondary winding in Figure 4.1 is in open circuit under no load conditions. A magnetizing current i_ϕ flows in the primary winding, which establishes the alternating flux ϕ in the magnetic core.The basic relationship between the applied voltage and the flux in the core follows from Ampere's law and Faraday's law:

$$e_1 = \frac{d\lambda_1}{dt} = N_1 \frac{d\phi_1}{dt} \tag{4.3}$$

By Lenz's law, e_1 is a counter-emf to v_1 and, in accordance with Kirchhoff's voltage law, $v_1 = e_1$.

At this point, we shall assume that we are dealing with sinusoidal excitation at frequency f ($\omega = 2\pi f$) and the amplitude of the flux is $\phi_{\max}$:

$$\phi(t) = \phi_{\max} \sin \omega t \tag{4.4}$$

$$e_1 = N_1 \frac{d\phi}{dt} = \omega N_1 \phi_{\max} \cos \omega t \tag{4.5}$$

or, in terms of the sine function:

$$e_1(t) = \omega N_1 \phi_{\max} \sin\left(\omega t + \frac{\pi}{2}\right) \tag{4.6}$$

which shows that the flux lags the applied voltage by 90°.

The amplitude of the primary EMF is

$$E_{1_{\max}} = 2\pi f N_1 \phi_{\max} \tag{4.7}$$

The magnetic flux may be expressed in terms of the flux density:

$$\phi_{\max} = B_{\max} A_c \tag{4.8}$$

and it follows that:

$$E_{1_{\text{rms}}} = \frac{E_{1_{\max}}}{\sqrt{2}} = 4.44\, f N_1 B_{\max} A_c \tag{4.9}$$

Since $V_{1\text{rms}} = E_{1\text{rms}}$, the rms value of the input voltage is related to the number of turns in the input or primary winding, the maximum flux density and core cross-sectional area by:

$$V_{1_{\text{rms}}} = 4.44\, f N_1 B_{\max} A_c \tag{4.10}$$

This is the celebrated transformer equation for sinusoidal excitation.

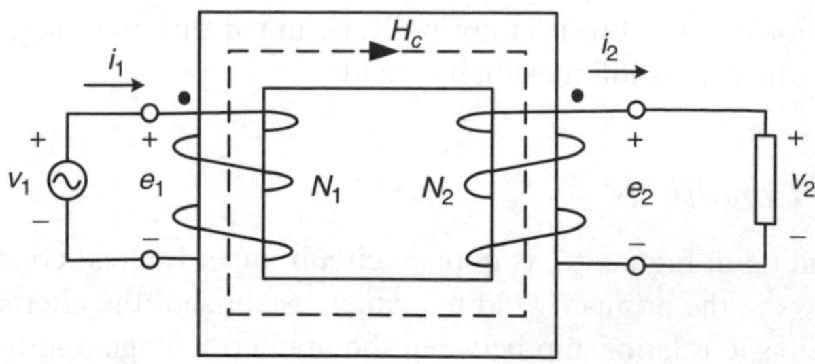

Figure 4.2 Two-winding transformer: load conditions.

4.1.2 Load Conditions

At this point, we shall apply a load to the secondary winding with N_2 turns, causing a current i_2 to flow, as shown in Figure 4.2.

The voltage v_l applied to the primary winding establishes the flux ϕ as before:

$$v_1 = e_1 = N_1 \frac{d\phi}{dt} \tag{4.11}$$

The common core flux links the secondary winding and induces an emf e_2 in the secondary winding and a voltage v_2 across the load:

$$v_2 = e_2 = N_2 \frac{d\phi}{dt} \tag{4.12}$$

Taking the ratio v_1/v_2 from Equations 4.11 and 4.12 yields:

$$\frac{v_1}{v_2} = \frac{N_1}{N_2} \tag{4.13}$$

In terms of the rms values of the voltages:

$$\frac{V_{1\text{rms}}}{V_{2\text{rms}}} = \frac{N_1}{N_2} = a \tag{4.14}$$

In summary, the voltage transformation ratio is directly proportional to the transformer turns ratio a.

The next step is to look at the effect of the load current. The mmf corresponding to the load current is $N_2 i_2$. Ampere's law dictates that the integral of the magnetic field intensity around a closed loop that links the primary and secondary windings is equal to the net mmf. With the direction of the mmf given by the right hand rule, then referring to Figure 4.2, we have:

$$H_c l_c = N_1 i_1 - N_2 i_2 \tag{4.15}$$

The negative sign of $N_2 i_2$ arises because the current in the secondary winding opposes the flux ϕ by the right hand screw rule convention.

The flux density B_c inside the core is related to H_c by the magnetic permeability:

$$B_c = \mu_r \mu_0 H_c \tag{4.16}$$

The flux inside the core is:

$$\phi_c = B_c A_c \tag{4.17}$$

Combining Equations 4.2, 4.15, 4.16 and 4.17 yields:

$$N_1 i_1 - N_2 i_2 = \phi_c \times \mathcal{R}_c \tag{4.18}$$

In the ideal transformer, we assume the core has infinite permeability ($\mu_r \to \infty$), that the resistance of the windings is negligible and that there is no core loss ($\sigma_c \to 0$). Infinite permeability in the core means that the magnetic reluctance is negligible, which, in turn, means that the mmf required to establish the flux is negligibly small. Thus, if the secondary mmf $N_2 i_2$ is established by the load, it must be countered by an mmf $N_1 i_1$ in the primary to satisfy Equation 4.18. In the hypothetical ideal transformer, Equation 4.18 becomes:

$$N_1 i_1 - N_2 i_2 = 0 \tag{4.19}$$

so:

$$N_1 i_1 = N_2 i_2 \tag{4.20}$$

or:

$$\frac{i_1}{i_2} = \frac{N_2}{N_1} \tag{4.21}$$

And in terms of rms values:

$$\frac{I_{1\mathrm{rms}}}{I_{2\mathrm{rms}}} = \frac{N_2}{N_1} = \frac{1}{a} \tag{4.22}$$

Thus, an ideal transformer changes currents in the inverse of the turns ratio of its windings.

There are no losses in an ideal transformer and the input power is equal to the output power:

$$v_1 i_1 = v_2 i_2 \tag{4.23}$$

Therefore:

$$\frac{i_1}{i_2} = \frac{v_2}{v_1} = \frac{N_2}{N_1} = \frac{1}{a} \tag{4.24}$$

4.1.3 Dot Convention

The windings in Figure 4.2 show the input current into the positive voltage terminal of the primary winding and the load current out of the positive voltage terminal of the secondary winding. This conveniently meets the conditions imposed on the mmf by Ampere's law. However, we could just as easily draw the windings as shown in Figure 4.3, and again we

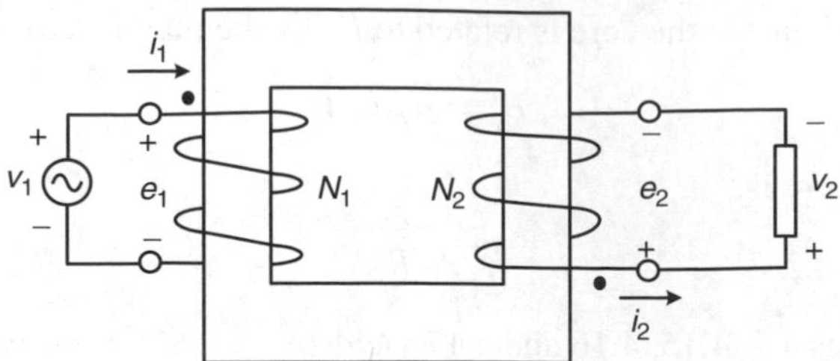

Figure 4.3 The ideal transformer: load conditions, alternative winding.

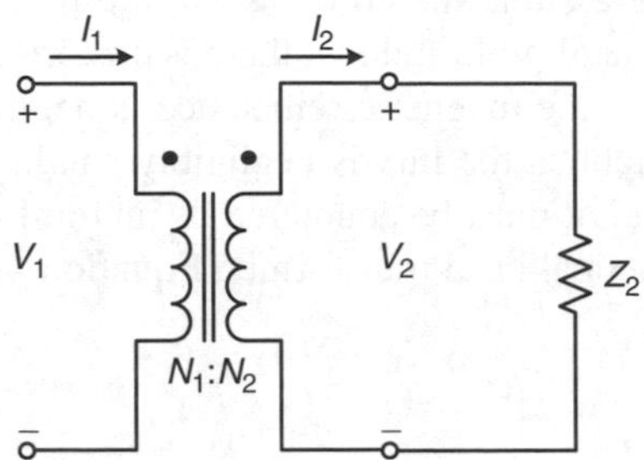

Figure 4.4 Electrical circuit symbol for a transformer.

can judiciously select the positive voltage terminal and the positive direction of current, so the relationship in Equation 4.15 holds. Obviously, great care must be taken in drawing the windings and in selecting the voltage and current polarities; adding more windings makes the situation more complex. In reality, most transformers are enclosed and it is not possible by inspection to tell the direction in which each coil is wound. In order to avoid any confusion, the dot convention is adopted.

The dot markings in Figures 4.2 and 4.3 indicate terminals of corresponding polarity. If one follows through either winding, beginning at the dotted terminal, both windings encircle the core in the same direction with respect to flux (in accordance with the right hand screw rule). Using this convention, the voltages at the dotted terminals are of the same instantaneous polarity for the primary and secondary windings. Similarly, the currents as shown are in phase. According to the convention, the instantaneous currents are in opposite directions through the windings, so therefore their mmfs cancel. In the electrical circuit symbol for a transformer, we can deduce the physical direction of the winding from the dot convention. Thus, the transformers of Figures 4.2 and 4.3 are represented by the electrical circuit symbol in Figure 4.4. The parallel bars between the two windings represent a common ferromagnetic core.

4.1.4 Reflected Impedance

When signals are transmitted in a circuit, the maximum power transfer theorem dictates that maximum power is transferred from a source to a load when the source impedance is equal to the load impedance. The transformer may be used to match the impedances between the source and load.

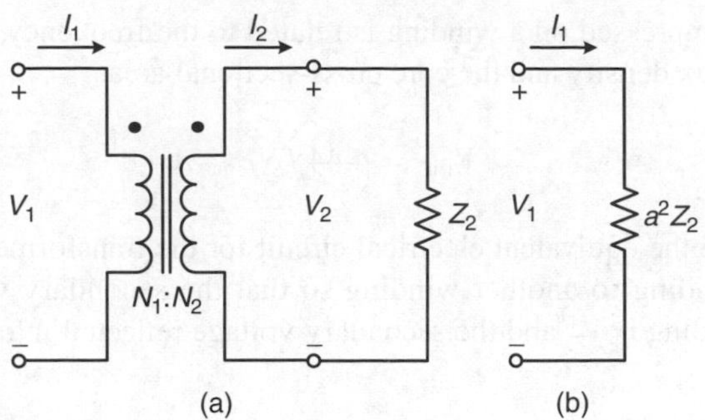

Figure 4.5 Reflected impedance in a transformer winding.

The ratio V_1/I_1 is the impedance seen by the input terminals of the transformer, recalling:

$$V_1 = aV_2 \tag{4.25}$$

$$I_1 = \frac{1}{a} I_2 \tag{4.26}$$

$$Z_2^1 = \frac{V_1}{I_1} = a^2 \frac{V_2}{I_2} = a^2 Z_2 \tag{4.27}$$

where Z_2 is the impedance of the load. Thus, the impedance Z_2 in the secondary may be replaced by the equivalent impedance Z_2^1, as seen from the primary terminals. The transformer equivalent circuit of Figure 4.5(a) is shown in Figure 4.5(b).

Reflected impedance is commonly used in electronic circuits to achieve maximum power transfer.

4.1.5 *Summary*

In an ideal transformer:

1. *Voltages* are transformed in the direct ratio of turns:

$$V_1 = \frac{N_1}{N_2} V_2 = aV_2 \tag{4.28}$$

2. *Currents* are transformed in the inverse ratio of turns:

$$I_1 = \frac{N_2}{N_1} I_2 = \frac{1}{a} I_2 \tag{4.29}$$

3. *Impedances* are transformed in the direct ratio squared:

$$Z_2^1 = \left(\frac{N_1}{N_2}\right)^2 Z_2 = a^2 Z_2 \tag{4.30}$$

The notation Z_2^1 means the secondary impedance Z_2 reflected in the primary.

The voltage impressed on a winding is related to the frequency, the number of turns, the maximum flux density and the core cross-sectional area.

$$V_{\text{rms}} = 4.44\, fNB_{\text{max}}A_c \tag{4.31}$$

In building up the equivalent electrical circuit for the transformer, we can refer quantities in one winding to another winding so that the secondary voltage reflected into the primary winding is V_2^1 and the secondary voltage reflected into the primary winding is V_1^2.

The relationships are:

$$V_2^1 = \frac{N_1}{N_2} V_2 = aV_2 \tag{4.32}$$

$$V_1^2 = \frac{N_2}{N_1} V_1 = \frac{1}{a} V_1 \tag{4.33}$$

4.2 Practical Transformer

So far, we have idealized the transformer to simplify its analysis. However, in a practical transformer, the following factors must be taken into account:

- magnetizing current and core loss;
- winding resistance;
- magnetic leakage flux.

In power electronics applications, winding capacitance may be an issue because a resonance condition can occur at high frequency. We will deal with winding capacitance in Chapter 8.

4.2.1 Magnetizing Current and Core Loss

The current in the primary winding of a transformer plays two roles:

1. It sets up the mutual flux in accordance with Ampere's law.
2. It balances the demagnetizing effect of the load current in the secondary winding.

The net mmf is $N_1I_1 - N_2I_2$ and, in terms of the magnetic circuit law, this may be related to the reluctance of the transformer core:

$$N_1I_1 - N_2I_2 = \phi_m \mathcal{R} \tag{4.34}$$

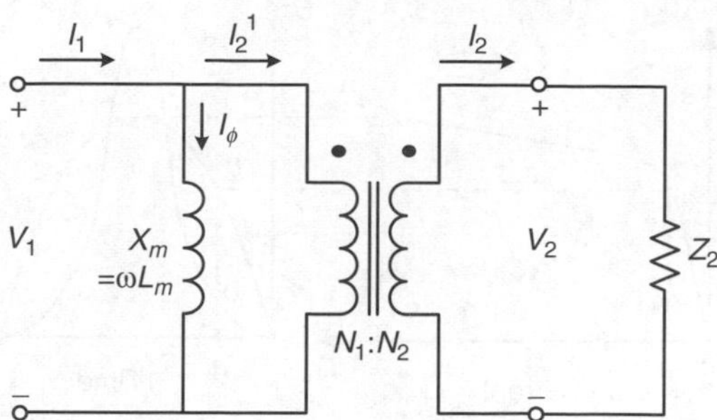

Figure 4.6 Electrical circuit for a transformer with the magnetizing inductance.

We had previously assumed that, in the ideal transformer, the core had infinite permeability. In reality, however, there is a finite permeability and there is an inductance associated with the reluctance of the core. We call this the magnetizing inductance L_m, as described in Section 4.1 and shown in Figure 4.6.

Thus, the primary current has two components: the magnetizing component I_ϕ and a load component reflected into the primary I_2^1:

$$I_2^1 = \frac{1}{a} I_2 \tag{4.35}$$

The instantaneous magnetizing current, i_ϕ, which establishes the flux in the ferromagnetic core, is determined by the magnetic properties of the core.

Let us examine the establishment of the core flux in more detail. Returning to no-load conditions and assuming as before that the winding resistances are negligible, according to Equations 4.4 and 4.6, the applied voltage leads the flux in the core by 90°.

At this point, we need to turn our attention to the magnetizing current. To simplify the construction of the magnetizing current curve, we will use a single value normal magnetization curve for flux versus current, as shown in Figure 4.7. This assumption, in effect, neglects hysteresis. The flux corresponding to the current *oa* is *ab* and the value of the current *oa* on the horizontal current axis of the magnetizing curve is drawn vertically at *b* in the time domain to give the vertical value of the current at that instant. In this manner, the magnetizing current on the time graph is generated and the resultant is shown, reflecting the effect of saturation. The first observation is that the magnetizing current and flux are in phase as expected, because mmf is the product of flux and reluctance.

The second observation is that the distorted magnetizing current contains harmonics, and Fourier analysis shows that these are odd harmonics. This further shows that the percentage of third and fifth harmonics will increase with increased distortion as the core goes further into saturation. Evidently, the peak value of the magnetizing current will increase rapidly as the transformer goes further into saturation. We assumed a single valued magnetization curve and neglected the hysteresis loss; the construction in Figure 4.7 may be repeated for hysteresis by noting the rising and falling values of flux,

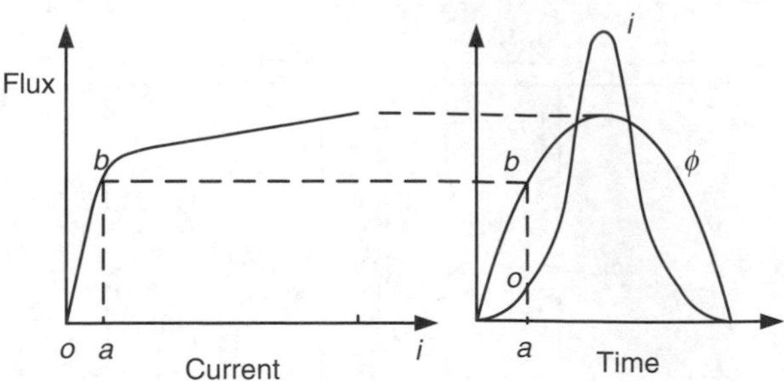

Figure 4.7 Magnetizing current wave shape.

which introduces further distortion of the magnetization curve, but the overall effect on harmonics is not radically altered. Hysteresis is a power loss in the core and, therefore, it will introduce a component of the magnetizing current that is in phase with the applied voltage. In power electronics applications, eddy current loss in the core will also add to the hysteresis loss, and the current representing these losses will also be in phase with the applied voltage.

As a first approximation, the magnetizing current can therefore be split into two components: one in phase with the applied voltage for the core loss I_c; and the other in phase with the flux I_m. This approach allows us to construct phasor diagrams for the transformer. The harmonic components of the magnetizing current could, in some circumstances, lead to resonant conditions with capacitive components of connected circuits. The magnetizing inductance L_m or the magnetizing reactance X_m may represent the flux in the core, and R_c may represent the core loss; the components of current through these circuit elements combine to form the magnetizing current. The magnetizing current is now represented by a shunt branch connected across V_1, consisting of R_c and X_m in parallel, as shown in Figure 4.8.

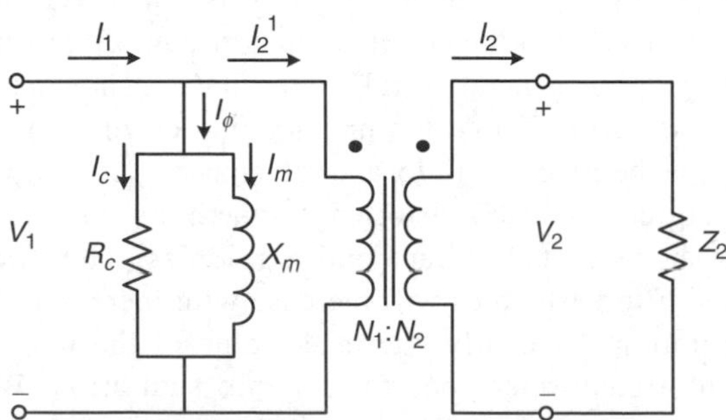

Figure 4.8 Electrical circuit for a transformer with the magnetizing branch.

4.2.2 Winding Resistance

Winding resistance can be represented by the resistances of the wires used in the windings, R_1 and R_2 for primary and secondary, respectively. The AC resistance due to the internal flux in the conductor may be approximated by

$$R_{ac} = R_{dc}\left[1 + \frac{\left(\frac{r_o}{\delta}\right)^4}{48 + 0.8\left(\frac{r_o}{\delta}\right)^4}\right] \tag{4.36}$$

Where δ is the skin depth in the conductor, as defined in Equation 1.21, and r_o is the radius of the conductor. For high-frequency operation, we have to take AC loss in the form of skin effect and proximity effects into account. We will deal with these effects in Chapter 6.

4.2.3 Magnetic Leakage

In the ideal transformer, the same flux links both the primary and secondary circuits. However, in practice, there is always some leakage flux which links only one winding.

Leakage inductance is a property of one winding relative to another. If there is a third winding on the transformer core, the primary-secondary leakage will be distinctly different from the primary-tertiary leakage and so on. Consider the two elementary coils in air presented in Figure 4.9, which constitute an elementary transformer. In Figure 4.9(a), coil 1 has an alternating current i_1 applied and coil 2 is open-circuited. This produces a magnetic field described by the flux lines in the diagram. Some of this flux links the second coil and is thus termed the mutual flux. The common or mutual flux is represented by dotted lines.

The remaining flux does not link the secondary, and is termed leakage flux (depicted by solid lines). In Figure 4.9(b), coil 1 is open-circuited and a current i_2 is applied to coil 2. Again, leakage flux is represented by solid flux lines and mutual flux is denoted by dotted lines. Clearly, the nature of the two leakage fields is quite different. Transformer action occurs

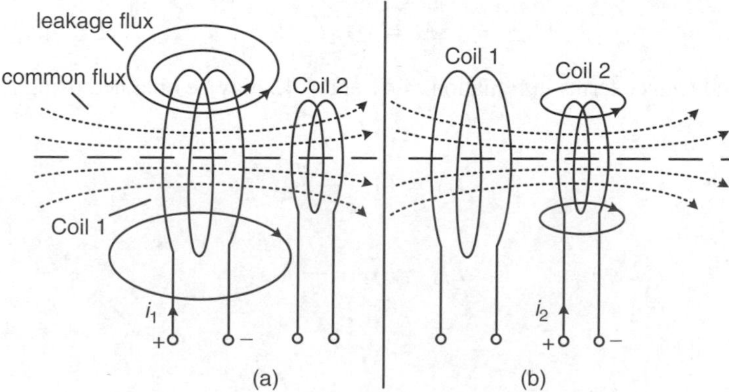

Figure 4.9 Leakage inductance in a transformer.

when current flows in both coils. In this case, the flux linking both coils has three components. If φ_{11} is the total flux linking coil 1 and φ_{22} is the total flux linking coil 2, then [1,2]:

$$\phi_{11} = \phi_{l1} + \phi_{21} + \phi_{12} \tag{4.37}$$

$$\phi_{22} = \phi_{l2} + \phi_{12} + \phi_{21} \tag{4.38}$$

Where φ_{l1} is the leakage flux associated with coil 1 due to i_1 in coil 1, φ_{21} is the flux linking both coils due to the current i_1, φ_{12} is the mutual flux due to the current i_2 in coil 2 and φ_{l2} is the leakage flux of coil 2. Each of these flux elements represents an inductance.From the flux equations, we can derive equations for the voltages on each coil:

$$V_1 = [L_{l1} + L_1]\frac{di_1}{dt} + M_{12}\frac{di_2}{dt} \tag{4.39}$$

$$V_2 = [L_{l2} + L_2]\frac{di_2}{dt} + M_{21}\frac{di_1}{dt} \tag{4.40}$$

From the form of these equations, we can extract the self inductances of coil 1 and coil 2 as L_{11} and L_{22} respectively, where:

$$L_{11} = L_{l1} + L_1 \tag{4.41}$$

$$L_{22} = L_{l2} + L_2 \tag{4.42}$$

Of course, as is always the case with mutual inductance, $M_{12} = M_{21} = M$.

Current in coil 1 sets up flux, some of which links coils 2. By definition, the mutual inductance is the ratio of the flux linking one coil due to the current in the other coil, so that:

$$L_1 = \frac{N_1}{N_2}M \tag{4.43}$$

$$L_2 = \frac{N_2}{N_1}M \tag{4.44}$$

Thus:

$$M = \sqrt{L_1 L_2} \tag{4.45}$$

The leakage inductance terms in Equations 4.41 and 4.42 may be obtained using Equations 4.43 and 4.44:

$$L_{l1} = L_{11} - \frac{N_1}{N_2}M \tag{4.46}$$

$$L_{l2} = L_{22} - \frac{N_2}{N_1}M \tag{4.47}$$

Define:

$$k_1 = \frac{L_1}{L_{11}} = 1 - \frac{L_{l1}}{L_{11}} \tag{4.48}$$

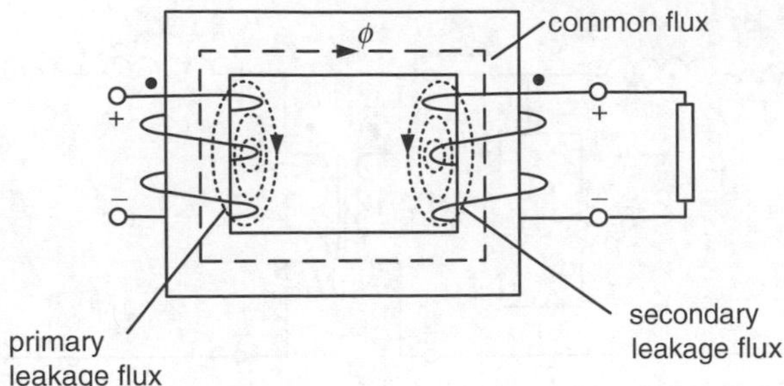

Figure 4.10 Leakage inductance in a transformer.

and:

$$k_2 = \frac{L_2}{L_{22}} = 1 - \frac{L_{l2}}{L_{22}} \tag{4.49}$$

It follows that:

$$k = \sqrt{k_1 k_2} \tag{4.50}$$

and:

$$M = k\sqrt{L_{11}L_{22}} \tag{4.51}$$

k is called the coupling coefficient. Taking this definition of k and the circuit relationships in Equations 4.39 and 4.40, with the appropriate dot convention, yields the classical equivalent electrical circuit representation of the transformer – the model that is normally used in circuit simulations of the transformer. Figure 4.10 shows the physical layout of the winding with the dot convention and the equivalent electrical circuit with coupled inductors. Leakage inductance is affected by high frequency operation; this will be discussed in Chapter 6.

The leakage affects can be represented by primary and secondary leakage inductors or primary and secondary leakage reactances X_{l1} and X_{l2}, respectively.

4.2.4 Equivalent Circuit

The equivalent circuit model is now complete and shown in Figure 4.11(a).

For an ideal transformer:

$$\frac{E_1}{E_2} = a \tag{4.52}$$

As a further step, we can refer all quantities in the secondary to primary in order to obtain the equivalent circuit of Figure 4.11(b).

The leakage reactance of the secondary winding referred to the primary winding is:

$$X_{l2}^1 = a^2 X_{l2} \tag{4.53}$$

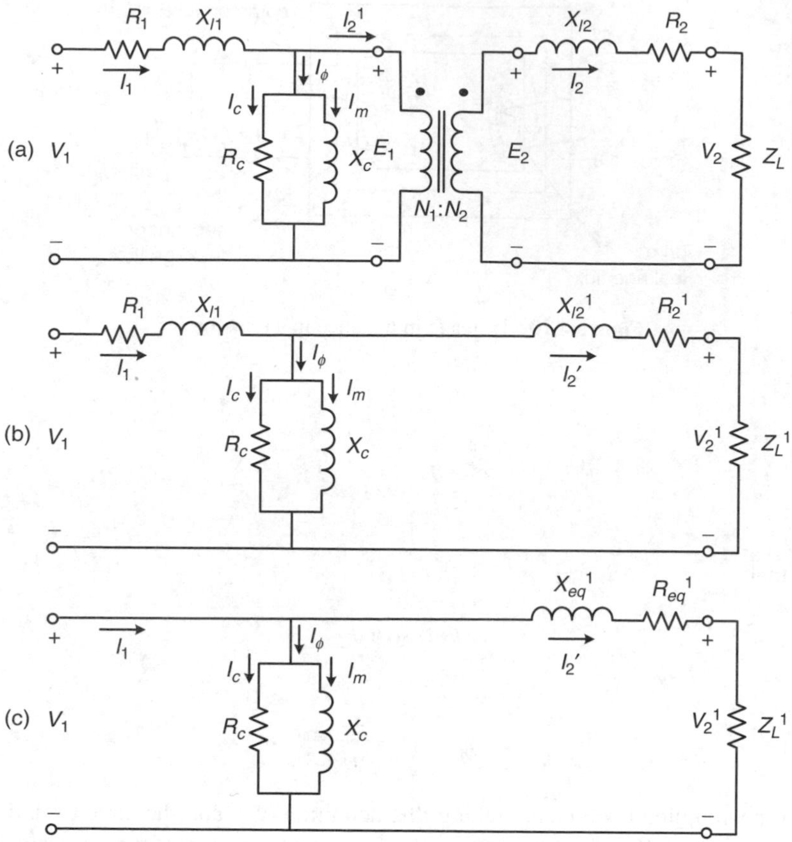

Figure 4.11 Transformer equivalent circuits.

The resistance of the secondary winding referred to the primary winding is:

$$R_2^1 = a^2 R_2 \tag{4.54}$$

The voltage across the secondary winding referred to the primary is given by Equation 4.32:

$$V_2^1 = aV_2 \tag{4.55}$$

And the current in the secondary referred to the primary is:

$$I_2^1 = \frac{1}{a} I_2 \tag{4.56}$$

Finally, we can combine corresponding quantities such as winding resistances and leakage reactances:

$$R_{\text{eq}} = R_1 + a^2 R_2 \tag{4.57}$$

$$X_{\text{eq}} = X_{l1} + a^2 X_{l2} \tag{4.58}$$

The shunt branch representing the core loss and the core magnetization may be moved to the input terminal with little loss of accuracy to obtain Figure 4.11(c). This is an approximation to Figure 4.11(b), since I_ϕ will change slightly, but it is very small, so the error is negligible. However, it greatly simplifies circuit calculations.

This is the usual equivalent electrical circuit that is used to represent transformers and, by traditional circuit analysis load voltage, regulation and transformer efficiency may be determined.

4.3 General Transformer Equations

So far, we have treated the transformer in its traditional role with sinusoidal excitation. For power electronics applications, we have to expand the analysis to include non-sinusoidal excitation and deal with frequencies well above the typical mains frequencies. We will begin by generalizing the equations for voltage, power and losses. The dissipation of the losses will determine the temperature rise in the windings, which will lead to an optimization of the transformer core size. In Chapter 6, we will show that further optimization will result from the analysis of high frequency loss in the windings.

4.3.1 The Voltage Equation

Faraday's law relates the impressed voltage on a winding v to the rate of change of flux density B, recalling Equation 4.11:

$$v = N\frac{d\phi}{dt} = NA_m\frac{dB}{dt} \tag{4.59}$$

where N is the number of turns and A_m is the effective cross-sectional area of the magnetic core. In the case of laminated and tape-wound cores, this is less than the physical area, A_c, due to interlamination space and insulation.

The layout of a typical transformer is shown in Figure 4.12 and the physical parameters are illustrated. The two areas are related by the core stacking factor, k_f ($A_m = k_f A_c$). Typically, k_f is 0.95 for laminated cores.

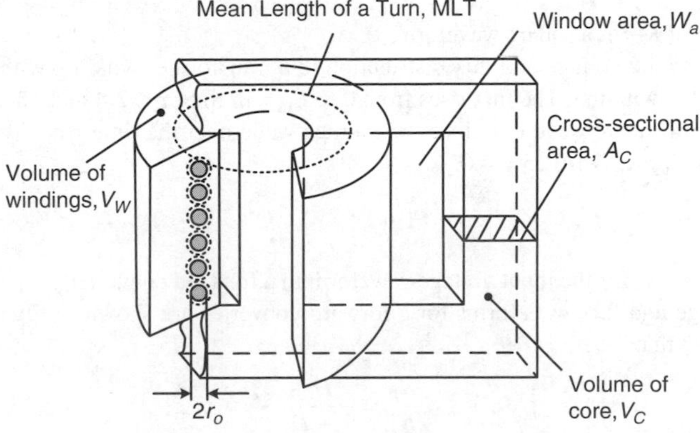

Figure 4.12 Typical layout of a transformer.

The average value of the applied voltage during the interval τ from the point where the flux density is zero to the point where it is at its maximum value ($B_{\max}$) is $\langle v \rangle$. This may be found by integrating Equation 4.59 [3]:

$$\begin{aligned}\langle v \rangle &= \frac{1}{\tau}\int_0^{\tau} v(t)dt \\ &= \frac{1}{\tau} N A_m \int_0^{B_{\max}} dB \\ &= \frac{1}{\tau} N A_m B_{\max}\end{aligned} \tag{4.60}$$

We want to relate this to the rms value of the applied voltage waveform. The form factor k is defined as the ratio of the rms value of the applied voltage waveform to the average value $\langle v \rangle$:

$$k = \frac{V_{\mathrm{rms}}}{\langle v \rangle} \tag{4.61}$$

Combining Equations 4.60 and 4.61 yields:

$$V_{\mathrm{rms}} = \frac{k}{\frac{\tau}{T}} f\, N\, B_{\max} A_m = K_v f\, N\, B_{\max} A_m \tag{4.62}$$

with:

$$K_v = \frac{k}{\frac{\tau}{T}} = \frac{k}{\tau f} \tag{4.63}$$

where $f = \dfrac{1}{T}$ is the frequency of the periodic applied voltage $v(t)$, and T is the period of $v(t)$.

Equation 4.62 has the same form as the classical transformer voltage equation, as given in Equation 4.9, with K, the waveform factor, defined by k, τ and T (or f). For a sinusoidal waveform, $K = 4.44$ and, for a square waveform, $K = 4.0$. The calculation of K for typical power electronic applications will be given in later examples.

Example 4.1

Establish the value of K_v for a square waveform.

Figure 4.13 shows the voltage and flux distribution in a transformer winding with a square wave of voltage applied to the winding. The flux rises from 0 to $B_{\max}$ in time $\tau = T/4$ and, therefore, $\tau/T = 0.25$. The form factor for a square wave is 1 since the average value over the time τ is V_{dc} and the rms value of the waveform is V_{dc}. K_v is $1/0.25 = 4.0$.

Example 4.2

Establish the value of K_v for the input voltage waveform in a forward converter.

The input voltage and flux waveforms for a forward converter are shown in Figure 4.14. The ratio (N_p/N_t) is fixed such that:

$$\frac{N_p}{N_t} = \frac{D}{1 - D}$$

so that the volt-seconds balance in the winding is maintained.

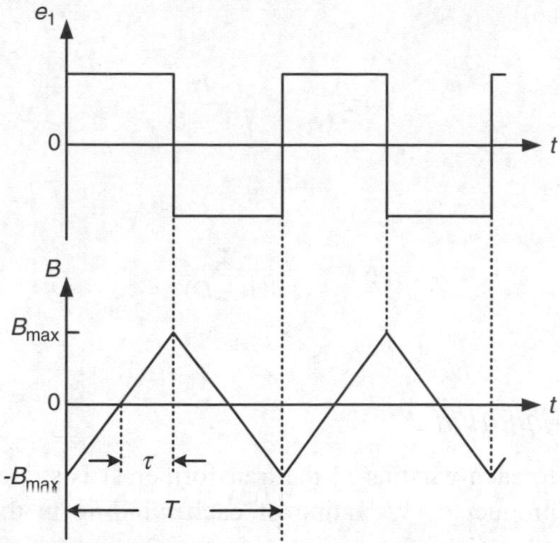

Figure 4.13 Square-wave voltage and flux waveforms.

The flux density increases from 0 to its maximum value in the time $\tau = DT$. The rms value of the voltage waveform in Figure 4.14 is:

$$V_{\text{rms}} = \sqrt{\frac{D}{1-D}} V_{\text{dc}}$$

The average value of the voltage waveform during the time τ is V_{dc}:

$$\langle v \rangle = V_{\text{dc}}$$

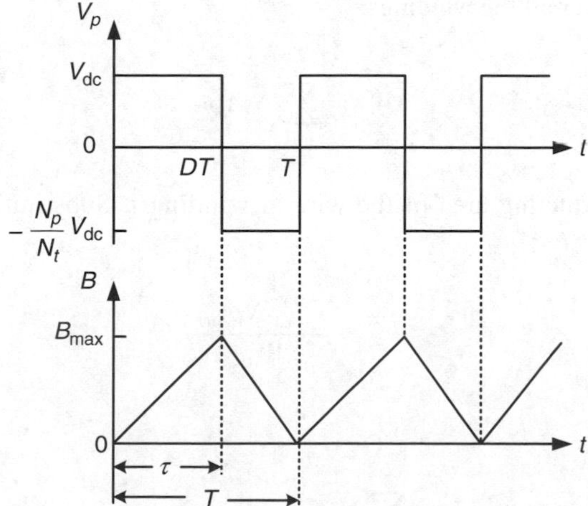

Figure 4.14 Voltage and flux waveforms with duty cycle D.

and thus, from Equation 4.61:

$$k = \frac{V_{\text{rms}}}{\langle v \rangle} = \sqrt{\frac{D}{1-D}}$$

and from Equation 4.63:

$$K_v = \frac{1}{\sqrt{D(1-D)}}$$

4.3.2 *The Power Equation*

Equation 4.62 applies to each winding of the transformer. It is straightforward to calculate the voltage × current product or VA rating of each winding in the transformer. Taking the sum of the VA products in an *n* winding transformer and taking the voltage given by Equation 4.62:

$$\sum \text{VA} = K_v f B_{\text{max}} A_m \sum_{i=1}^{n} N_i I_i \tag{4.64}$$

N_i is the number of turns in winding *i* that carries a current with rms value I_i.

The window utilization factor k_u is the ratio of the total conduction area W_c for all conductors in all windings to the total window winding area W_a of the core:

$$k_u = \frac{W_c}{W_a} \tag{4.65}$$

The total conduction area is related to the number of conductors (turns) and the area of each conductor summed over all the windings:

$$W_c = \sum_{i=1}^{n} N_i A_{wi} \tag{4.66}$$

where A_{wi} is the conducting area of the wire in winding *i*. Substituting Equation 4.66 in Equation 4.65:

$$k_u = \frac{\sum_{i=1}^{n} N_i A_{wi}}{W_a} \tag{4.67}$$

Thus:

$$\sum_{i=1}^{n} N_i A_{wi} = k_u W_a \tag{4.68}$$

The current density in each winding is $J_i = I_i/A_{wi}$. Normally, the wire area and the conduction area are taken as the area of bare conductor. However, we can account for skin effect in a conductor and proximity effect between conductors by noting that the increase in resistance due to these effects is manifested by reducing the effective conduction area. The skin effect factor, k_s, is the increase in resistance (or decrease in conduction area) due to skin effect, and likewise for the proximity effect factor, k_x:

$$k_s = \frac{R_{\text{ac}}}{R_{\text{dc}}} \tag{4.69}$$

$$k_x = \frac{R'_{\text{ac}}}{R_{\text{dc}}} \tag{4.70}$$

Incorporating these definitions into the window utilization factor:

$$k_u = \frac{k_b}{k_s k_x} \tag{4.71}$$

where k_b is the ratio of bare conductor total area to the window winding area.

The definition in Equation 4.71 makes allowance in the window utilization factor for skin and proximity effects. At this point, we do not have analytical expression for skin and proximity effects; we will deal with these effects in detail in Chapter 6. Typically, $k_b = 0.7$, $k_s = 1.3$ and $k_x = 1.3$, giving $k_u = 0.4$.

Combining Equations 4.64 and 4.68, with the same current density J_o in each winding, yields the total VA for all the windings. In Section 3.3, we showed that the optimum distribution of current between multiple windings is achieved when the same current density is applied to each winding:

$$\sum \text{VA} = K_v f B_{\max} k_f A_c J_o k_u W_a \tag{4.72}$$

The product of the core cross-sectional area and the window winding area $A_c W_a$ appears in Equation 4.72 and is an indication of the core size, and is designated window-cross-section product A_p. Rearranging Equation 4.72 relates the summation of the VA ratings of all the windings to the physical, electrical and magnetic properties of the transformer:

$$\sum \text{VA} = K_v f B_{\max} J_o k_f k_u A_p \tag{4.73}$$

4.3.3 Winding Loss

The ohmic or I^2R loss in any of the windings is:

$$I^2R = \rho_w \frac{l_{wi}}{A_{wi}} I_i^2 = \rho_w \frac{N_i \text{MLT}(J_o A_{wi})^2}{A_{wi}} \tag{4.74}$$

The electrical resistivity of the conductor is ρ_w and the length of the conductor in the winding is l_{wi}, i.e. the product of the number of turns N_i and the mean length of a turn (MLT).

The current in the winding is expressed in terms of the current density.The total resistive loss for all the windings is:

$$P_{\text{cu}} = \sum RI^2 = \rho_w \sum_{i=1}^{n} \frac{N_i \text{MLT}(J_o A_{wi})^2}{A_{wi}} \tag{4.75}$$

Incorporating the definition of window utilization factor, k_u (Equation 4.67), and noting that the volume of the windings (fully wound $k_u = 1$) is $V_w = \text{MLT} \times W_a$, then:

$$P_{\text{cu}} = \rho_w V_w k_u J_o^2 \tag{4.76}$$

4.3.4 Core Loss

In general, the core loss per unit volume are given in W/m^3 in accordance with the general Steinmetz equation (Equation 1.29, reproduced below):

$$P_{\text{fe}} = K_c f^{\alpha} B_{\text{max}}^{\beta}$$

where K_c, α and β are constants. Typical values are given in Table 1.1. The core loss includes hysteresis and eddy current losses. The manufacturer's data is normally measured for sinusoidal excitation. In the absence of test data on the design core, the manufacturer's data must be used in establishing the constants in Equation 1.29. The constants may also be deduced from measurements of the core loss; this will be dealt with in Chapter 8.

4.3.5 Optimization

Eliminating the current density in Equation 4.76 using Equation 4.73 yields an expression for the copper or winding loss:

$$P_{\text{cu}} = \rho_w V_w k_u \left[\frac{\sum \text{VA}}{K_v f B_{\text{max}} k_f k_u A_p} \right]^2 = \frac{a}{f^2 B_{\text{max}}^2} \tag{4.77}$$

It is written in this form to show that the copper loss is inversely proportional to both the frequency squared and the flux density squared.

Rewriting Equation 1.29 shows that the core loss is also dependent on the frequency and the flux density:

$$P_{\text{fe}} = V_c K_c f^{\alpha} B_{\text{max}}^{\beta} = b f^{\alpha} B_{\text{max}}^{\beta} \tag{4.78}$$

The total loss is made up of the combined core and winding losses:

$$P = \frac{a}{f^2 B_{\text{max}}^2} + b f^{\alpha} B_{\text{max}}^{\beta} \tag{4.79}$$

The domain of P is in the first quadrant of the f-$B_{\max}$ plane. P is positive everywhere and it is singular along the axes. If $\alpha = \beta$, P has a global minimum $\left\{ \dfrac{dP}{d(fB_m)} = 0 \right\}$ at the frequency flux density product:

$$f_o B_o = \left[\frac{2a}{\beta b} \right]^{\frac{1}{\beta + 2}} \tag{4.80}$$

For $\alpha = \beta = 2$, with Equations 4.77 and 4.78, the frequency flux density product is:

$$f_o B_o = \sqrt[4]{\frac{\rho_w V_w k_u}{\rho_c V_c K_c}} \sqrt{\frac{\sum \mathrm{VA}}{K_v k_f k_u A_p}} \tag{4.81}$$

Given that B_o must be less than the saturation flux density B_{sat}, there is a critical frequency, given by Equation 4.81 above, by which the total loss may be minimized by selecting an optimum value of flux density which is less than the saturation value ($B_o < B_{\mathrm{sat}}$). Equation 4.81 shows that $f_o B_o$ is related to power density in the transformer, since A_p is related to core size.

In the more general case ($\alpha \neq \beta$), there is no global minimum. The minimum of P at any given frequency is obtained by taking the partial derivative with respect to $B_{\max}$ and setting it to zero:

$$\frac{\partial P}{\partial B_{\max}} = -\frac{2a}{f^2 B_{\max}^3} + \beta b f^\alpha B_{\max}^{\beta - 1} = 0 \tag{4.82}$$

The minimum loss occurs when:

$$P_{\mathrm{cu}} = \frac{\beta}{2} P_{\mathrm{fe}} \tag{4.83}$$

for a fixed frequency f.

The minimum value of P at any given flux density is obtained by taking the partial derivative with respect to f and setting it to zero. The minimum loss occurs when:

$$P_{\mathrm{cu}} = \frac{\alpha}{2} P_{\mathrm{fe}} \tag{4.84}$$

for a fixed flux density $B_{\max}$.

Evaluation of Equation 4.84 at $B_o = B_{\mathrm{sat}}$ gives the critical frequency above which the total loss is minimized by operating at an optimum value of flux density that is less than the saturation value ($B_o < B_{\mathrm{sat}}$):

$$f_o^{\alpha+2} B_o^{\beta+2} = \frac{2}{\alpha} \frac{\rho_w V_w k_u}{\rho_c V_c K_c} \left[\frac{\sum \mathrm{VA}}{K_v k_f k_u A_p} \right]^2 \tag{4.85}$$

The nature of Equation 4.85 is illustrated in Figure 4.15. The two sets of curves shown are for low frequency (say, 50 Hz) and high frequency (say, 50 kHz).

At 50 Hz, the optimum flux density (at point B) is greater than the saturation flux density, so therefore the minimum loss achievable is at point A. However, the winding and core losses are not equal. At 50 kHz, the optimum flux density is less than the saturation flux density and the

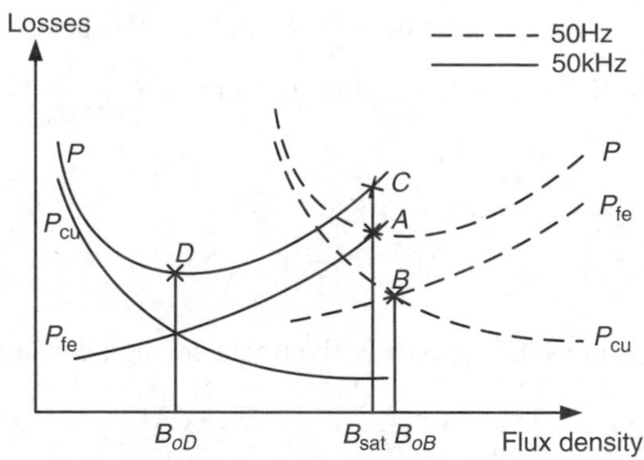

Figure 4.15 Winding, core, and total losses at different frequencies.

core and winding losses are equal at the optimum point *D*. The first step in a design is to establish whether the optimum flux density given by the optimization criterion in Equation 4.83 is greater or less than the saturation flux density. We will return to this in Chapter 5, when we will set out the design methodology based on the analysis presented above.

4.4 Power Factor

The VA rating of each winding is required in order to proceed with the transformer design, as indicated by Equation 4.73. Traditionally, the concept of power factor was applied to sinusoidal waveforms of current and voltage, and the power factor is simply the cosine of the phase angle θ between the waveforms. In power electronics, waveforms are very often a mixture of sinusoids and square waves. The definition of power factor is:

$$k_p = \frac{\langle p \rangle}{V_{\text{rms}} I_{\text{rms}}} \tag{4.86}$$

where $\langle p \rangle$ is the average power delivered at the terminals where V_{rms} and I_{rms} are the rms values of voltage and current respectively.

Consider a case where the voltage is a sinusoid and the current is a square wave. The average power is:

$$\langle p \rangle = \left(\frac{1}{T}\right) \int_0^T v(t)i(t) = V_{\text{rms}} I_{1\text{rms}} \cos(\theta) \tag{4.87}$$

Since $v(t)$ is a pure sinusoid, only the fundamental of the current waveform will yield a DC or average value of the product $v(t)i(t)$. V_{rms} is the rms value of voltage and $I_{1\text{rms}}$ is the rms value of the fundamental of the current waveform. θ is the phase angle between the voltage waveform and the fundamental of the current waveform.

Equation 4.87 can be rewritten:

$$\langle p \rangle = V_{\text{rms}} I_{\text{rms}} \frac{I_{1\text{rms}}}{I_{\text{rms}}} \cos(\theta) = V_{\text{rms}} I_{\text{rms}} k_d k_\theta \tag{4.88}$$

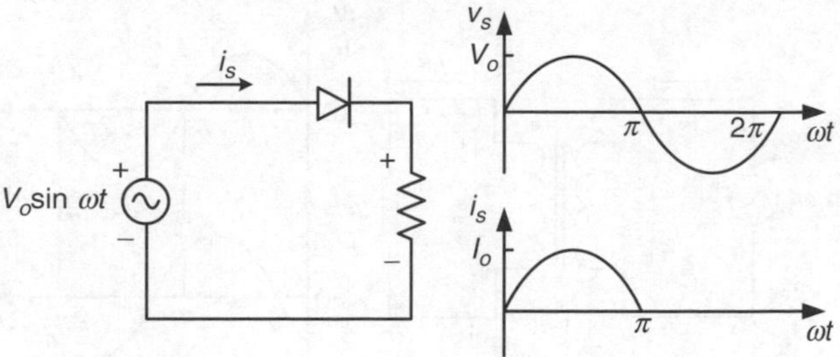

Figure 4.16 Half-wave rectifier with a resistive load.

Where k_d is called the distortion factor and k_θ is called the displacement factor. Note that I_{rms} is the rms value of the total waveform including all the harmonics. Clearly, for pure sinusoids, k_d is unity and k_θ is equal to cos θ, the classical definition of power factor.

Example 4.3

Calculate the power factor of the source in a half-wave rectifier with a resistive load. The circuit and associated voltage v_s and current i_s waveforms are shown in Figure 4.16.

We can write down the rms values of the source voltage and current using the relationships:

$$V_{s\text{rms}} = \frac{V}{\sqrt{2}}$$

$$I_{s\text{rms}} = \frac{I}{2}$$

where V is the peak value of the voltage waveform and I is the peak value of the current waveform.

The average power delivered to the load is:

$$\langle p \rangle = \frac{VI}{2\pi}\int_0^{\pi} \sin(\omega t)^2 d(\omega t) = \frac{VI}{4}$$

The power factor is:

$$k_p = \frac{\langle p \rangle}{V_{\text{rms}} I_{\text{rms}}} = \frac{1}{\sqrt{2}}$$

Example 4.4

Calculate the $\sum$VA rating for a transformer with a sinusoidal input. The transformer has a 1 : 1 turns ratio, and associated waveforms for voltage and current are shown in Figure 4.17.

The load power factor is denoted by k_p and the output power by P_o, where $P_o = V_s I_s k_p$. The power factor is simply $k_p = \cos\theta$, because the voltage and current waveforms are sinusoidal.

For the overall efficiency of a power electronics converter, transformer losses play a minor role in most cases and, for practical designs, we shall assume that the efficiency is 100%.

The VA rating of the primary winding is:

$$\text{VA}_p = V_p I_p = \frac{P_o}{k_p}$$

k_p is the power factor of the primary winding and is the same as the power factor of the load, since the reflected voltage and current from the secondary to the primary maintain the phase angle θ.

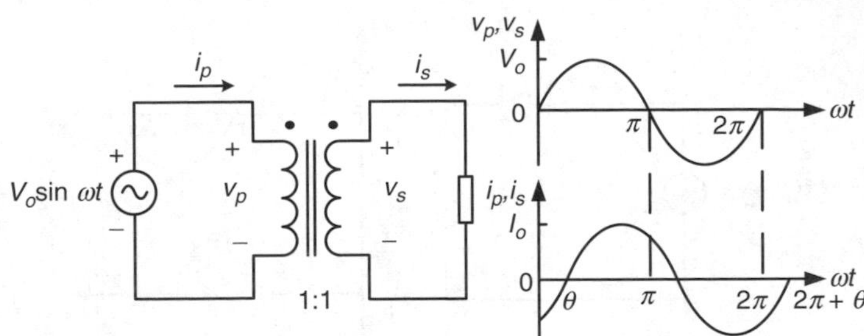

Figure 4.17 Transformer and associated waveforms.

In the secondary winding, the VA rating is:

$$\mathrm{VA}_s = V_s I_s = \frac{P_o}{k_p}$$

The total VA rating for both the primary and secondary windings is now:

$$\sum \mathrm{VA} = (1+1)\frac{P_o}{k_p} = \frac{2}{k_p} P_o$$

Example 4.5 Centre-tapped rectifier

Calculate the $\sum$ VA rating for a centre tapped rectifier shown in Figure 4.18. The waveforms are shown for voltage in each of the output windings and for the input winding with the notation shown in the transformer diagram. The waveforms are shown for both a resistive load and an inductive load, where the inductance is assumed to be very large.

The simplest approach is to calculate the power factor of each winding, recognizing that each of the secondary windings handles half the throughput of power.

The average power delivered through each secondary winding is:

$$\langle p \rangle = \frac{P_o}{2}.$$

(a) Resistive load

The average output power is defined as:

$$P_o = \frac{V_o}{\sqrt{2}} \frac{I_o}{\sqrt{2}} = \frac{V_o I_o}{2}$$

Secondary windings

The rms value of the secondary voltage is:

$$V_{s1} = V_{s2} = \frac{V_o}{\sqrt{2}}$$

$$I_{s1} = I_{s2} = \frac{I_o}{2}$$

The average power through each of the secondary windings is:

$$\langle p_s \rangle = \frac{P_o}{2}$$

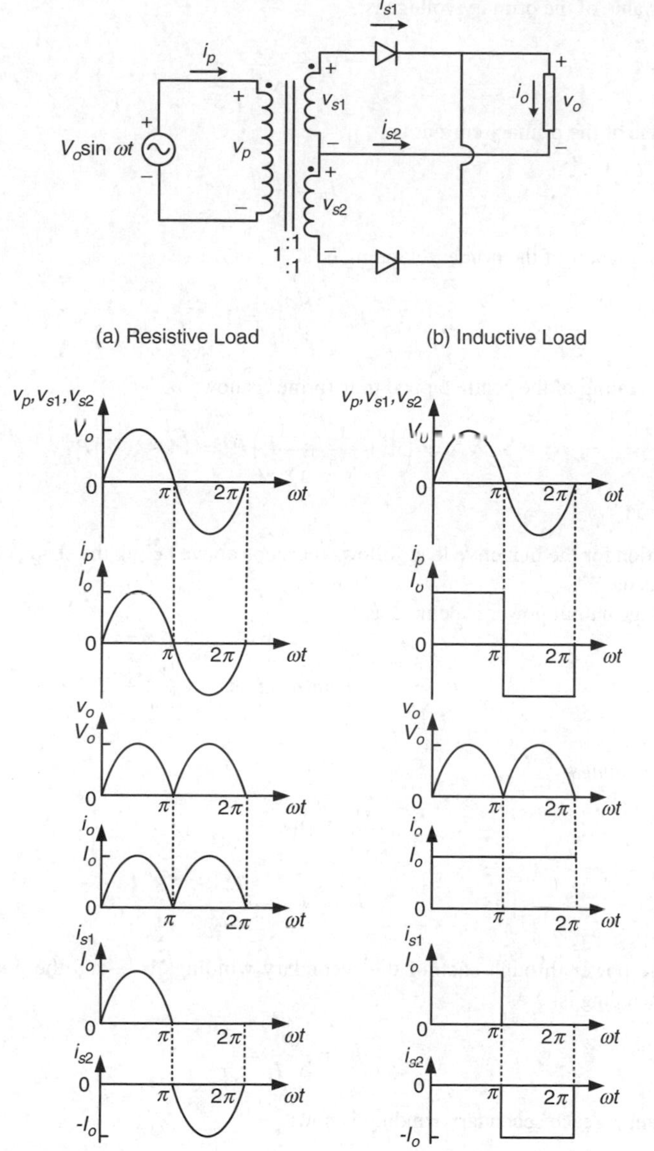

Figure 4.18 Centre-tapped rectifier and associated waveforms for (a) resistive and (b) inductive loads.

and the power factor of each secondary winding is:

$$k_p = \frac{\langle p_s \rangle}{V_s I_s} = \frac{1}{\sqrt{2}}$$

The VA rating of each secondary winding is now:

$$\mathrm{VA}_s = \frac{\langle p_s \rangle}{k_p} = \frac{P_o}{\sqrt{2}}$$

Primary winding

The rms value of the primary voltage is:

$$V_p = \frac{V_o}{\sqrt{2}}$$

The rms value of the primary current is:

$$I_p = \frac{I_o}{\sqrt{2}}$$

and the power factor of the primary winding is:

$$k_p = \frac{\langle p \rangle}{V_p I_p} = 1$$

The total VA rating of the centre tapped transformer is now:

$$\sum \text{VA} = \left(1 + \frac{1}{\sqrt{2}} + \frac{1}{\sqrt{2}}\right) P_o = \left(1 + \sqrt{2}\right) P_o$$

(b) Inductive load

The solution for the inductive load follows the steps above taking the shape of the current waveform into account.

The average output power is defined as:

$$P_o = \frac{V_o I_o}{\pi} \int_0^{\pi} \sin(\omega t) d(\omega t) = \frac{2 V_o I_o}{\pi}$$

Secondary windings

$$V_{s1} = V_{s2} = \frac{V_o}{\sqrt{2}}$$

$$I_{s1} = I_{s2} = \frac{I_o}{\sqrt{2}}$$

The average power through each of the secondary windings is $\frac{P_o}{2}$ and the power factor of each secondary winding is:

$$k_p = \frac{\langle p \rangle}{V_s I_s} = \frac{2}{\pi}$$

The VA rating of each secondary winding is now:

$$\text{VA} = \frac{\langle p \rangle}{k_p} = \frac{\pi}{4} P_o$$

Primary winding

$$V_p - \frac{V_o}{\sqrt{2}}$$

$$I_p = I_o$$

$$k_p = \frac{\langle p \rangle}{V_p I_p} = \frac{2\sqrt{2}}{\pi} = 0.9$$

The total VA rating of the centre tapped transformer with an inductive load is now:

$$\sum \text{VA} = \left(\frac{1}{0.9} + \frac{\pi}{4} + \frac{\pi}{4}\right) P_o = \left(\frac{1}{0.9} + \frac{\pi}{2}\right) P_o$$

4.5 Problems

4.1 List five different types of transformers and describe their applications.

4.2 Explain why the rms value of the primary current of a transformer is not zero when the secondary current is zero.

4.3 Derive k and K_v for a sine wave input voltage waveform in a transformer.

4.4 Derive k and K_v for a pulsed waveform with a duty cycle D input voltage waveform in a transformer.

4.5 Derive k and K_v for a triangular waveform with a duty cycle D input voltage waveform in a transformer.

4.6 Calculate the power factor for the full-wave bridge rectifier with an inductive load shown in Figure 4.19.

4.7 Calculate the input power factor for the full-wave bridge rectifier with capacitive filter shown in Figure 4.20.

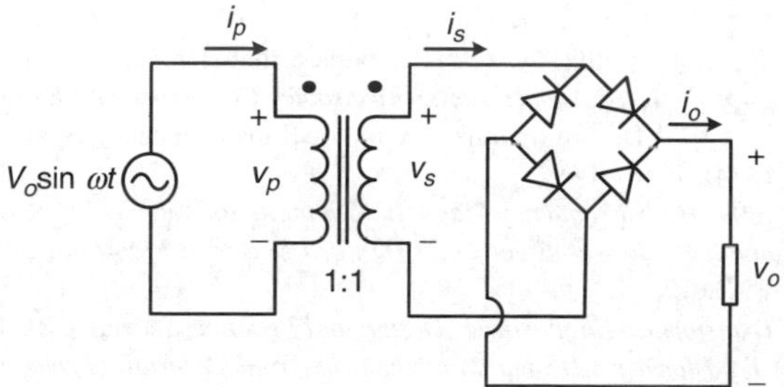

Figure 4.19 Full-wave bridge rectifier.

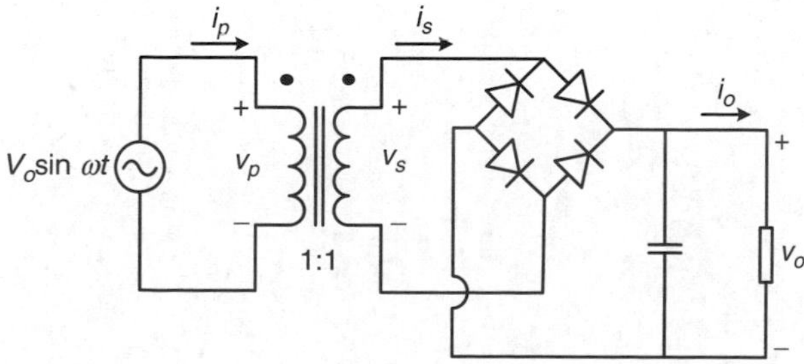

Figure 4.20 Full-wave bridge rectifier with a capacitive filter.

References

1. Hurley, W.G., Wilcox, D.J., and McNamara, P.S. (1991) Calculation of short circuit impedance and leakage impedance in transformer windings. *Proceedings of the IEEE Power Electronics Specialists Conference, PESC*, pp. 651–658.
2. Hurley, W.G. and Wilcox, D.J. (1994) Calculation of leakage inductance in transformer windings. *IEEE Transactions on Power Electronics* **9** (1), 121–126.
3. Hurley, W.G., Wolfle, W.H., and Breslin, J.G. (1998) Optimized transformer design: inclusive of high-frequency effects. *IEEE Transactions on Power Electronics* **13** (4), 651–659.

Further Reading

1. Blume, L.F. (1982) *Transformer Engineering*, John Wiley & Sons, New York.
2. Bueno, M.D.A. (2001) *Inductance and Force Calculations in Electrical Circuits*, Nova Science Publishers, Huntington.
3. Del Vecchio, R.M., Poulin, B., Feghali, P.T. *et al.* (2001) *Transformer Design Principles: With Applications to Core-Form Power Transformers*, 1st edn, CRC Press, Boca Raton, FL.
4. Dowell, P.L. (1966) Effects of eddy currents in transformer windings. *Proceedings of the Institute of Electrical and Electronic Engineers* **113** (8), 1387–1394.
5. Erickson, R.W. (2001) *Fundamentals of Power Electronics*, 2nd edn, Springer, Norwell, MA.
6. Fitzgerald, A.E., Kingsley, C. Jr, and Umans, S.D. (2002) *Electric Machinery*, 6th edn, McGraw-Hill, New York.
7. Flanagan, W.M. (1992) *Handbook of Transformer Design and Application*, 2nd edn, McGraw-Hill, New York.
8. Georgilakis, P.S. (2009) *Spotlight on Modern Transformer Design (Power Systems)*, 1st edn, Springer, New York.
9. Hoke, A.F. and Sullivan, C.R. (2002) An improved two-dimensional numerical modeling method for E-core transformers. *Proceedings of the IEEE Applied Power Electronics Conference and Exposition, APEC*, pp. 151–157.
10. Jieli, L., Sullivan, C.R., and Schultz, A. (2002) Coupled-inductor design optimization for fast-response low-voltage. *Proceedings of the IEEE Applied Power Electronics Conference and Exposition, APEC*, pp. 817–823.
11. Judd, F. and Kressler, D. (1977) Design optimization of small low-frequency power transformers. *IEEE Transactions on Magnetics* **13** (4), 1058–1069.
12. Kazimierczuk, M.K. (2009) *High-Frequency Magnetic Components*, John Wiley & Sons, Chichester.
13. Krein, P.T. (1997) *Elements of Power Electronics (Oxford Series in Electrical and Computer Engineering)*, Oxford University Press, Oxford.
14. Kulkarni, S.V. (2004) *Transformer Engineering: Design and Practice*, 1st edn, CRC Press, New York.
15. B.H.E. Limited (2004) *Transformers: Design, Manufacturing, and Materials (Professional Engineering)*, 1st edn, McGraw-Hill, New York.
16. McAdams, W.H. (1954) *Heat Transmission*, 3rd edn, McGraw-Hill, New York.
17. McLyman, C.W.T. (1997) *Magnetic Core Selection for Transformers and Inductors*, 2nd edn, Marcel Dekker Inc., New York.
18. McLyman, C.W.T. (2002) *High Reliability Magnetic Devices*, 1st edn, Marcel Dekker Inc., New York.

5

Transformer Design

Transformers used in power electronics applications normally serve to provide isolation from the input mains and to reduce voltage stress on switching components by more closely matching the operating voltage to the switch voltage ratings. We saw in Chapter 4 that the size of the transformer is reduced as the frequency of operation is increased. These three objectives may be simultaneously achieved in a DC-DC converter.

While it is true to say that the design of conventional power transformers is well documented, the additional issues that arise in high-frequency operation need special attention. One approach is to develop empirical design rules, but these tend to lead to conservative designs that often mean core sizes that are too large for the intended application. The switching waveforms encountered in modern converters mean that non-sinusoidal excitation occurs at high frequencies. We have seen in Chapters 1 and 4 that the losses are frequency dependant: hysteresis loss and eddy current loss in ferromagnetic cores and skin and proximity effects in windings.

Traditionally, the starting point for transformer design is that winding loss is approximately equal to the core loss. However, in a typical power frequency transformer, the ratio may be as high as 6 : 1. This was explained in Section 4.3.5 by virtue of the fact that flux density is limited by its saturation value. When the optimization is not limited by the saturation flux density, then as the frequency increases, the actual flux density decreases and, in many cases, the operating flux density may be a fraction of the saturation flux density.

It is not sufficient to confine the design to electrical issues, because the heat generated in the transformer must be dissipated through the surface of the transformer (consisting of the exposed core surface and the winding surface). The dissipation of the losses will ultimately determine the maximum temperature inside the transformer. The maximum temperature may be limited by the Curie temperature in the core material, or by the temperature rating of the insulation used in the conductors.

The design of a modern transformer must incorporate elements of circuit analysis, magnetic circuit laws and heat transfer. In Chapter 4, we restated the traditional equations for transformers to take high frequency, power factor and non-sinusoidal excitation into account – the fundamental relationships as presented form the bedrock of the approach to the design methodology. Dimensional analysis of the various relationships is invoked to establish a set of

Transformers and Inductors for Power Electronics: Theory, Design and Applications, First Edition.
W. G. Hurley and W. H. Wölfle.

robust design rules. Finally, the overall approach is informed by the avoidance of unnecessary design factors that often tend to cloud the main design criteria.

The purpose of a transformer is to transfer energy from the input winding to the output winding through electromagnetic induction. The overall aim of the design methodology aim is to optimize this energy transfer in a given application, in terms of minimizing the core and winding losses. Optimization can be based on cost, weight or volume; all of these criteria lead to the conclusion that the core loss is approximately equal to the winding loss in an optimized design.

The main factors effecting transformer operation are the operating temperature, the electrical frequency and the maximum flux density. We demonstrated in Chapter 4 that there is a critical frequency above which the total loss can be minimized by selecting a value of maximum flux density that is less than the saturation flux density. Below the critical frequency, the transfer of energy through the transformer is restricted by the limitation that maximum operating flux density cannot be greater than the saturation value for the ferromagnetic material used in the core construction.

In this chapter, we will establish the design rules to select the core and winding in a transformer. Several examples from power electronics are presented to illustrate the robust nature of the methodology. The approach is based on design rules that are derived from first principles, thus ensuring the general applicability of the design algorithms. The main specifications are power rating, frequency and temperature rise. When the core and winding are selected, the overall transformer is evaluated for temperature rise and efficiency. Several examples are included to illustrate the design methodology, including a centre-tapped rectifier transformer, a forward converter and a push-pull converter.

5.1 The Design Equations

In Chapter 4, we established the basic electrical relationships in the transformer. In the case of the inductor design, we established an expression for A_p in terms of the stored energy, and it is possible to establish an expression for A_p in the case of the transformers in terms of the power transferred.

5.1.1 Current Density in the Windings

The optimum value of current density in the windings may be found from the optimum criterion (Equation 4.83) using the equations for copper loss (Equation 3.22), core loss (Equation 1.29) and thermal heat transfer (Equation 3.14).

From Equation 4.83, for a fixed frequency:

$$P_{\text{cu}} + P_{\text{fe}} = \frac{\beta + 2}{\beta} P_{\text{cu}} \tag{5.1}$$

and:

$$P_{\text{cu}} + P_{\text{fe}} = \frac{\beta+2}{\beta}\left[\rho_w V_w k_u J_o^2\right] = h_c A_t \Delta T \tag{5.2}$$

Extracting the current density:

$$J_o = \sqrt{\frac{\beta}{\beta+2}\frac{h_c A_t \Delta T}{\rho_w V_w k_u}}. \tag{5.3}$$

Employing the dimensional analysis equations (Equations 3.25 to 3.27) and taking $\beta = 2$ gives an expression for the current density in terms of the temperature rise in the windings and the core-window winding area product:

$$J_o = K_t \sqrt{\frac{\Delta T}{2k_u}} \frac{1}{\sqrt[8]{A_p}} \tag{5.4}$$

where K_t is defined in Equation 3.29.

5.1.2 Optimum Flux Density unlimited by Saturation

The optimum design is at point A in Figure 5.1, and the optimum flux density in the core is not limited by saturation. The optimum conditions established by Equation 5.1 may be exploited to establish a formula for A_p in terms of the design specifications: output power, frequency and temperature rise.

Taking B_o as the flux density at the optimum operating point and J_0 as the corresponding value of the current density given by Equation 5.4, and substituting these values in the power equation (Equation 4.73), gives us an expression for the core window winding area product as:

$$A_p = \left[\frac{\sqrt{2}\sum \text{VA}}{K_v f B_o k_f K_t \sqrt{k_u \Delta T}}\right]^{8/7} \tag{5.5}$$

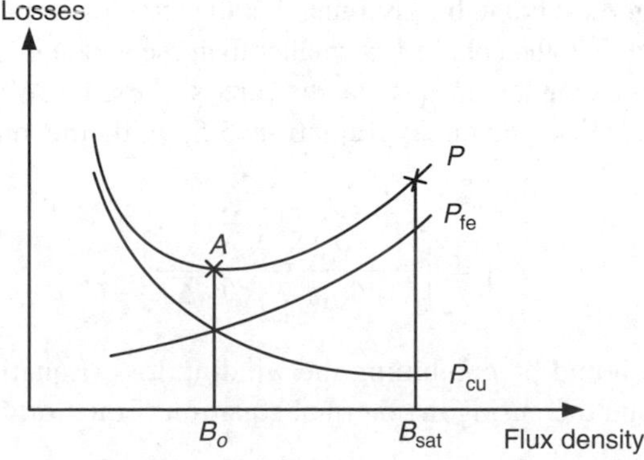

Figure 5.1 Winding, core, and total losses unlimited by saturation.

At this point, we do not know the value of B_o, but this is required to size the core. We can find B_o by taking a closer look at the optimum conditions.

Taking $\beta = 2$ as before means that the core and winding losses are equal and the copper loss is equal to half the total loss. Therefore, with P as the total of the copper and iron losses (the same as Q in Equation 3.14):

$$\frac{\left[\frac{P}{2}\right]^{\frac{2}{3}}}{P_{cu}^{\frac{1}{12}} P_{fe}^{\frac{7}{12}}} = 1 \tag{5.6}$$

The rather unusual format of Equation 5.6 is designed to extract B_o.

Substituting for P with Equation 5.2, P_{cu} with Equation 4.83 and P_{fe} with Equation 1.29, and incorporating the dimensional analysis given by Equations 3.25 to 3.27, yields the following equation

$$\frac{[h_c k_a \Delta T]^{\frac{2}{3}}}{2^{\frac{2}{3}} [\rho_w k_w k_u]^{\frac{1}{12}} \left[k_c K_c f^{\alpha} B_o^2\right]^{\frac{7}{12}} (J_o A_p)^{\frac{1}{6}}} = 1 \tag{5.7}$$

Extracting $(JA_p)^{1/6}$ from the power equation (Equation 4.73) yields an expression for B_o:

$$B_o = \frac{[h_c k_a \Delta T]^{\frac{2}{3}}}{2^{\frac{2}{3}} [\rho_w k_w k_u]^{\frac{1}{12}} [k_c K_c f^{\alpha}]^{\frac{7}{12}}} \left[\frac{K_v f k_f k_u}{\Sigma \mathrm{VA}}\right]^{\frac{1}{6}} \tag{5.8}$$

The optimum flux density may be found from the specifications of the application and the material constants.

5.1.3 Optimum Flux Density limited by Saturation

The optimum design is at point B in Figure 5.2, and the optimum flux density in the core is limited by saturation. The value of B_{max} in the voltage equation is fixed by B_{sat}.

The initial estimate of A_p is found by assuming that the total loss is equal to twice the copper loss (at point C in Figure 5.2, the core loss is smaller than the winding loss). Assuming that the total loss is double the winding loss means that the core is oversized. We will refine this later.

The initial value of A_p is given by Equation 5.5, with the maximum flux density given by B_{sat}:

$$A_{p1} = \left[\frac{\sqrt{2} \sum \mathrm{VA}}{K_v f B_{sat} k_f K_t \sqrt{k_u \Delta T}}\right]^{\frac{8}{7}} \tag{5.9}$$

The current density is found by combining the winding loss (Equation 4.76) and the core loss (Equation 1.29), and then using the thermal equation (Equation 5.2):

$$J_o^2 = \frac{h_c A_t \Delta T}{\rho_w V_w k_u} - \frac{V_c K_c f^{\alpha} B_{max}^{\beta}}{\rho_w V_w k_u} \tag{5.10}$$

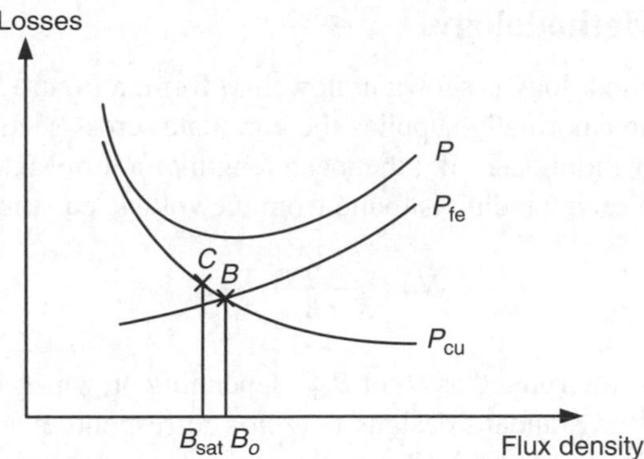

Figure 5.2 Winding, core and total losses unlimited by saturation.

Extracting J_o^2 from the power equation (Equation 4.73) and invoking the dimensional analysis for A_t, V_c and V_w in Equations 3.24 to 3.26 gives an expression for A_p:

$$\frac{k_c K_c f^\alpha B_{\max}^\beta}{\rho_w k_w k_u} A_p^2 - \frac{h_c k_a \Delta T}{\rho_w k_w k_u} A_p^{7/4} + \left[\frac{\Sigma \mathrm{VA}}{K_v f B_{\max} k_f k_u}\right]^2 = 0 \tag{5.11}$$

We now have an expression for A_p in the form:

$$f(A_p) = a_0 {A_p}^2 - a_1 A_p^{\frac{7}{4}} + a_2 = 0 \tag{5.12}$$

The roots of $f(A_p)$ are found numerically using the Newton Raphson Method:

$$\begin{aligned} A_{p_{i+1}} &= A_{p_i} - \frac{f(A_{p_i})}{f'(A_{p_i})} \\ &= A_{p_i} - \frac{a_0 A_{p_i}^2 - a_1 {A_{p_i}}^{\frac{7}{4}} + a_2}{2a_0 A_{p_i} - \frac{7}{4} a_1 A_{p_i}^{\frac{3}{4}}} \end{aligned} \tag{5.13}$$

A_{pi} is given by Equation 5.9, and one iteration should be sufficient.

Finally, we need to calculate the corresponding value of the current density. Proceeding as before, with dimensional analysis for A_t in Equation 5.10, gives us an expression for J_o in terms of the transformer specifications and material properties:

$$J_o = \sqrt{\frac{h_c k_a \sqrt{A_p} \Delta T - V_c K_c f^\alpha B_{\max}^\beta}{\rho_w V_w k_u}} \tag{5.14}$$

Note that the volume of the winding (fully wound $k_u = 1$) is given by $V_w = \mathrm{MLT} \times W_a$, and the volume of the core is $V_c = l_c \times A_c$.

5.2 The Design Methodology

The overall design methodology is shown in flowchart form in Figure 5.3.

The core manufacturer normally supplies the core data: cross-section, A_m (or A_c with no laminations), window winding area W_a, the mean length of a turn MLT and the core volume. The number of turns in each winding is found from the voltage equation (Equation 4.62):

$$N = \frac{V_{\text{rms}}}{K_v f B_{\max} A_m}. \tag{5.15}$$

In this equation, $B_{\max}$ is interpreted as B_o or B_{sat}, depending on which is lower.

The selected core from standard designs may not correspond exactly to the value of A_p given by Equation 5.5 or Equation 5.12 and, therefore, the current density should be calculated using Equation 5.14. The calculation of the core and winding losses follows the procedures established in Chapter 3.

The design procedure may be summarized as follows:

Specifications:

- Input voltage and current
- Output voltage and current/power
- Frequency of operation
- Maximum core temperature or temperature rise
- Ambient temperature

Circuit parameters:

- Waveform factor
- Power factor

Core Selection:

- Core material
- Maximum flux density
- Core window winding area product

Winding design:

- Number of turns
- Current density
- Wire selection

Losses:

- Copper loss
- Winding loss
- Efficiency

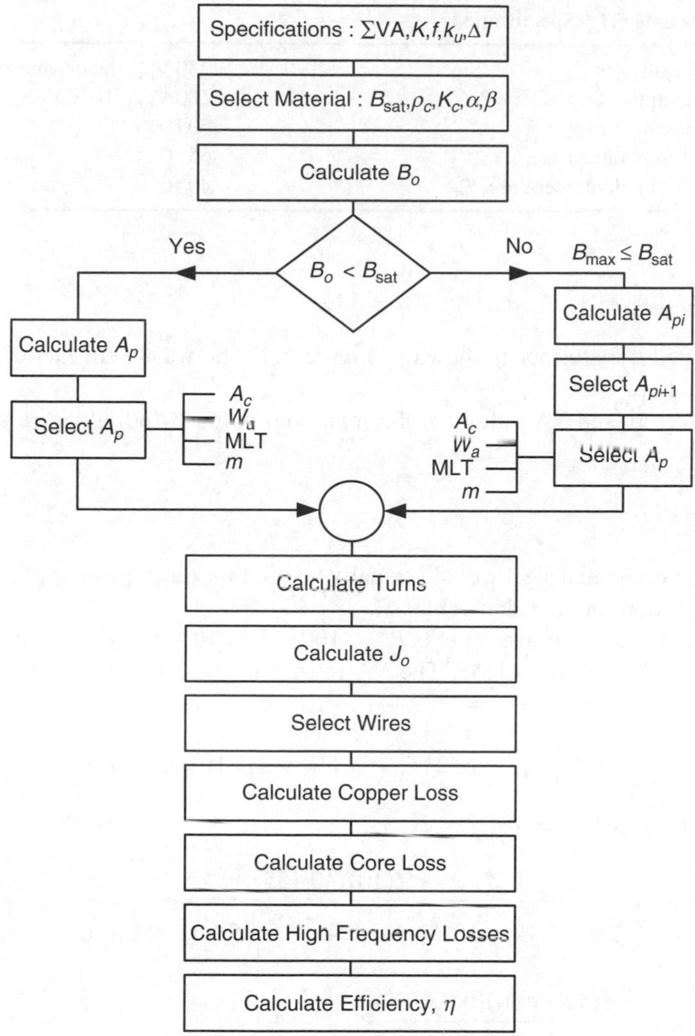

Figure 5.3 Flow chart of design process.

In Chapter 6, we will refine the design to take high-frequency skin and proximity effects into account.

5.3 Design Examples

5.3.1 Example 5.1: Centre-Tapped Rectifier Transformer

Specifications

The design specifications for the transformer are listed in Table 5.1.

Table 5.1 Specifications

Input	230 V_{rms}, sine wave
Output	100 V_{rms}, 10 A
Frequency, f	50 Hz
Temperature rise, ΔT	55 °C
Ambient temperature, T_a	40 °C

Circuit Parameters

The centre-tapped transformer is shown in Figure 5.4. The waveform factor $K_v = 4.44$ for sine wave excitation.

The power factors and VA ratings of the input and output windings were established in Example 4.5.

Core Selection

Laminated grain orientated steel would normally be used for this type of application. Typical material specifications are listed in Table 5.2.

The output power of the transformer is $P_o = (100 + 1) \times 10 = 1010$ W, assuming a forward voltage drop of 1 V for each diode. The VA ratings of the windings were established in Example 4.5, giving:

$$\sum \mathrm{VA} = \left(1 + \sqrt{2}\right) P_o = \left(1 + \sqrt{2}\right) 1010 = 2438 \ \mathrm{VA}$$

The optimum flux density (Equation 5.8) is:

$$B_o = \frac{[(10)(40)(55)]^{2/3}}{2^{2/3}\left[(1.72 \times 10^{-8})(10)(0.4)\right]^{1/12}\left[(5.6)(3.388)(50)^{1.7}\right]^{7/12}} \cdot \left[\frac{(4.44)(50)(0.95)(0.4)}{2438}\right]^{1/6} = 4.1 \ \mathrm{T}$$

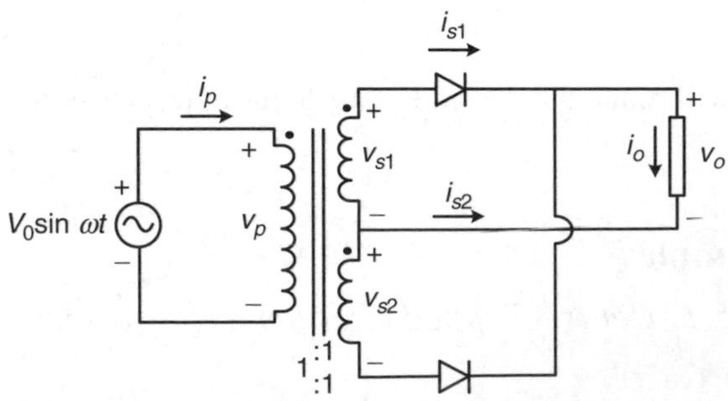

Figure 5.4 Centre-tapped rectifier with a resistive load.

Table 5.2 Material specifications

K_c	3.388
α	1.7
β	1.9
B_{sat}	1.5 T

Since $B_o > B_{sat}$, the design is saturation-limited.
1st iteration: $B_{max} = 1.5$ T and, from Equation 5.9, A_{p1} is:

$$A_{p1} = \left[\frac{\sqrt{2}(2438)}{(4.44)(50)(1.5)(0.95)(48.2\times 10^3)\sqrt{(0.4)(55)}}\right]^{8/7} \times 10^8 = 1166\ \text{cm}^4$$

2nd iteration:
From Equation 5.11:

$$a_0 = \frac{(5.6)(3.388)(50)^{1.7}(1.5)^{1.9}}{(1.72\times 10^{-8})(10)(0.4)} = 4.606\times 10^{11}$$

$$a_1 = \frac{(10)(40)(55)}{(1.72\times 10^{-8})(10)(0.4)} = 3.198\times 10^{11}$$

$$a_2 = \left[\frac{2438}{(4.44)(50)(1.5)(0.95)(0.4)}\right]^2 = 371.3$$

and from Equation 5.13:

$$\begin{aligned} A_{p2} &= 1166\times 10^{-8} \\ &\quad - \frac{(4.606\times 10^{11})(1166\times 10^{-8})^2 - (3.198\times 10^{11})(1166\times 10^{-8})^{7/4} + 371.3}{2(4.606\times 10^{11})(1166\times 10^{-8}) - \frac{7}{4}(3.198\times 10^{11})(1166\times 10^{-8})^{3/4}} \\ &= 859\ \text{cm}^4 \end{aligned}$$

A third iteration is not necessary.

A tape-wound toroidal core with 0.23 mm laminations is suitable. The manufacturer's data for this core is summarized in Table 5.3

Table 5.3 Core and winding specifications

A_c	19.5 cm^2
W_a	50.2 cm^2
A_p	979 cm^4
V_c	693 cm^3
k_f	0.95
k_u	0.4
MLT	28 cm
ρ_{20}	1.72 μΩ-cm
α_{20}	0.00393

Winding Design

Primary turns

$$N_p = \frac{V_p}{K_v B_{\max} A_c f} = \frac{230}{(4.44)(1.5)(19.5 \times 10^{-4})(50)} = 354 \text{ turns}$$

Secondary turns

The rms value of each secondary voltage is $(100 + 1) = 101$ V, which includes a 1 V forward voltage drop in the diode.

$$N_s = N_p \frac{V_s}{V_p} = 354 \frac{101}{230} = 155 \text{ turns}$$

Wire size

The resistivity of the copper must be adjusted for temperature using Equation 3.37:

$$T_{\max} = 40 + 55 = 95\,^{\circ}\mathrm{C}$$

and the resistivity is:

$$\rho_w = (1.72 \times 10^{-8})[1 + (0.00393)(95 - 20)] = 2.23 \times 10^{-8}\ \Omega\mathrm{m}$$

The current density (Equation 5.14) for the chosen core is:

$$J_o = \sqrt{\frac{hk_a\sqrt{A_p}\Delta T - V_c K_c f^{\alpha} B_m^{\beta}}{\rho_w V_w k_u}}$$

$$= \sqrt{\frac{(10)(40)\sqrt{979 \times 10^{-8}}(55) - (693 \times 10^{-6})(3.388)(50)^{1.7}(1.5)^{1.9}}{(2.23 \times 10^{-8})(28 \times 10^{-2})(50.2 \times 10^{-4})(0.4)(0.4)}} = 2.277 \times 10^6\ \mathrm{A/m^2}$$

Remember that $V_w = \mathrm{MLT} \times W_a$.

Primary copper loss

$$I_p = \frac{P_o}{k_{pp} V_p} = \frac{1010}{(1)(230)} = 4.39\ \mathrm{A}$$

The cross-sectional area of the conductor is:

$$A_w = I_p / J_o = 4.39/2.277 = 1.929\ \mathrm{mm}^2$$

This is equivalent to a 1.57 mm bare diameter wire. Select a 1.6 mm diameter wire with a DC resistance of 8.50 mΩ/m at 20 °C (Table A.1).

Use Equation 3.37 to correct the winding resistance for temperature:

$$R_{dc} = (28 \times 10^{-2})(354)(8.50 \times 10^{-3})[1 + (0.00393)(95 - 20)] = 1.091\ \Omega$$

The primary copper loss is:

$$P_{\text{cu}} = I_{\text{rms}}^2 R_{\text{dc}} = (4.39)^2(1.091) = 21.04\,\text{W}$$

Secondary copper loss

The rms value of the load current, which is a full wave-rectified sine wave, is 10 A. The current appears in each secondary winding as a half-wave-rectified sine wave, as shown in Figure 5.5.

From Table 6.1: $I_{s1\text{rms}} = I_{s2\text{rms}} = 10/2 = 5\,\text{A}$.

$I_s = 5\,\text{A}$.

$A_w = I_s/J_o = 5/2.277 = 2.196\,\text{mm}^2$

This is equivalent to a 1.67 mm bare diameter wire. Select a 1.8 mm diameter wire with a DC resistance of 6.72 mΩ/m at 20 °C (Table A.1).

Use Equation 3.37 to correct the winding resistance for temperature:

$$R_\theta = (28 \times 10^{-2})(155)\left(6.72 \times 10^{-3}\right)[1 + (0.00393)(95 - 20)] = 0.378\,\Omega$$

The secondary copper loss (two windings) is:

$$P_{\text{cu}} = I_{\text{rms}}^2 R_{\text{dc}} = (5)^2(0.378)(2) = 18.88\,\text{W}$$

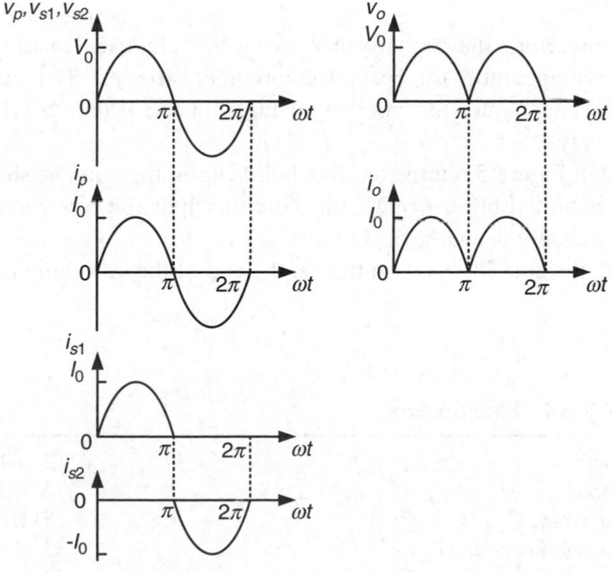

Figure 5.5 Centre-tapped rectifier voltage and current waveforms.

The skin depth at 50 Hz in copper is $\delta = 66/\sqrt{50} = 9.3$ mm, and therefore does not create additional loss, since the radii of both primary and secondary conductors are less than the skin depth (see Figure 6.3).

Core loss

$$P_{\text{fe}} = V_c K_c f^{\alpha} B_{\max}^{\beta} = (693 \times 10^{-6})(3.388)(50)^{1.7}(1.5)^{1.9} = 3.92\,\text{W}$$

Total losses:	Primary copper	21.04 W
	Secondary copper	18.88 W
		3.92 W
		43.84 W

Efficiency:

$$\text{Efficiency} = \frac{1010}{1010 + 43.84} 100 = 95.8\,\%$$

5.3.2 *Example 5.2: Forward Converter*

Specifications

The design specifications for the transformer are listed in Table 5.4.

Circuit Parameters

In a forward converter, the transformer provides electrical isolation and adjusts the input/output voltage ratio for correct component stresses. The circuit is shown in Figure 5.6. N_p, N_s and N_t are the number of turns in the primary, secondary and reset windings respectively.

When switch Q in Figure 5.6 turns on, flux builds up in the core, as shown in Figure 5.7. When the switch is turned off at $t = DT$, this core flux must be reset, otherwise core creep takes place and eventually the core will saturate.

Assume at $t = 0$, the core flux is 0, so that, at the end of the switching period, DT, the flux is $\phi_{\max}$.

Table 5.4 Specifications

Input	$12 \rightarrow 36$ V
Output	9 V, 7.5 A
Frequency, f	25 kHz
Temperature rise, ΔT	35 °C
Ambient temperature, T_a	40 °C

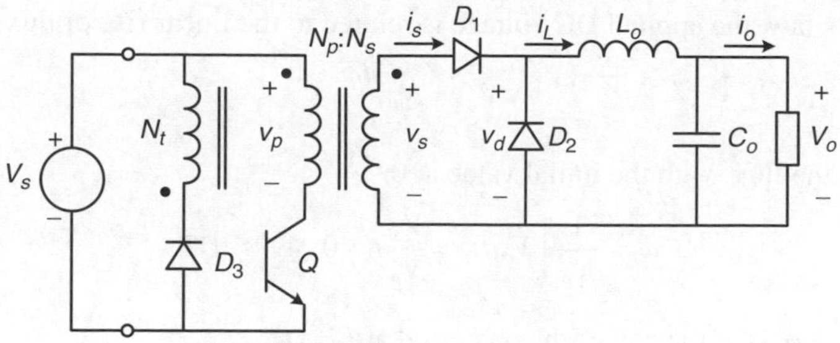

Figure 5.6 Forward converter.

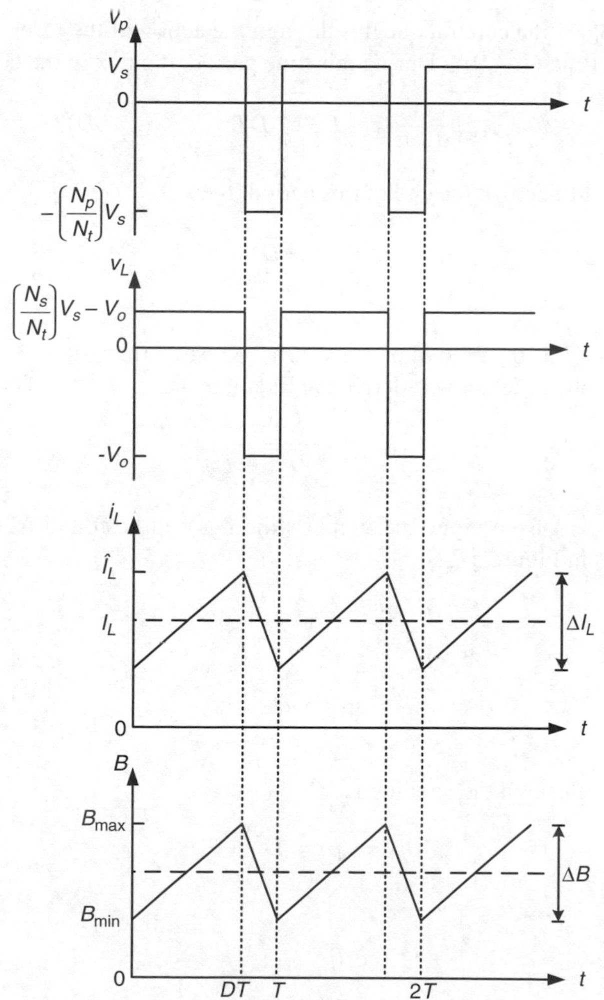

Figure 5.7 Forward converter voltage and flux waveforms.

By Faraday's law, the applied DC voltage is related to the linear rise of flux in the core:

$$V_s = N_p \frac{d\phi}{dt}$$

Integrating for the flux, with the initial value at 0:

$$\phi = \frac{1}{N_p}\int V_s dt = \frac{V_s}{N_p} t \quad 0 \le t \le DT$$

The maximum flux is at the end of the on period at $t = DT$:

$$\phi_{\max} = \frac{V_s}{N_p} DT$$

At $t = DT$, resetting of the core flux begins through the action of the reset winding, this must be achieved in the time $(1 - D)T$. During this time period, the flux in the core is:

$$\phi = \phi_{\max} - \frac{V_s}{N_t}(t - DT) \quad DT \le t \le (1 - D)T$$

If the flux is to reset to zero at the end of this period, then:

$$\frac{N_p}{N_t} = \frac{D}{1 - D}$$

For a duty cycle of 75%, the ratio of primary turns to reset turns ratio is 3.

The voltage waveform factor was derived in Example 4.2:

$$K_v = \frac{1}{\sqrt{D(1 - D)}}$$

The power factor, k_p of the input and output windings can be found from the voltage and current waveforms in Figure 5.7:

$$V_{\text{rms}} = \sqrt{\frac{D}{1 - D}} V_s$$

$$I_{\text{rms}} = \sqrt{D}\frac{N_s}{N_p} I_o$$

The average power through the winding is:

$$P_{\text{av}} = \langle p \rangle = D \frac{N_s}{N_p} V_s I_o$$

The power factor k_p is:

$$k_p = \frac{\langle p \rangle}{V_{\text{rms}} I_{\text{rms}}} = \sqrt{1 - D}$$

in both the primary and secondary windings.

Table 5.5 Material specifications

K_c	37.2
α	1.13
β	2.07
B_{sat}	0.4T

Core Selection

Ferrite would normally be used for this type of application at the specified frequency. The material specifications for Mn-Zn are listed in Table 5.5.

The output power of the transformer is $P_o = (9+1) \times 7.5 = 75$ W, assuming a forward voltage drop of 1 V for the diode.

The duty cycle is $D = 9/12 = 0.75$ and the power factor is then $\sqrt{1-D} = 0.5$:

$$K_v = \frac{1}{\sqrt{0.75(1-0.75)}} = 2.31$$

The power factor and VA ratings of the windings are established above giving

$$\sum \text{VA} = \left(\frac{1}{k_{pp}} + \frac{1}{k_{ps}}\right) P_o = (2+2)(75) = 300 \text{ VA}$$

Adding 5% for the reset winding gives $\sum$ VA = 315 VA. Set $k_u = 0.4$.

The optimum flux density is found using Equation 5.8. However, B_o is the amplitude of the flux waveform. On the other hand, B_{max} in Equation 4.63 is the maximum flux density, and in the case of the forward converter, this is at least equal to the flux ripple ΔB (in continuous conduction) and the amplitude of the flux used in the core loss calculation Equation 1.29 is $\Delta B/2$. In calculating B_o using Equation 5.8, 2 K_v may be used to properly account for these effects.

$$B_o = \frac{[(10)(40)(35)]^{2/3}}{2^{2/3}\left[(1.72 \times 10^{-8})(10)(0.4)\right]^{1/12}\left[(5.6)(37.2)(25\,000)^{1.13}\right]^{7/12}} \cdot \left[\frac{(2 \times 2.31)(25\,000)(1.0)(0.4)}{315}\right]^{1/6} = 0.186 \text{ T}$$

$B_{max} = 2B_o = 0.372$ T. This is less than B_{sat}, and A_p from Equation 5.5 is:

$$A_p = \left[\frac{\sqrt{2}(315)}{(2.31)(25\,000)(0.372)(1.0)(48.2 \times 10^3)\sqrt{(0.4)(35)}}\right]^{8/7} \times 10^8 = 1.173 \text{ cm}^4$$

Select the ETD39 core. The core specifications are given in Table 5.6.

Table 5.6 Core and winding specifications

A_c	$1.25\,\text{cm}^2$
W_a	$1.78\,\text{cm}^2$
A_p	$2.225\,\text{cm}^4$
V_c	$11.5\,\text{cm}^3$
k_f	1.0
k_u	0.4
MLT	6.9 cm
ρ_{20}	1.72 μΩ-cm
α_{20}	0.00393

Winding Design

The ratio V_{rms}/K is given by DV_s from the analysis above. Thus, in calculating the number of turns, take D at its maximum value, i.e. $D = 0.75$ and $K_v = 2.31$. The RMS value of the input voltage waveform is $V_{\text{rms}} = \sqrt{\dfrac{D}{1-D}}V_{\text{dc}} = \sqrt{3}V_{\text{dc}}$.

Primary turns

$$N_p = \frac{V_p}{K_v B_{\max} A_c f} = \frac{\sqrt{3}(12)}{(2.31)(2 \times 0.186)(1.25 \times 10^{-4})(25\,000)} = 7.7 \text{ turns}$$

rounded up to 9 turns.

Secondary turns

In this design, the number of secondary turns is equal to the number of primary turns, i.e. 9 turns.

The number of turns in the reset winding is:

$$N_t = \frac{1-D}{D}N_p = \frac{1-0.75}{0.75}(9) = 3 \text{ turns}$$

Wire size

The current density (Equation 5.4) is:

$$J_o = (48.2 \times 10^3)\frac{35}{\sqrt{2(0.4)}}\frac{1}{\sqrt[8]{(2.225 \times 10^{-8})}} = 2.885 \times 10^6 \text{ A/m}^2$$

Primary current

$$I_p = \frac{P_o}{k_{pp}V_p} = \frac{75}{(0.5)\sqrt{3}(12)} = 7.22 \text{ A}$$

$$A_w = I_p/J_o = 7.22/2.885 = 2.502 \text{ mm}^2$$

This corresponds to a 1.79 mm diameter. A 1.8 mm diameter wire with a dc resistance of 6.72 mΩ/m at 20 °C will suffice (Table A.1).

Primary copper loss

Use Equation 3.37 to correct the winding resistance for temperature:

$$T_{\max} = 40 + 35 = 75\,^{\circ}\mathrm{C}.$$

$$R_{\mathrm{dc}} = 9 \times 6.9 \times 10^{-2} \times 6.72 \times 10^{-3} \times [1 + 0.00393 \times (75 - 20)] \times 10^{3} = 5.08\ \mathrm{m}\Omega$$

The copper loss in the primary winding is:

$$P_{\mathrm{cu}} = R_{\mathrm{dc}} I_{\mathrm{rms}}^{2} = 5.08 \times 10^{-3} \times (7.22)^{2} = 0.264\ \mathrm{W}$$

Secondary current

$$I_s = \sqrt{D} I_o = \sqrt{0.75}(7.5) = 6.50\ \mathrm{A}$$

$$A_w = I_s / J_o = 6.50/2.885 = 2.252\ \mathrm{mm}^2$$

This corresponds to a 1.69 mm diameter. A 1.8 mm diameter wire with a DC resistance of 6.72 mΩ/m at 20 °C will suffice (Table A.1).

Secondary copper loss

Use Equation 3.37 to correct the winding resistance for temperature:

$$R_{\mathrm{dc}} = 9 \times 6.9 \times 10^{-2} \times 6.72 \times 10^{-3} \times [1 + 0.00393 \times (75 - 20)] \times 10^{3} = 5.08\ \mathrm{m}\Omega$$

The copper loss in the secondary winding is:

$$P_{\mathrm{cu}} = R_{\mathrm{dc}} I_{\mathrm{rms}}^{2} = 5.08 \times 10^{-3} \times (6.50)^{2} = 0.214\ \mathrm{W}$$

High frequency effects

These are dc values. At 25 kHz, the skin depth $\delta = 66/\sqrt{25\,000} = 0.42$ mm. This is less than the radius of either primary or secondary conductor, so it increases the resistance.

The correction factor to account for skin effect is given by Equation 1.22.

Primary AC resistance:

$$R_{pac} = 5.08 \left[1 + \frac{(0.9/0.42)^4}{48 + 0.8(0.9/0.42)^4} \right] = 6.73\ \mathrm{m}\Omega$$

$$I_p^2 R_{pac} = (7.22)^2 (6.73 \times 10^{-3}) = 0.350\ \mathrm{W}$$

Secondary AC resistance:

$$R_{sac} = 5.08 \left[1 + \frac{(0.9/0.42)^4}{48 + 0.8(0.9/0.42)^4} \right] = 6.73\ \mathrm{m}\Omega$$

$$I_s^2 R_{sac} = (6.50)^2 (6.73 \times 10^{-3}) = 0.284\ \mathrm{W}$$

Core loss

The flux density ripple ΔB can be calculated by using Faraday's law:

$$\Delta B = \frac{V_s DT}{N_P A_c} = \frac{(12)(0.75)}{(9)(1.25 \times 10^{-4})(25\,000)} = 0.320\,\mathrm{T}$$

$$P_{\mathrm{fe}} = V_c K_c f^{\alpha} (\Delta B/2)^{\beta} = (11.5 \times 10^{-6})(37.2)(25\,000)^{1.13}(0.16)^{2.07} = 0.898\,\mathrm{W}$$

Total losses:

Primary copper	0.350 W
Secondary copper	0.284 W
Core	0.898 W
	1.532 W

Efficiency

$$\text{Efficiency} = \frac{75}{75 + 1.532} = 98\%$$

5.3.3 Example 5.3: Push-Pull Converter

Specifications

The specifications for the push-pull converter are given in Table 5.7.

Circuit Parameters

The circuit diagram for the push-pull converter is shown in Figure 5.8 and its associated voltage and current waveforms are shown in Figure 5.9. We assume for simplicity that the turns ratio is 1 : 1.

In Figure 5.8, switch 1 turns on at $t = 0$ and turns off at time DT'. By defining the duty cycle in this manner, the combined on-time of the two switches is DT and the output voltage is DV_s. The switching period is T and each switch controls the voltage waveform for $T' = T/2$.

The flux density increases from 0 to its maximum value in the time $\tau = DT'/2 = DT/4$.

The RMS value of the input voltage waveform in Figure 5.9 is found by:

$$V_{\mathrm{rms}} = \sqrt{D} V_s$$

Table 5.7 Specifications

Input	36 → 72 V
Output	24 V, 12.5 A
Frequency, f	50 kHz
Temperature rise, ΔT	35 °C
Ambient temperature, T_a	45 °C

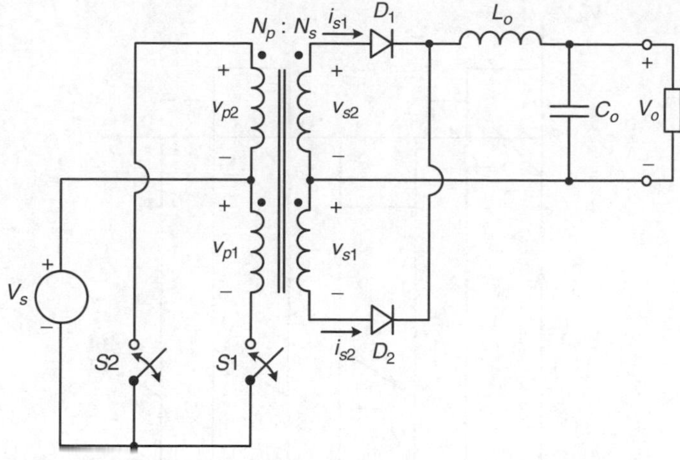

Figure 5.8 Push-pull converter circuit.

The average value of the input voltage waveform during the time τ is V_s:

$$\langle v \rangle = V_s$$

and thus, from Equation 4.61:

$$k = \frac{V_{\mathrm{rms}}}{\langle v \rangle} = \frac{\sqrt{D}V_s}{V_s} = \sqrt{D}$$

$$\frac{\tau}{T} = \frac{DT'/2}{T} = \frac{DT/4}{T} = \frac{D}{4}$$

$$K_v = \frac{k}{\tau/T} = \frac{\sqrt{D}}{D/4} = \frac{4}{\sqrt{D}}.$$

For $D = 1.0$, $K = 4.0$, as expected for a square waveform.

When both switches are off in the circuit of Figure 5.8, half the output current circulates in each of the secondary windings. This circulating current contributes to heating, but there is no transfer of power through the transformer. We can include this effect by correctly defining the power factor of the windings.

The RMS value of the secondary current (neglecting the ripple), as shown in Figure 5.9, is given by:

$$I_{s\mathrm{rms}} = \frac{I_o}{2}\sqrt{(1+D)}.$$

The RMS value of the secondary voltage is:

$$V_{s\mathrm{rms}} = \sqrt{D}V_s = \frac{V_o}{\sqrt{D}}.$$

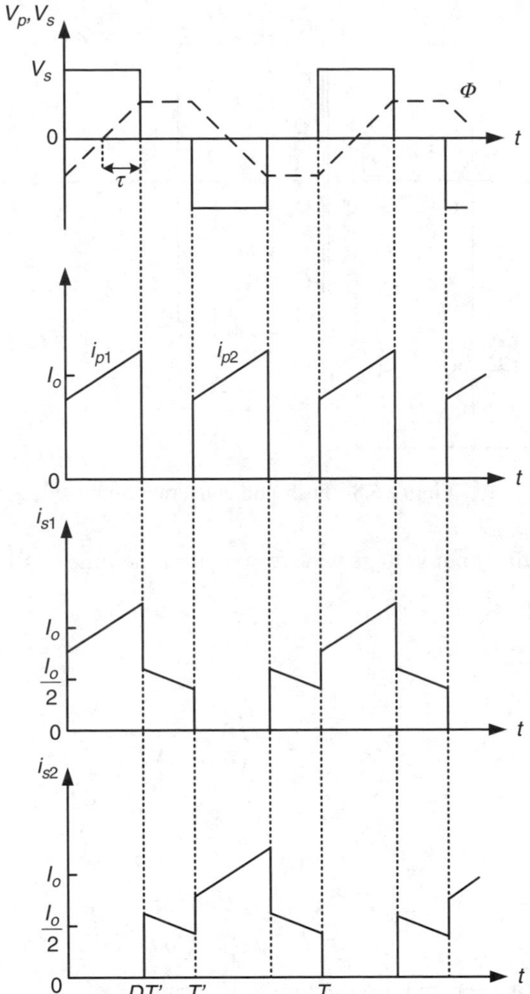

Figure 5.9 Push-pull converter voltage and current waveforms.

The VA rating of each secondary winding is now:

$$V_{srms} I_{srms} = \frac{1}{2} \frac{\sqrt{1+D}}{\sqrt{D}} V_o I_o = \frac{\sqrt{1+D}}{\sqrt{D}} \frac{P_o}{2}.$$

Recalling the definition of power factor and noting that, for each winding, the average power $\langle p \rangle = P_o/2$, where P_o is the total output power, the power factor of each secondary winding is:

$$k_{ps} = \frac{\langle p \rangle}{V_{srms} I_{srms}} = \sqrt{\frac{D}{1+D}}.$$

For $D = 1$, $k_{ps} = 1/\sqrt{2}$ as expected.

The RMS values of the input voltage and current are:

$$V_{\text{prms}} = \sqrt{D}V_s,$$

$$I_{\text{prms}} = \sqrt{(D/2)}I_s.$$

The average power delivered by each of the primary windings is:

$$\langle p \rangle = \int_0^{DT'} v(t)i(t)dt = \frac{1}{T}V_sI_sDT' = \frac{D}{2}V_sI_s$$

For a 1 : 1 turns ratio, $I_{\text{dc}} = I_0$ and the power factor in each primary winding is then given by:

$$k_{pp} = \frac{\langle p \rangle}{V_{\text{prms}}I_{\text{prms}}} = \frac{1}{\sqrt{2}}.$$

We can now sum the VA ratings over the two input windings and the two output windings. The average power through each secondary winding is $P_o/2$ and the average power through each primary winding is $P_o/2$. Thus we have:

$$\begin{aligned}\sum \text{VA} &= \left(\frac{1}{k_{pp}}\left(\frac{P_o}{2} + \frac{P_o}{2}\right) + \frac{1}{k_{ps}}\left(\frac{P_o}{2} + \frac{P_o}{2}\right)\right) \\ &= \left(\sqrt{2} + \sqrt{\frac{1+D}{D}}\right)P_o\end{aligned}$$

For the input voltage range 36 V to 72 V, the duty cycle can vary between 33% and 67%. For an input voltage of 36 V, the duty cycle is 24/36 = 67%. The waveform factor is $K_v = 4/\sqrt{D} = 4.88$.

Core Selection

Ferrite would normally be used for this type of application at the specified frequency. The material specifications for EPCOS N67 Mn-Zn ferrite are listed in Table 5.8.

The output power of the transformer is $P_o = (24 + 1.0) \times 12.5 = 312.5$ W, assuming a forward voltage drop of 1.5 V for the diode. In terms of core selection, the maximum dissipation occurs at maximum duty cycle, that is $D = 0.67$:

$$\sum \text{VA} = \left(\sqrt{2} + \sqrt{\frac{1+0.67}{0.67}}\right)(312.5) = 935 \text{ VA}$$

Table 5.8 Material specifications

K_c	9.12
α	1.24
β	2.0
B_{sat}	0.4 T

Table 5.9 Core and winding specifications

A_c	$1.73\,\text{cm}^2$
W_a	$2.78\,\text{cm}^2$
A_p	$4.81\,\text{cm}^4$
V_c	$17.70\,\text{cm}^3$
k_f	1.0
k_u	0.4
MLT	7.77 cm
ρ_{20}	1.72 μΩ-cm
α_{20}	0.00393

The optimum flux density Equation 5.8 is:

$$B_o = \frac{[(10)(40)(35)]^{2/3}}{2^{2/3}\left[(1.72\times10^{-8})(10)(0.4)\right]^{1/12}\left[(5.6)(9.12)(50000)^{1.24}\right]^{7/12}} \cdot \left[\frac{(4.88)(50000)(1.0)(0.4)}{935}\right]^{1/6} = 0.126\,\text{T}$$

The optimum flux density is less than B_{sat} and A_p from Equation 5.5 and is:

$$A_p = \left[\frac{\sqrt{2}(935)}{(4.88)(50\,000)(0.126)(1.0)(0.4)(48.2\times10^3)\sqrt{(0.4)(35)}}\right]^{8/7} \times 10^8 = 2.693\,\text{cm}^4$$

The EPCOS ETD44 core is suitable. The core specifications are given in Table 5.9.

Winding Design

Primary turns

$$N_p = \frac{V_p}{K_v B_{\max} A_c f} = \frac{\sqrt{0.67}(36)}{(4.88)(0.126)(1.73\times10^{-4})(50000)} = 5.5\text{ turns}$$

Choose six turns.

Secondary turns

We assumed a 1 : 1 turns ratio. so the number of secondary turns is 6.

Wire size

The current density (Equation 5.4) is:

$$J_o = K_t\sqrt{\frac{\Delta T}{2k_u}}\frac{1}{\sqrt[8]{A_p}} = (48.2\times10^3)\sqrt{\frac{35}{2(0.4)}}\frac{1}{\sqrt[8]{(4.81\times10^{-8})}} = 2.620\times10^6\ \text{A/m}^2$$

Primary current

$$I_p = \frac{P_o/2}{k_{pp}V_p} = \frac{312.5/2}{(0.707)(29.5)} = 7.5\ \text{A}$$

$$A_w = I_p/J_o = 2.863\ \text{mm}^2$$

Standard 0.1 × 30 mm copper foil with a DC resistance of 5.8 mΩ/m at 20 °C meets this requirement.

Primary copper loss

Use Equation 3.37 to correct the winding resistance for temperature:

$$T_{\max} = 45 + 30 = 75\ ^\circ\text{C}.$$

$$R_{\text{dc}} = (6)(7.77 \times 10^{-2})(5.80 \times 10^{-3})[1 + (0.00393)(75 - 20)] \times 10^3 = 3.29\ \text{m}\Omega$$

The copper loss in each the primary winding is:

$$P_{\text{cu}} = R_{\text{dc}} I_{\text{rms}}^2 = (3.29 \times 10^{-3})(7.5)^2 = 0.185\ \text{W}$$

Secondary current

$$I_s = \frac{I_o}{2}\sqrt{1+D} = \frac{12.5}{2}\sqrt{1+0.67} = 8.08\ \text{A}$$

$$A_w = I_s/J_o = 3.083\ \text{mm}^2$$

Again, standard 0.1 × 30 mm copper foil meets this requirement.

Secondary copper loss

Use Equation 3.37 to correct the winding resistance for temperature:

$$R_{\text{dc}} = (6)(7.77 \times 10^{-2})(5.80 \times 10^{-3})[1 + (0.00393)(75 - 20)] \times 10^3 = 3.29\ \text{m}\Omega$$

The copper loss in each of the secondary winding is:

$$P_{\text{cu}} = R_{\text{dc}} I_{\text{rms}}^2 = (3.29 \times 10^{-3})(8.08)^2 = 0.215\ \text{W}$$

High frequency effects

The skin depth in copper at 50 kHz is 0.295 mm, which is greater than the thickness of the foil and therefore does not present a problem. See Example 7.6.

Core loss

The peak value of the flux density in the selected core is:

$$B_{\max} = \frac{\sqrt{D}V_{dc}}{K_v f N_p A_c} = \frac{\sqrt{0.67}(36)}{(4.88)(50\,000)(6)(1.73 \times 10^{-4})} = 0.116\,\text{T}$$

$$P_{\text{fe}} = V_c K_c f^{\alpha} B_{\max}^{\beta} = (17.7 \times 10^{-6})(9.12)(50\,000)^{1.24}(0.116)^{2.0} = 1.466\,\text{W}$$

Total losses:		
	Primary copper	0.185 W
		0.185 W
	Secondary copper	0.215 W
		0.215 W
	Core	1.466 W
		2.266 W

Efficiency

$$\text{Efficiency} = \frac{312.5}{312.5 + 2.266} = 99.3\%$$

5.4 Transformer Insulation

The window utilization factor in a transformer is typically 40%, which arises due to the requirements to isolate windings from each other. In general, there is a distinction between the insulation required for the proper functioning of a transformer (operational insulation) and the safe isolation of a circuit from a hazardous voltage (double insulation or reinforced insulation).

These are three basic means of providing insulation:

- By insulator; this consists of a dielectric material (e.g. mylar tape) separating two conductors.
- By creepage; this is the distance along an insulating surface (e.g. from one solder pin to another on a bobbin) between two conductors on the surface.
- By clearance, which is normally understood to be the air gap between two conducting bodies.

The insulation capability achieved by dielectric materials is rated in kV/mm or (kV/sec)/mm. In the case of creepage, the roughness of the surface and the degree of pollution defines the insulation capability. Clearance normally applies to air, and the breakdown voltage is a function of ionization and conduction of the air in question, depending on pressure, temperature, humidity and pollution.

To quantify the quality of the insulation, the voltages that are applied to the insulation barriers need to be defined clearly. There are three main voltage definitions required for insulation barriers:

- The *rms working voltage* is the rms value of the highest voltage to which the insulation can be subjected when the transformer is operating under conditions of normal use.
- The *peak working voltage* is the peak value of the working voltage and it may be an order of magnitude greater than the rms working voltage, depending on the shape of the voltage waveform. The rms and peak working voltages are steady state voltages that occur in normal operation.

- *Transient voltages* must also be considered. These are irregular and represent abnormal operation. There are different categories of overvoltage, and the applicable values depend on the types of equipment and the location of the installation.

5.4.1 Insulation Principles

The wire used in the winding of a transformer has single or double temperature-resistant lacquer insulation on its surface, which meets the usual turn-to-turn voltage insulation requirements. In a multilayer transformer, the additive effect of the turns means that the wire insulation might not be sufficient to insulate all the turns from each other. In this case, additional insulation is required between the layers (intra-winding insulation).

Basic insulation meets the operating insulation requirements in a transformer. The basic insulation should be compatible with the rms working voltage, the peak working voltage and the relevant overvoltage category of the applied circuits.

Transformers are often used for isolating hazardous high-voltage circuits from low-voltage circuits that may be exposed to human contact. The hazard to humans means that additional insulation is required to meet the safe extra low voltage (SELV) requirements. The additional insulation requirement results from the tolerance of a single failure and therefore requires double basic insulation. The additional insulation, combined with the basic insulation, is referred to as *double insulation* or *reinforced insulation* in the relevant standards.

In addition to the insulation requirements, the insulation materials in a transformer must withstand the highest possible operating temperatures. During abnormal operations, such as a fault inside the transformer or a short-circuit condition, the internal temperature within the coil could rise to a sufficiently high level to cause ignition of the insulation materials. Insulation materials are therefore classified under different flammability categories, depending on the maximum temperature to which the material will be exposed.

5.4.2 Practical Implementation

The main insulation elements of a transformer are illustrated in Figure 5.10. The bobbin isolates the windings from the core; there is lacquer insulation on the wire; there are

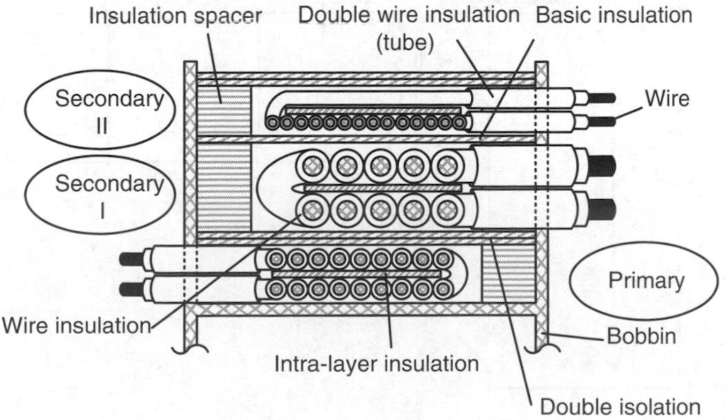

Figure 5.10 Transformer insulation.

insulation tapes between the winding layers; there are sleeves over the wires; and the insulation spacer achieves the required creepage distances.

The cross-section of the bobbin assembly for an E-type core, with the various types of insulation, is shown in Figure 5.10. The example shows one primary (hazardous voltage) winding with two layers and two secondary windings. The intra-layer insulation (basic insulation), consisting of tapes, isolates the working voltages between the wires, while the inter-layer insulation (double insulation) tape provides the required isolation between the windings. Tape that provides basic insulation is always required between windings.

The low-voltage secondary winding is separated from the primary winding by a double layer of basic insulation tape which achieves double insulation, or reinforced insulation. This achieves the safe separation of the hazardous voltage on the primary to the low voltage on the secondary side (SELV). Space holders (or insulation spacers) on each side of a winding increase the required creepage distance by increasing the distance from the end wires in the primary windings to the end wires of the secondary windings. These space holders are often made up of several layers of insulation tape. The wire sleeves provide the double insulation for coil wires entering and exciting the bobbin. The final assembly is impregnated with an insulating resin to prevent the ingress of pollution and moisture.

The insulation requirements ultimately depend on the application and the relevant safety standards. The widely used standards that are relevant to the insulation requirements in power supplies and its transformers for office and household applications are IEC60950 and IEC60335. Special transformers need to comply with specific standards, for example IEC6155-8, IEC6155-2 and IEC6155-16. These international standards form the basis for nationally adopted standards in various jurisdictions.

5.5 Problems

5.1 Rework Example 5.1 for a centre-tapped rectifier with a highly inductive load.

5.2 Calculate the core size in a full-bridge converter with the specifications of Example 5.3. The circuit is shown in Figure 5.11 and the voltage and current waveforms are shown in Figure 5.12.

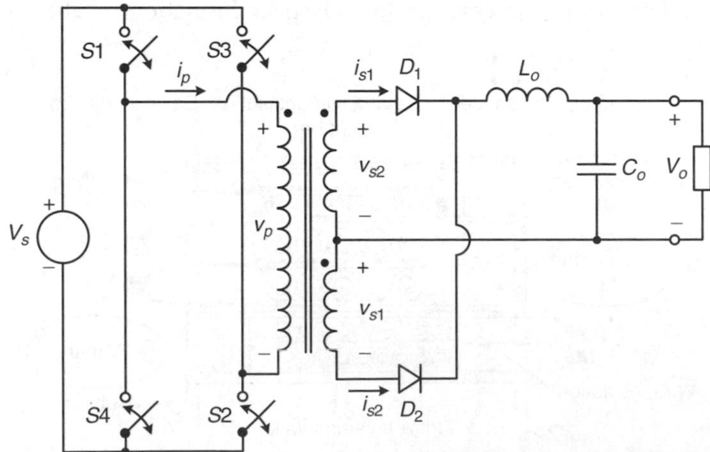

Figure 5.11 Full-bridge converter circuit.

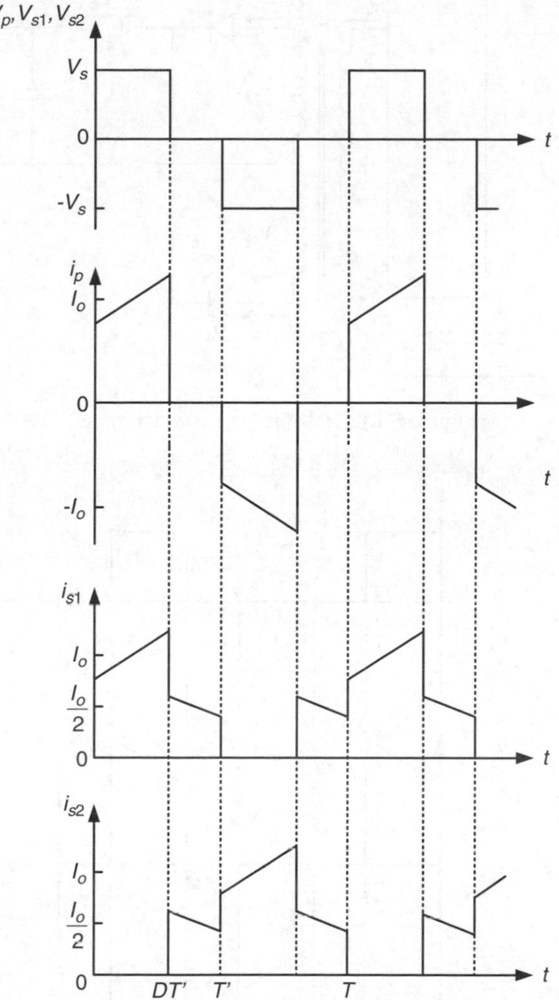

Figure 5.12 Full-bridge converter voltage and current waveforms.

5.3 Calculate the core size in a half-bridge converter with the specifications of Example 5.3. The circuit is shown in Figure 5.13 and the voltage and current waveforms are shown in Figure 5.14.

5.4 A 600 W forward converter operates at 80 kHz with a nominal DC input voltage of 325 V and a nominal DC output voltage of 24 V. Using an EE ferrite core (N87 material in Table 1.1), select the transformer core and calculate the copper loss in the windings and the core loss. The maximum allowed temperature on the transformer isolation materials is 110 °C and the maximum temperature rise allowed in the winding is 35 °C.

5.5 A 600 W push-pull transformer operates at 80 kHz with a rectified AC input voltage of 230 V and a DC output voltage of 24 V. Using an EE ferrite core (N87 material in

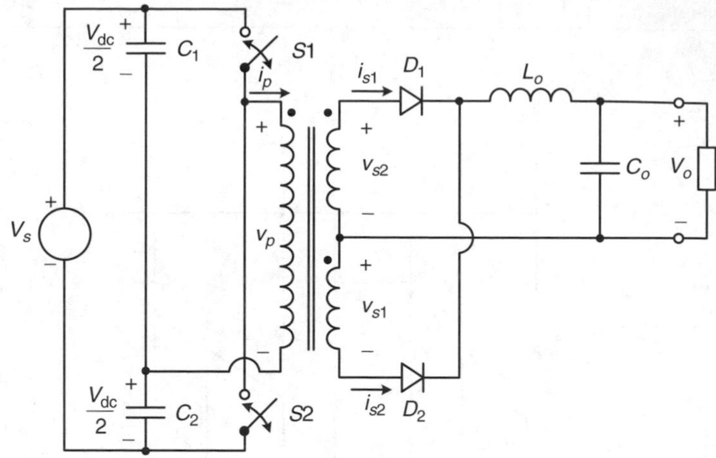

Figure 5.13 Half-bridge converter circuit.

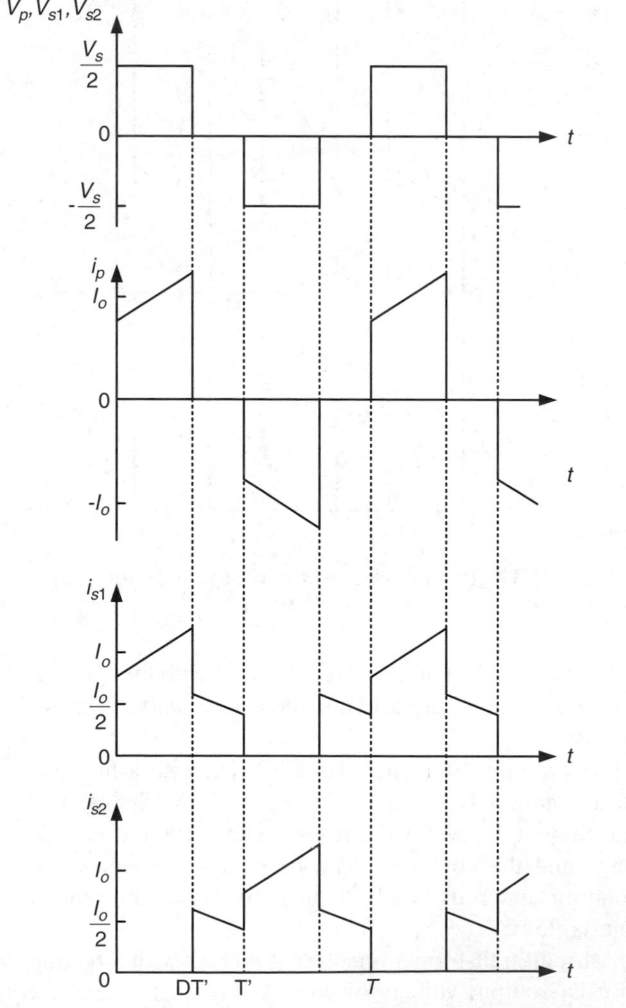

Figure 5.14 Half-bridge converter voltage and current waveforms.

Table 1.1), select the transformer core and calculate the copper loss in the windings and the core loss. The maximum ambient temperature of the transformer is 70 °C and the maximum temperature rise allowed in the winding is 25 °C.

5.6 A power supply with an AC input voltage of 110 V and active power factor correction boosts the input voltage to 400 V DC. This voltage is supplied to a half bridge converter with 200 W output power and DC output voltage of 24 V. Using a ferrite core (material similar to N87 in Table 1.1), select the core for a transformer with a centre-tapped output winding. Calculate the copper loss in the windings and the core loss for a nominal operating frequency of 90 kHz. The maximum allowed operating temperature of the transformer materials is 130 °C, with a maximum temperature inside the power supply of 100 °C. Hint: Use the waveforms in Example 4.5 to find the power factor of the windings.

MATLAB Program for Example 5.1

```
%example 5.1 Centre-Tapped Rectifier Transformer

alpha=1.7;
alpha20=0.00393;
beta=1.9;
deltaT=55;
ro20=1.72e-8;
row=1.72e-8;

Ac=19.5e-4;
Ap=979e-8;
Bm=1.5;
Bsat=1.5;
f=50;
h=10;
Is=5;
ka=40;
Kc=3.388;
kc=5.6;
kf=0.95;
kpp=1;
Kt=48.2e3;
ku=0.4;
Kv=4.44;
kw=10;
MLT=28e-2;
N=354;
N2=155;
Po=1010;
Tmax=95;
```

```
Vc=693e-6;
Vp=230;
Vs=101;
Vw =0.28e-2*50.2e-4;
wire_Rdc1=8.50e-3;
wire_Rdc2=6.72e-3;

sumVA=(1+sqrt(2))*Po
Bo=([(h*ka*deltaT)^(2/3)]/[2^(2/3)*(row*kw*ku)^(1/12)*
(kc*Kc*f^alpha)^(7/12
)])*[(Kv*f*kf*ku)/(sumVA)]^(1/6)
Ap1=[(sqrt(2)*sumVA)/(Kv*f*Bsat*kf*Kt*sqrt(ku*deltaT))]^(8/7)
a0 =(kc*Kc*f^alpha*Bsat^beta)/(row*kw*ku)
a1 =(h*ka*deltaT)/(row*kw*ku)
a2 =((sumVA)/(Kv*f*Bsat*kf*ku))^2
Ap2 = Ap1-(a0*Ap1^2-a1*Ap1^(7/4)+a2)/(2*a0*Ap1-(7/4)*a1*Ap1^(3/4))
Np=Vp/(Kv*Bm*Ac*f)
Ns=Np*Vs/Vp
Tmax=40+55;
row=ro20*[1+alpha20*(Tmax-20)]
Jo = sqrt((h*ka*sqrt(Ap)*deltaT-Vc*Kc*f^alpha*Bsat^beta)/
(row*Vw*ku))
Ip=Po/(kpp*Vp)
Aw1=Ip/Jo
Rdc1=MLT*N*wire_Rdc1*(1+alpha20*(Tmax-20))
Pcu1=((Ip^2)*Rdc1)
Aw2=Is/Jo
Rdc2=MLT*N2*wire_Rdc2*(1+alpha20*(Tmax-20))
Pcu2=(Is^2*Rdc2*2)
Pfe=Vc*Kc*f^alpha*Bm^beta
Ptot=Pfe+Pcu1+Pcu2
efficiency=Po/(Po+Ptot)
```

MATLAB Program for Example 5.2

```
%example 5.2 Forward Converter

Vs=12;
alpha=1.13;
alpha20=0.00393;
beta=2.07;
delta=0.42;
deltaT = 35;
Tamb = 40;
row = 1.72e-8;
```

```
Bsat=0.4;
D=0.75;
f = 25e3;
h = 10;
Io=7.5;
ka = 40;
kc = 5.6;
kw = 10;
Kc = 37.2;
kf = 1.0;
Kt = 48.2e3;
ku = 0.4;
Po=75;
rop=0.9;
ros=0.9;
Vp=sqrt(3)*Vs;
kpp=sqrt(1-D);
kps=sqrt(1-D);
Kv=1/(sqrt(D*(1-D)))
sumVA=(1/kpp+1/kps)*Po
sumVA2=1.05*sumVA
Bo=(((h*ka*deltaT)^(2/3))/((2^(2/3))*((row*kw*ku)^(1/12))*
((kc*Kc*f^alpha)^
(7/12))))*([(2*Kv*f*kf*ku)/sumVA2]^(1/6))
Ap1=([(sqrt(2)*sumVA2)/(Kv*f*2*Bo*kf*Kt*sqrt(ku*deltaT))]^(8/7))
*10^8
Ap2=2.2251e-8;
Ac=1.25e-4;
Vc=11.5e-6;
MLT=6.9e-2;
Np=Vp/(Kv*2*Bo*Ac*f)
Np=9;
Nt=(1-D)/D*Np
Tmax=Tamb+deltaT;
Jo=Kt*((sqrt(deltaT))/(sqrt(2*ku)))*(1/(Ap2)^(1/8))
Ip=Po/(kpp*Vp)
Awp=Ip/Jo
wire_Rdcp=6.72e-3;
Rdcp=MLT*Np*wire_Rdcp*(1+alpha20*(Tmax-20))
Pcup=Rdcp*Ip^2
Is=sqrt(D)*Io
Aws=Is/Jo
wire_Rdcs=6.72e-3;
Rdcs=MLT*Np*wire_Rdcs*(1+alpha20*(Tmax-20))
Pcus=Rdcs*Is^2
Rpac=Rdcp*[1+(rop/delta)^4/(48+0.8*(rop/delta)^4)]
```

```
Ip2Rpac=Ip^2*Rpac
Rsac=Rdcs*[1+(ros/delta)^4/(48+0.8*(ros/delta)^4)]
Is2Rsac=Is^2*Rsac
deltaB=(Vs*D)/(Np*Ac*f)
Pfe=Vc*Kc*(f^alpha)*((deltaB/2)^beta)
Ptot=Pfe+Ip2Rpac+Is2Rsac
Efficiency=Po/(Po+Ptot)
```

MATLAB Program for Example 5.3

```
%example 5.3 Push-Pull Converter

alpha=1.24;
alpha20=0.00393;
beta=2.0;
deltaT=35;
row=1.72e-8;

Ac=1.73e-4;
D=0.67;
f=50e3;
h=10;
Io=12.5;
ka=40;
kc=5.6;
Kc=9.12;
Kc2=9.12;
kf=1;
kpp=0.707;
Kt=48.2e3;
ku=0.4;
Kv=4.88;
kw=10;
MLT=7.77e-2;
Po=312.5;
Tmax=45+35;
Vc=17.7e-6;
Vdc=36;
wire_Rdc=5.80e-3;
wire_Rdc2=5.80e-3;

sumVA=round((sqrt(2)+sqrt((1+D)/D))*Po)
Bo=(((h*ka*deltaT)^(2/3))/((2^(2/3))*((row*kw*ku)^(1/12))*
((kc*Kc*f^alpha)^
(7/12))))*([(Kv*f*kf*ku)/sumVA]^(1/6))
```

```
Ap1=([(sqrt(2)*sumVA)/(Kv*f*Bo*kf*Kt*sqrt(ku*deltaT))]^(8/7))*10^8
Vp=sqrt(0.67)*36
Np=round(Vp/(Kv*Bo*Ac*f))

Kt=48.2e3;
Ap2=4.81e-8;
Jo=Kt*((sqrt(deltaT))/(sqrt(0.8)))*(1/(Ap2)^(1/8))

Ip=(Po/2)/(kpp*Vp)
Aw1=Ip/Jo

Rdc=MLT*Np*wire_Rdc*(1+alpha20*(Tmax-20))
Pcu1=Rdc*Ip^2

Is=(Io/2)*sqrt(1+D)
Aw2=Is/Jo
Rdc2=MLT*Np*wire_Rdc2*(1+alpha20*(Tmax-20))
Pcu2=Rdc*Is^2
Bmax=(sqrt(D)*Vdc)/(Kv*f*Np*Ac)
Pfe=Vc*Kc2*(f^alpha)*(Bmax^beta)
Ptot=Pfe+2*Pcu1+2*Pcu2
efficiency=Po/(Po+Ptot)
```

Further Reading

1. Bartoli, M., Reatti, A., and Kazimierczuk, M.K. (1994) High-frequency models of ferrite core inductors. *Proceedings of the IEEE Industrial Electronics, Control and Instrumentation, IECON*, pp. 1670–1675.
2. Bartoli, M., Reatti, A., and Kazimierczuk, M.K. (1994) Modelling iron-powder inductors at high frequencies. *Proceedings of the IEEE Industry Applications Conference, IAS*, pp. 1225–1232.
3. Bennett, E. and Larson, S.C. (1940) Effective resistance to alternating currents of multilayer windings. *Transactions of the American Institute of Electrical Engineers* **59** (12), 1010–1017.
4. Blume, L.F. (1982) *Transformer Engineering*, John Wiley & Sons, New York.
5. Bueno, M.D.A. (2001) *Inductance and Force Calculations in Electrical Circuits*, Nova Science Publishers, Huntington.
6. Carsten, B. (1986) High frequency conductor losses in switchmode magnetics. *Proceedings of the High Frequency Power Converter Conference*, pp. 155–176.
7. Cheng, K.W.E. and Evans, P.D. (1994) Calculation of winding losses in high-frequency toroidal inductors using single strand conductors. *IEE Proceedings-Electric Power Applications B* **141** (2), 52–62.
8. Cheng, K.W.E. and Evans, P.D. (1995) Calculation of winding losses in high frequency toroidal inductors using multistrand conductors. *IEE Proceedings-Electric Power Applications B* **142** (5), 313–322.
9. Del Vecchio, R.M., Poulin, B., Feghali, P.T. *et al.* (2001) *Transformer Design Principles: With Applications to Core-Form Power Transformers*, 1st edn, CRC Press, Boca Raton, FL.
10. Dowell, P.L. (1966) Effects of eddy currents in transformer windings. *Proceedings of the Institution of Electrical Engineers* **113** (8), 1387–1394.
11. Erickson, R.W. (2001) *Fundamentals of Power Electronics*, 2nd edn, Springer, Norwell, MA.
12. Evans, P.D. and Chew, W.M. (1991) Reduction of proximity losses in coupled inductors. *IEE Proceedings-Electric Power Applications B* **138** (2), 51–58.
13. Ferreira, J.A. (2010) *Electromagnetic Modelling of Power Electronic Converters (Power Electronics and Power Systems)*, 1st edn, Springer, Norwell, MA.

14. Fitzgerald, A.E., Kingsley, C. Jr, and Umans, S.D. (2002) *Electric Machinery*, 6th edn, McGraw-Hill, New York.
15. Flanagan, W.M. (1992) *Handbook of Transformer Design and Application*, 2nd edn, McGraw-Hill, New York.
16. Georgilakis, P.S. (2009) *Spotlight on Modern Transformer Design (Power Systems)*, 1st edn, Springer, New York.
17. Hanselman, D.C. and Peake, W.H. (1995) Eddy-current effects in slot-bound conductors. *IEE Proceedings-Electric Power Applications B* **142** (2), 131–136.
18. Hoke, A.F. and Sullivan, C.R. (2002) An improved two-dimensional numerical modeling method for E-core transformers. *Proceedings of the IEEE Applied Power Electronics Conference and Exposition, APEC*, pp. 151–157.
19. Hurley, W.G. and Wilcox, D.J. (1994) Calculation of leakage inductance in transformer windings. *IEEE Transactions on Power Electronics* **9** (1), 121–126.
20. Hurley, W.G., Wilcox, D.J., and McNamara, P.S. (1991) Calculation of short circuit impedance and leakage impedance in transformer windings. *Proceedings of the IEEE Power Electronics Specialists Conference, PESC*, pp. 651–658.
21. Hurley, W.G., Wolfle, W.H., and Breslin, J.G. (1998) Optimized transformer design: inclusive of high-frequency effects. *IEEE Transactions on Power Electronics* **13** (4), 651–659.
22. Jieli, L., Sullivan, C.R., and Schultz, A. (2002) Coupled-inductor design optimization for fast-response low-voltage. *Proceedings of the IEEE Applied Power Electronics Conference and Exposition, APEC*, pp. 817–823.
23. Judd, F. and Kressler, D. (1977) Design optimization of small low-frequency power transformers. *IEEE Transactions on Magnetics* **13** (4), 1058–1069.
24. Kassakian, J.G. and Schlecht, M.F. (1988) High-frequency high-density converters for distributed power supply systems. *Proceedings of the IEEE* **76** (4), 362–376.
25. Kassakian, J.G., Schlecht, M.F., and Verghese, G.C. (1991) *Principles of Power Electronics (Addison-Wesley Series in Electrical Engineering)*, Prentice Hall, Reading, MA.
26. Kazimierczuk, M.K. (2009) *High-Frequency Magnetic Components*, John Wiley & Sons, Chichester.
27. Krein, P.T. (1997) *Elements of Power Electronics (Oxford Series in Electrical and Computer Engineering)*, Oxford University Press, Oxford.
28. Kulkarni, S.V. (2004) *Transformer Engineering: Design and Practice*, 1st edn, CRC Press, New York.
29. B. H. E. Limited (2004) *Transformers: Design, Manufacturing, and Materials (Professional Engineering)*, 1st edn, McGraw-Hill, New York.
30. McAdams, W.H. (1954) *Heat Transmission*, 3rd edn, McGraw-Hill, New York.
31. McLyman, C.W.T. (1997) *Magnetic Core Selection for Transformers and Inductors*, 2nd edn, Marcel Dekker Inc., New York.
32. McLyman, C.W.T. (2002) *High Reliability Magnetic Devices*, 1st edn, Marcel Dekker Inc., New York.
33. McLyman, C.W.T. (2004) *Transformer and Inductor Design Handbook*, 3rd edn, Marcel Dekker Inc., New York.
34. E. S. MIT (1943) *Magnetic Circuits and Transformers (MIT Electrical Engineering and Computer Science)*, The MIT Press, Cambridge, MA.
35. Muldoon, W.J. (1978) Analytical design optimization of electronic power transformers. *Proceedings of Power Electronics Specialists Conference, PESC*, pp. 216–225.
36. Pentz, D.C. and Hofsajer, I.W. (2008) Improved AC-resistance of multiple foil windings by varying foil thickness of successive layers. *The International Journal for Computation and Mathematics in Electrical and Electronic Engineering,* **27** (1), 181–195.
37. Perry, M.P. (1979) Multiple layer series connected winding design for minimum losses. *IEEE Transactions on Power Apparatus and Systems* **PAS-98** (1), 116–123.
38. Petkov, R. (1996) Optimum design of a high-power, high-frequency transformer. *IEEE Transactions on Power Electronics* **11** (1), 33–42.
39. Pollock, J.D., Lundquist, W., and Sullivan, C.R. (2011) Predicting inductance roll-off with dc excitations. *Proceedings of the IEEE Energy Conversion Congress and Exposition, ECCE*, pp. 2139–2145.
40. Pressman, A.I., Bellings, K., and Morey, T. (2009) *Switching Power Supply Design*, 3rd edn, McGraw-Hill, New York.
41. Ramo, S., Whinnery, J.R., and Van Duzer, T. (1984) *Fields and Waves in Communication Electronics*, 2nd edn, John Wiley & Sons, New York.

42. Sagneri, A.D., Anderson, D.I., and Perreault, D.J. (2010) Transformer synthesis for VHF converters. *Proceedings of International Power Electronics Conference, IPEC*, pp. 2347–2353.
43. Smith, B. (2009) *Capacitors, Inductors and Transformers in Electronic Circuits (Analog Electronics Series)*, Wexford College Press, Wexford.
44. Snelling, E.C. (1988) *Soft Ferrites: Properties and Applications*, 2nd edn, Butterworths, London.
45. Sullivan, C.R. (1999) Optimal choice for number of strands in a litz-wire transformer winding. *IEEE Transactions on Power Electronics* **14** (2), 283–291.
46. Sullivan, C.R. and Sanders, S.R. (1996) Design of microfabricated transformers and inductors for high-frequency power conversion. *IEEE Transactions on Power Electronics* **11** (2), 228–238.
47. Urling, A.M., Niemela, V.A., Skutt, G.R., and Wilson, T.G. (1989) Characterizing high-frequency effects in transformer windings-a guide to several significant articles. *Proceedings of the IEEE Applied Power Electronics Conference and Exposition, APEC*, pp. 373–385.
48. Van den Bossche, A. (2005) *Inductors and Transformers for Power Electronics*, 1st edn, CRC Press, New York.
49. Vandelac, J.P. and Ziogas, P.D. (1988) A novel approach for minimizing high-frequency transformer copper losses. *IEEE Transactions on Power Electronics* **3** (3), 266–277.
50. Venkatachalam, K., Sullivan, C.R., Abdallah, T., and Tacca, H. (2002) Accurate prediction of ferrite core loss with nonsinusoidal waveforms using only Steinmetz parameters. *Proceedings of IEEE Workshop on Computers in Power Electronics, COMPEL*, pp. 36–41.
51. Venkatraman, P.S. (1984) Winding eddy current losses in switch mode power transformers due to rectangular wave currents. *Proceedings of the 11th National Solid-State Power Conversion Conference, Powercon* 11, pp. A1.1–A1.11.
52. Williams, R., Grant, D.A., and Gowar, J. (1993) Multielement transformers for switched-mode power supplies: toroidal designs. *IEE Proceedings-Electric Power Applications B* **140** (2), 152–160.
53. Ziwei, O., Thomsen, O.C., and Andersen, M. (2009) The analysis and comparison of leakage inductance in different winding arrangements for planar transformer. *Proceedings of the IEEE Power Electronics and Drive Systems, PEDS*, pp. 1143–1148.

6

High Frequency Effects in the Windings[1]

It is clear from Chapter 2 that operating the transformer at high frequency reduces its size. However, there are additional loss mechanisms that come into play at high frequencies. These contribute to temperature rise, and a larger core may be required to increase the surface area to improve heat dissipation. A proper understanding of high-frequency effects is required to ensure a trade-off between reduced size and increased loss.

It was shown in Section 4.3.5 that the optimum design has the core loss approximately equal to the copper (winding) loss. The methodology for transformer design developed in Chapter 5 assumed that the winding loss was due to the DC resistance of the windings. In reality, and in particular for power electronics applications, high-frequency operation leads to increased AC loss due to skin and proximity effects.

The skin effect and the proximity effect give rise to increased loss in conductors, due to the non-uniform distribution of current in the conductors. These effects are a direct result of Faraday's law, whereby eddy currents are induced to oppose the flux created in the windings by the AC currents. Both of these effects are examined in detail in the sections to follow, and straightforward formulae are given to calculate them.

Traditionally, Dowell's celebrated formula is used to calculate the high-frequency proximity effect loss with sinusoidal excitation. We will expand Dowell's formula [1] to take account of arbitrary current waveforms that are encountered in switch mode power supplies by calculating the loss at the individual frequencies in the Fourier series. The resultant loss can be calculated for a range of winding layer thicknesses, and an optimum layer thickness can be established graphically. We will also use a more convenient approach, based on the derivative of the current waveform, to simplify the calculations. This will lead to a simple expression for the optimum layer thickness in a multilayer winding.

[1] Parts of this chapter are reproduced with permission from [2] Hurley, W.G., Gath, E., and Breslin, J.G. (2000) Optimizing the AC resistance of multilayer transformer windings with arbitrary current waveforms. *IEEE Transactions on Power Electronics* **15** (2), 369–376.

Transformers and Inductors for Power Electronics: Theory, Design and Applications, First Edition.
W. G. Hurley and W. H. Wölfle.

6.1 Skin Effect Factor

An isolated round conductor carrying AC current generates a concentric alternating magnetic field which, in turn, induces eddy currents (Faraday's law). Figure 6.1 shows the distribution of the magnetic field at low frequencies. The nature of the flux density inside and outside the conductor was analyzed in Section 1.2.

Eddy currents oppose the flux, and the resulting distribution of current means that eddy currents cancel some of the current at the centre of the conductor, while increasing the current near the surface, as shown in Figure 6.1. The overall effect is that the total current flows in a smaller annular area. At high frequencies, the current flows in an equivalent annular cylinder at the surface, with thickness δ, called the skin depth. This field problem can be solved from Maxwell's equations. In a linear homogeneous isotropic medium, Maxwell's equations take the following form:

$$\nabla \times \mathbf{H} = \sigma \mathbf{E} \tag{6.1}$$

$$\nabla \times \mathbf{E} = -\mu_0 \frac{\partial \mathbf{H}}{\partial t} \tag{6.2}$$

For a sinusoidal current, the **H** field takes the form $H = H_\phi e^{j\omega t}$. This gives rise to current in the z direction that varies with radius only. Combining Equations 6.1 and 6.2 with sinusoidal excitation yields a general expression for the current density J in a conductor such as copper with relative permeability $\mu_r = 1$:

$$\nabla \times \mathbf{H} = \sigma \mathbf{E} = J \tag{6.3}$$

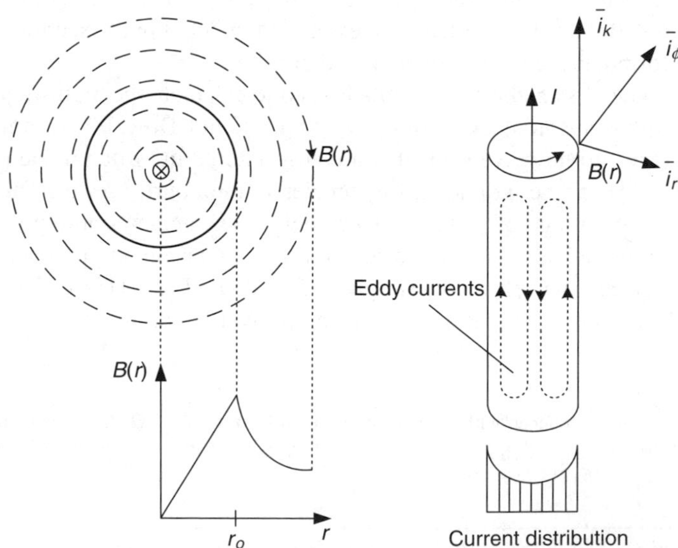

Figure 6.1 Eddy currents in a circular conductor.

σ is the conductivity of the conductor material and:

$$\frac{d^2J}{dr^2}+\frac{1}{r}\frac{dJ}{dr}-j\omega\mu_0\sigma J=0 \tag{6.4}$$

This is a modified Bessel's equation. The general solution is:

$$J(r)=AI_0(mr)+BK_0(mr) \tag{6.5}$$

where I_0 and K_0 are modified Bessel functions of the first and second kind, of order 0, and $m=\sqrt{j\omega\mu_0\sigma}$, so that the argument of I_0 and K_0 is complex. We take the principal value of the square root, that is, $\sqrt{j}=e^{\frac{j\pi}{4}}$.

The coefficients A and B are determined from the boundary conditions and will be complex. It is worth noting that the solutions of Equation 6.4 are in fact combinations of the Kelvin functions with real argument, *viz.* $ber(r\sqrt{\omega\mu_0\sigma})$, $bei(r\sqrt{\omega\mu_0\sigma})$, $ker(r\sqrt{\omega\mu_0\sigma})$ and $kei(r\sqrt{\omega\mu_0\sigma})$, though in our analysis we find it more convenient to use the modified Bessel functions with complex arguments. MATLAB can calculate these functions directly without resorting to calculation of the real and imaginary parts individually. The quantity m is a function of the frequency and the conductivity of the conductor; this may be related to the skin depth δ_0 of the conducting material:

$$mr_o=(1+j)\frac{r_o}{\delta_0} \tag{6.6}$$

$$\delta_0=\frac{1}{\sqrt{\pi f\mu_0\sigma}} \tag{6.7}$$

The solution of Equation 6.5, taking the boundary conditions into account, means that $B=0$, since the current is finite at the centre of the core at $r=0$. Taking the current density as $J(r_o)$ at the outer surface, the solution is:

$$\frac{J(r)}{J(r_o)}=\frac{I_0(mr)}{I_0(mr_o)} \tag{6.8}$$

A plot of the normalized magnitude of $J(r)$ is shown in Figure 6.2 for different values of r_o/δ_0, corresponding to 1 kHz, 10 kHz, 100 kHz and 1 MHz in a 2.5 mm diameter copper conductor.

It is clear that, even at 100 kHz, there is a significant reduction of current in the centre of the copper wire. The MATLAB program to generate the plots in Figure 6.2 is given at the end of this chapter, and it may be used to generate plots for other conductors or for different frequencies.

Assuming cylindrical symmetry, the various components of the electric field intensity **E** and the magnetic field intensity **H** inside the cylinder, in cylindrical coordinates (r, ϕ, z), satisfy the following identities [2]:

$$E_r=0,\quad E_z=0,\quad \frac{\partial E_\phi}{\partial z}=0 \tag{6.9}$$

$$H_r=0,\quad H_\phi=0,\quad \frac{\partial H_z}{\partial \phi}=0 \tag{6.10}$$

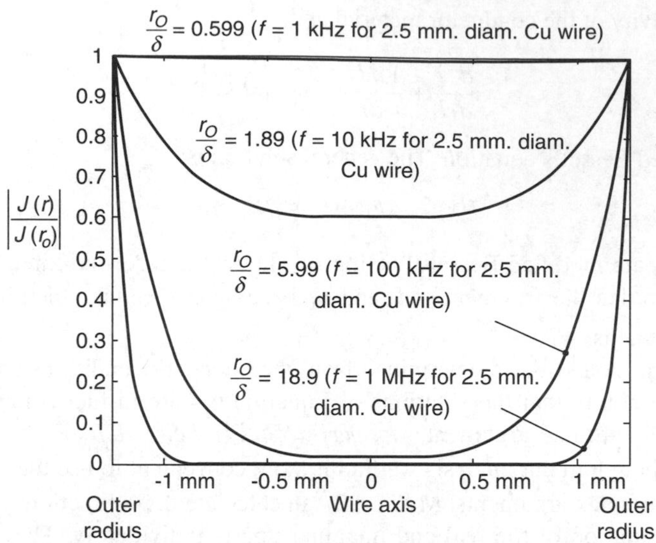

Figure 6.2 Current distribution in a circular conductor.

The two Maxwell's equations then reduce to:

$$E_z = E(r_o)\frac{I_0(mr)}{I_0(mr_o)} \tag{6.11}$$

and:

$$H_\varphi = \frac{1}{j\omega\mu_0}\frac{dE_z}{dr} = \frac{\sigma E(r_o)}{m^2}\frac{I_0'(mr)}{I_0(mr_o)} \tag{6.12}$$

Ampere's law relates the magnetic field intensity to the current (Equation 1.8):

$$\oint_C \mathbf{H} \cdot d\mathbf{l} = i \tag{1.8}$$

Taking the contour at the outer surface:

$$H_\varphi(r_o) = \frac{I}{2\pi r_o} \tag{6.13}$$

Substituting back into Equation 6.12 and noting that:

$$\frac{d}{dr}I_0(mr) = mI_1(mr) \tag{6.14}$$

yields the internal impedance of the conductor per unit length:

$$Z_i = \frac{E(r_o)}{I} = R_{\mathrm{dc}}\frac{mr_oI_0(mr)}{2I_1(mr_o)} \tag{6.15}$$

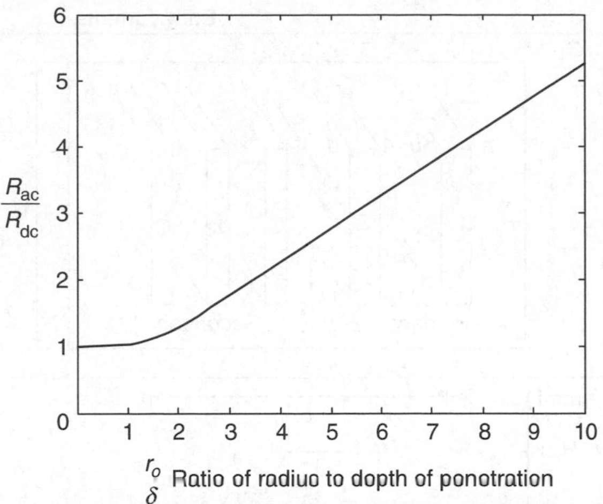

Figure 6.3 R_{ac}/R_{dc} due to skin effect.

R_{dc} is simply the DC resistance of the conductor per unit length. We now have a very compact form of the internal impedance of the conductor. The AC resistance is given by the real part of Z_i and internal inductive reactance is given by the imaginary part of Z_i.

This expression is easily evaluated with MATLAB, but the following approximations are useful, defining $k_s = R_{ac}/R_{dc}$:

$$k_s = 1 + \frac{\left(\frac{r_o}{\delta_0}\right)^4}{48 + 0.8\left(\frac{r_o}{\delta_0}\right)^4} \qquad \frac{r_o}{\delta_0} < 1.7 \tag{6.16}$$

$$= 0.25 + 0.5\left(\frac{r_o}{\delta_0}\right) + \frac{3}{32}\left(\frac{\delta_0}{r_o}\right) \qquad \frac{r_o}{\delta_0} > 1.7 \tag{6.17}$$

At very high frequencies, $\delta_0 << r_o$ and:

$$k_s \approx 0.5\frac{r_o}{\delta_0} = 0.5 r_o \sqrt{\pi f \mu_0 \sigma} \tag{6.18}$$

showing that the AC resistance is proportional to the square root of frequency. A plot of R_{ac}/R_{dc} calculated with Equation 6.15 is shown in Figure 6.3.

6.2 Proximity Effect Factor

The proximity effect arises when the distribution of current in one layer of a winding influences the distribution in another layer. Consider the transformer depicted in Figure 6.4, which has a two-layer primary winding, each layer carrying current I.

Assuming ideal magnetic material ($\mu_r \to \infty$, $\sigma \to 0$), the magnetic field intensity H goes to zero inside the core. With the aid of Ampere's law, we can plot the field intensity across the window of the transformer.

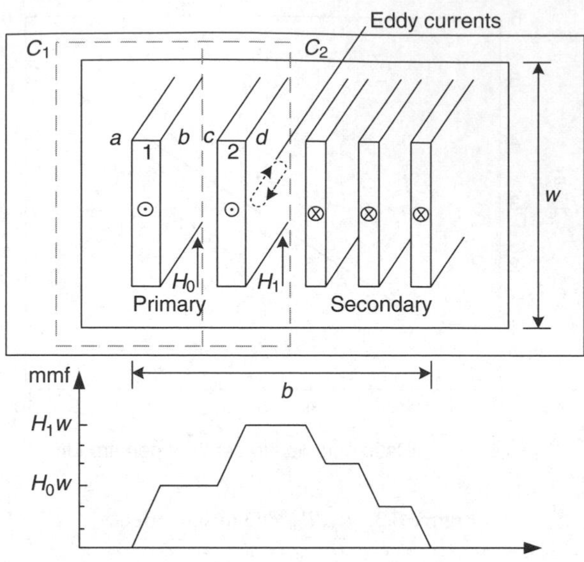

Figure 6.4 mmf in a transformer winding.

Consider N turns carrying current I within a layer. For contour C_1 in Figure 6.4:

$$H_0 w = NI \tag{6.19}$$

and:

$$H_0 = \frac{NI}{w} \tag{6.20}$$

For contour C_2 in Figure 6.4:

$$H_1 w = 2NI \tag{6.21}$$

and

$$H_1 = \frac{2NI}{w} = 2H_0 \tag{6.22}$$

The situation is best illustrated by an mmf plot. Recall from Chapter 4 that the mmf in the secondary winding will cancel the mmf in the primary, restoring the field intensity to zero at the far side of the window.

The field intensity is alternating at the operating frequency of the transformer. The presence of the field intensity adjacent to the winding layers will induce eddy currents in the conductors, thus opposing the magnetic field in accordance with Faraday's law, as shown in Figure 6.4.

The field intensity to the right of layer 1 in Figure 6.4 is H_0 and it is zero to the left. The direction of the induced eddy currents is such that the overall current distribution is increased on the right hand side of the current layer and reduced on the left. At sufficiently high frequency (where the skin depth is less than the thickness d of the layer), the current distribution will be zero on the left hand side of the layer and the current may be considered to flow in a layer of thickness δ; the overall resistance is increased because there is a smaller conduction area. It must be remembered that the net current flowing is not changed from its DC value.

In the next section, the proximity effect factor is derived from first principles. The derivation is based on foil layers which extend the full height of the window. Refinements for layers that consist of round conductors or foils that do not extend the full length of the window can be accommodated by the concept of 'porosity', in which an equivalent area of copper is spread along the length w.

The general formula for the proximity effect factor can be derived by solving Maxwell's equations for an annular cylindrical layer of thickness d and with field intensities H_+ and H_- on either side.

6.2.1 AC Resistance in a Cylindrical Conductor

Maxwell's equations (Equations 6.1 and 6.2) apply to the annular cylindrical conducting layer, shown in Figure 6.5, which carries a sinusoidal current $i_\phi(t) = I_\phi e^{j\omega t}$. The conductivity of the conducting medium is σ and the physical dimensions are shown in Figure 6.5. H_- and H_+ are the magnetic fields parallel to the inside and outside surfaces of the cylinder, respectively. We shall see shortly that H_- and H_+ are independent of z.

Assuming cylindrical symmetry, the various components of the electric field intensity **E** and the magnetic field intensity **H** inside the cylinder, in cylindrical co-ordinates (r, ϕ, z), satisfy the identities in Equations 6.9 and 6.10. The two equations (6.1 and 6.2) then reduce to:

$$-\frac{\partial H_z}{\partial r} = \sigma E_\phi \tag{6.23}$$

$$\frac{1}{r}\frac{\partial}{\partial r}(rE_\phi) = -j\omega\mu_0 H_z \tag{6.24}$$

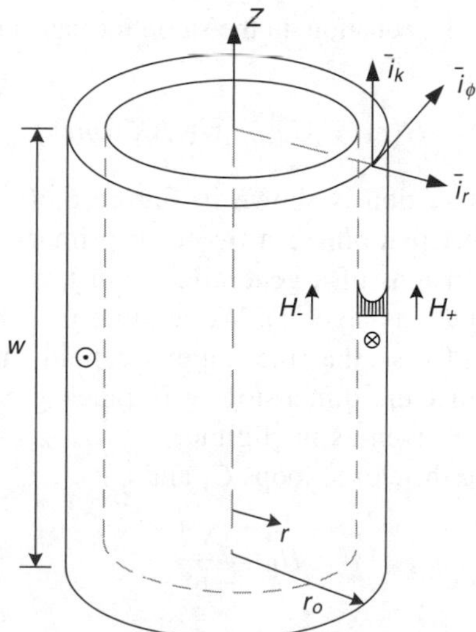

Figure 6.5 Conducting cylinder.

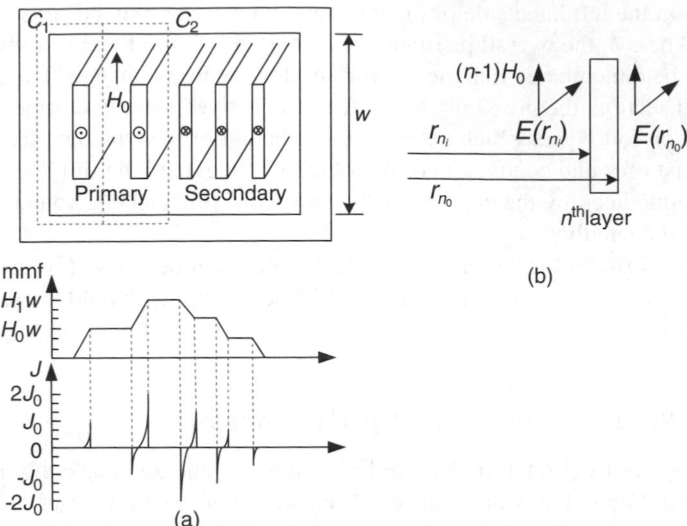

Figure 6.6 Transformer cross-section with (a) associated mmf diagram and current density at high frequency, (b) generalized *n*th layer.

Since **H** has only a *z*-component and **E** has only a ϕ-component, we drop the subscripts without ambiguity. Furthermore, the electric and magnetic field intensities are divergence-free, so it follows that E and H are functions of r only. Substituting the expression for E given by Equation 6.23 into Equation 6.24 then yields the ordinary differential equation:

$$\frac{d^2H}{dr^2} + \frac{1}{r}\frac{dH}{dr} - j\omega\mu_0\sigma H = 0 \tag{6.25}$$

This is another modified Bessel's equation, in the same format as Equation 6.4, and the general solution is:

$$H(r) = AI_0(mr) + BK_0(mr) \tag{6.26}$$

A typical transformer cross-section is shown in Figure 6.6(a), with associated mmf diagram and current density distribution for a two-turn primary and a three-turn secondary winding. The physical dimensions of a generalized *n*th layer are shown in Figure 6.6(b) (the innermost layer is counted as layer 1). We assume that the magnetic material in the core is ideal ($\mu_r \to \infty$, $\sigma \to 0$), so that the magnetic field intensity goes to zero inside the core. We also assume that the dimension w is much greater than the radial dimensions, so that end effects are taken as negligible.

Invoking Ampere's law for the closed loops C_1 and C_2:

$$H_0 = \frac{NI}{w} \tag{6.27}$$

where N is the number of turns in layer n, each carrying constant current I. n here refers to the layer number and should not be confused with the harmonic number n to be used later.

Applying the inner and outer boundary conditions for layer n, that is:

$$H(r_{n_i}) = (n-1)H_0 \tag{6.28}$$

and

$$H(r_{n_o}) = nH_0 \tag{6.29}$$

to the general solution (Equation 6.26), we obtain the coefficients:

$$A = \frac{[(nK_0(mr_{n_i}) - (n-1)K_0(mr_{n_o})]H_0}{I_0(mr_{n_o})K_0(mr_{n_i}) - K_0(mr_{n_o})I_0(mr_{n_i})} \tag{6.30}$$

$$B = \frac{[(-nI_0(mr_{n_i}) + (n-1)I_0(mr_{n_o})]H_0}{I_0(mr_{n_o})K_0(mr_{n_i}) - K_0(mr_{n_o})I_0(mr_{n_i})} \tag{6.31}$$

The corresponding value of $E(r)$ is found from Equation 6.23, that is:

$$E(r) = -\frac{1}{\sigma}\frac{dH(r)}{dr} \tag{6.32}$$

Using the modified Bessel function identity (Equation 6.14) and:

$$\frac{d}{dr}K_0(mr) = -m\,K_1(mr) \tag{6.33}$$

the electric field intensity is then given by:

$$E(r) = -\frac{m}{\sigma}[AI_1(mr) - BK_1(mr)] \tag{6.34}$$

The Poynting vector E × H represents the energy flux density per unit area crossing the surface per unit time. In the cylindrical coordinate system in Figure 6.5, the power per unit area into the cylinder is given by E × H on the inside surface and –E × H on the outside surface. Since **E** and **H** are orthogonal, the magnitude of the Poynting vector is simply the product $E(r)H(r)$, and its direction is radially outwards.

The power per unit length (around the core) of the inside surface of layer n is:

$$\begin{aligned} P_{n_i} &= E(r_{n_i})H(r_{n_i})w \\ &= -\frac{m}{\sigma}(n-1)H_0 w[AI_1(mr_{n_i}) - BK_1(mr_{n_i})] \end{aligned} \tag{6.35}$$

A and B are given by Equations 6.30 and 6.31 respectively, H_0 is given by Equation 6.27, so:

$$\begin{aligned} P_{n_i} = \frac{N^2I^2m}{\sigma w \Psi}&\{(n-1)^2[I_0(mr_{n_o})K_1(mr_{n_i}) + K_0(mr_{n_o})I_1(mr_{ni})] \\ &-n(n-1)[I_0(mr_{n_i})K_1(mr_{n_i}) + K_0(mr_{n_i})I_1(mr_{n_i})]\} \end{aligned} \tag{6.36}$$

where we define:

$$\Psi = I_0(mr_{n_o})K_0(mr_{n_i}) - K_0(mr_{n_o})I_0(mr_{n_i}) \tag{6.37}$$

In a similar fashion, we find the power per unit length (around the core) of the outside surface of layer n is:

$$\begin{aligned} P_{n_o} &= -E(r_{n_o})H(r_{n_o})w \\ &= \frac{m}{\sigma}nH_o w[AI_1(mr_{n_o}) - BK(mr_{n_o})] \\ &= \frac{N^2I^2m}{\sigma w\Psi}\{n^2[I_0(mr_{n_i})K_1(mr_{n_o}) + K_0(mr_{n_i})I_1(mr_{n_o})] \\ &\quad -n(n-1)[I_0(mr_{n_o})K_1(mr_{n_o}) + K_0(mr_{n_o})I_1(mr_{n_o})]\} \end{aligned} \tag{6.38}$$

The minus sign is required to find the power into the outer surface.

We now assume that $mr >> 1$ and use the leading terms in the asymptotic approximations for the modified Bessel functions I_0, I_1, K_0, K_1 (noting for purposes of validity $arg(mr) = \frac{\pi}{4} < \frac{\pi}{2}$):

$$I_0(mr) \approx I_1(mr) \approx \frac{e^{mr}}{\sqrt{2\pi mr}}, \qquad K_0(mr) \approx K_1(mr) \approx \sqrt{\frac{\pi}{2mr}}e^{-mr} \tag{6.39}$$

Substituting these into Equations 6.36 and 6.38 and rearranging, yields the total power dissipation for layer n as:

$$P_{n_i} + P_{n_o} = \frac{N^2I^2m}{\sigma w}\left[(2n^2 - 2n + 1)\coth(md_n) - \frac{n^2 - n}{\sinh(md_n)}\left(\sqrt{\frac{r_{n_o}}{r_{n_i}}} + \sqrt{\frac{r_{n_i}}{r_{n_o}}}\right)\right] \tag{6.40}$$

where $d_n \equiv r_{n_o} - r_{n_i}$ is the thickness of layer n. This result was obtained from the Poynting vector for the complex field intensities, so the real part represents the actual power dissipation.

We now assume that each layer has constant thickness d, so that $d_n = d$ (independent of n). Furthermore, we assume that $d << r_{n_i}$. Then, using the Taylor expansion:

$$\sqrt{1+\epsilon} + \frac{1}{\sqrt{1+\epsilon}} = 2 + \frac{\epsilon^2}{4} + O(\epsilon^3) \tag{6.41}$$

it follows that if $\frac{d}{r_{n_i}} < 10\%$, the error incurred by approximating the sum of the square roots in Equation 6.40 by 2 is in the order of 0.1%. The total power dissipation in layer n then becomes:

$$\begin{aligned} \Re(P_{n_i} + P_{n_o}) &\approx \Re\left(\frac{N^2I^2m}{\sigma w}\left[(2n^2 - 2n + 1)\coth(md) - \frac{2(n^2 - n)}{\sinh(md)}\right]\right) \\ &= \Re\left(\frac{N^2I^2m}{\sigma w}\left[\coth(md) + 2(n^2 - n)\tanh\left(\frac{md}{2}\right)\right]\right) \end{aligned} \tag{6.42}$$

The DC power per unit length of layer n is:

$$P_{\text{dc}} = R_{\text{dc}}(NI)^2 = \frac{(NI)^2}{\sigma w d}. \tag{6.43}$$

The AC resistance factor is the ratio of the AC resistance to the DC resistance:

$$\frac{R_{\text{ac}}}{R_{\text{dc}}} = \frac{\Re(P_{n_i} + P_{n_o})}{P_{\text{dc}}} = \Re\left(md\left[\coth(md) + 2(n^2 - n)\tanh\left(\frac{md}{2}\right)\right]\right) \tag{6.44}$$

Finally, the general result for p layers is:

$$\begin{aligned}\frac{R_{\text{ac}}}{R_{\text{dc}}} &= \frac{\Re\sum_{n=1}^{p}(P_{n_i} + P_{n_o})}{\sum_{n=1}^{p} P_{\text{dc}}}\\ &= \frac{1}{p}\Re\left(md\sum_{n=1}^{p}\left[\coth\,(md) + 2(n^2 - n)\tanh\left(\frac{md}{2}\right)\right]\right)\\ &= \Re\left(md\left[\coth\,(md) + \frac{2(p^2 - 1)}{3}\tanh\left(\frac{md}{2}\right)\right]\right)\end{aligned} \tag{6.45}$$

From the definition of m:

$$md = \sqrt{\omega\mu_0\sigma}\left(\frac{1+j}{\sqrt{2}}\right)d = (1+j)\frac{d}{\delta_0} = (1+j)\Delta \tag{6.46}$$

where Δ is the ratio of the layer thickness d to the skin depth δ_0. The AC resistance factor for p layers is then:

$$\begin{aligned}\frac{R_{\text{ac}}}{R_{\text{dc}}} &= \Re\left(\Delta(1+j)\left[\coth\,(\Delta(1+j)) + \frac{2(p^2 - 1)}{3}\tanh\left(\frac{\Delta}{2}(1+j)\right)\right]\right)\\ &= \Delta\left[\frac{\sinh 2\Delta + \sin 2\Delta}{\cosh 2\Delta - \cos 2\Delta} + \frac{2(p^2 - 1)}{3}\,\frac{\sinh \Delta - \sin \Delta}{\cosh \Delta + \cos \Delta}\right]\end{aligned} \tag{6.47}$$

This is Dowell's formula [1].

This is the ratio of AC resistance to DC resistance for proximity effect with sinusoidal excitation, designated k_p, and it is plotted in Figure 6.7. It is clear that, as the number of layers increases, there is a substantial increase in the AC resistance for a given layer thickness and frequency.

This is a very good approximation to the original cylindrical solution, particularly if the layer thickness is less than 10% of the radius of curvature. Windings that consist of round conductors, or foils that do not extend the full winding window, may be treated as foils with equivalent thickness d and effective conductivity $\sigma_w = \eta\sigma$. η is called the porosity factor.

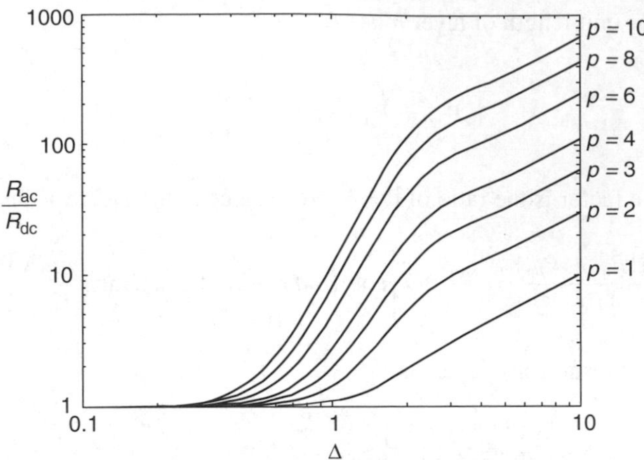

Figure 6.7 Proximity effect factor for sinusoidal excitation.

A round conductor of diameter D is equivalent to a square conductor of side length:

$$d = \sqrt{\frac{\pi}{4}}D \tag{6.48}$$

And the porosity factor is:

$$\eta = \frac{Nd}{w} \tag{6.49}$$

The effective conductivity is then:

$$\sigma_w = \eta\sigma \tag{6.50}$$

When the foil extends a distance w_f across a window w, then:

$$\eta = \frac{w_f}{w} \tag{6.51}$$

These calculations are shown graphically in Figure 6.8, and a detailed treatment of wire conductors is given by Ferreira [3] and Snelling [4].

The trigonometric and hyperbolic functions in Equation 6.47 may be represented by the series expansions:

$$\frac{\sinh(2\Delta) + \sin(2\Delta)}{\cosh(2\Delta) - \cos(2\Delta)} \approx \frac{1}{\Delta} + \frac{4}{45}\Delta^3 - \frac{16}{4725}\Delta^7 + O(\Delta^{11}) \tag{6.52}$$

$$\frac{\sinh(\Delta) - \sin(\Delta)}{\cosh(\Delta) + \cos(\Delta)} \approx \frac{1}{6}\Delta^3 - \frac{17}{2520}\Delta^7 + O(\Delta^{11}) \tag{6.53}$$

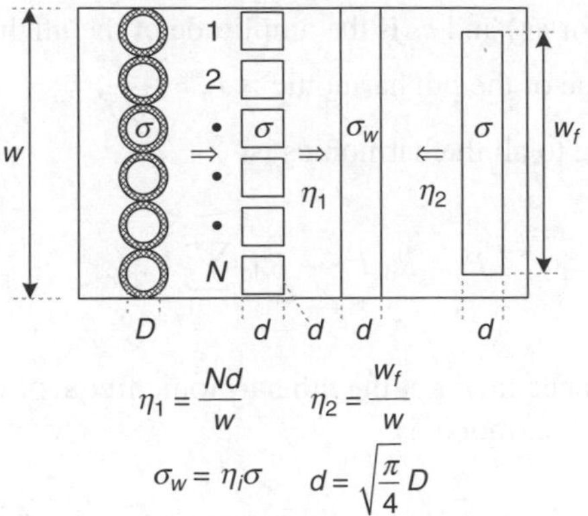

Figure 6.8 Porosity factor for foils and round conductors.

If only terms up to the order of Δ^3 are used, the relative error incurred in Equation 6.52 is less than 1.2% for $\Delta < 1.2$ and the relative error in Equation 6.53 is less than 4.1% for $\Delta < 1$ and less than 8.4% if $\Delta < 1.2$. The asymptotic values of the functions on the left hand side of Equations 6.52 and 6.53 are 1 for $\Delta > 2.5$.

Terms up to the order of Δ^3 in Equations 6.48 and 6.49 are sufficiently accurate for our purposes. Thus, Equation 6.47, with the aid of Equations 6.48 and 6.49, becomes:

$$\frac{R_{\mathrm{ac}}}{R_{\mathrm{dc}}} = 1 + \frac{\Psi}{3}\Delta^4 \tag{6.54}$$

where:

$$\Psi = \frac{5p^2 - 1}{15} \tag{6.55}$$

We now have a simple and straightforward formula to calculate the AC resistance for sinusoidal excitation in a winding with p layers.

6.3 Proximity Effect Factor for an Arbitrary Waveform

So far, we have the proximity effect factor for sinusoidal excitation. For power electronic applications, we need to extend the analysis to include higher frequencies as they appear in the Fourier series of a non-sinusoidal waveform.

An arbitrary periodic current waveform may be represented by its Fourier series:

$$i(t) = I_{\mathrm{dc}} + \sum_{n=1}^{\infty} a_n \cos n\omega t + b_n \sin n\omega t. \tag{6.56}$$

The sine and cosine terms may be combined to give an alternative form

$$i(t) = I_{\mathrm{dc}} + \sum_{n=1}^{\infty} c_n \cos(n\omega t + \varphi_n) \tag{6.57}$$

where I_{dc} is the DC value of $i(t)$ and c_n is the amplitude of the nth harmonic with corresponding phase φ_n. The rms value of the nth harmonic is $I_n = \frac{c_n}{\sqrt{2}}$.

The total power loss due to all the harmonics is:

$$P = R_{dc} I_{dc}^2 + R_{dc} \sum_{n=1}^{\infty} k_{p_n} I_n^2 \tag{6.58}$$

where k_{pn} is the AC resistance factor at the nth harmonic frequency.

The skin depth at the nth harmonic is:

$$\delta_n = \frac{1}{\sqrt{\pi n f \mu_r \mu_o \sigma}} = \frac{\delta_o}{\sqrt{n}} \tag{6.59}$$

and defining Δ_n as:

$$\Delta_n = \frac{d}{\delta_n} = \sqrt{n}\frac{d}{\delta_o} = \sqrt{n}\Delta \tag{6.60}$$

k_{pn} may be found from Equation 6.47.

$$\begin{aligned} k_{p_n} &= \sqrt{n}\Delta \left[\frac{\sinh(2\sqrt{n}\Delta) + \sin(2\sqrt{n}\Delta)}{\cosh(2\sqrt{n}\Delta) - \cos(2\sqrt{n}\Delta)} + \frac{2(p^2-1)}{3} \frac{\sinh(\sqrt{n}\Delta) - \sin(\sqrt{n}\Delta)}{\cosh(\sqrt{n}\Delta) + \cos(\sqrt{n}\Delta)} \right]. \\ &= \Delta_n \left[\frac{\sinh(2\Delta_n) + \sin(2\Delta_n)}{\cosh(2\Delta_n) - \cos(2\Delta_n)} + \frac{2(p^2-1)}{3} \frac{\sinh(\Delta_n) - \sin(\Delta_n)}{\cosh(\Delta_n) + \cos(\Delta_n)} \right] \end{aligned} \tag{6.61}$$

R_{eff} is the AC resistance due to $i(t)$ so that the total power loss is $P = R_{eff} I_{rms}^2$, I_{rms} being the rms value of $i(t)$. Thus, the ratio of effective AC resistance to DC resistance is:

$$\frac{R_{eff}}{R_{dc}} = \frac{I_{dc}^2 + \sum_{n=1}^{\infty} k_{p_n} I_n^2}{{I_{rms}}^2}. \tag{6.62}$$

R_{dc} is the DC resistance of a foil of thickness d. Define R_δ as the DC resistance of a foil of thickness δ_0, recalling that δ_0 is the skin depth at the fundamental frequency of the periodic waveform. Therefore:

$$\frac{R_\delta}{R_{dc}} = \frac{d}{\delta_0} = \Delta \tag{6.63}$$

So:

$$\frac{R_{eff}}{R_{dc}} = \frac{R_{eff}}{R_\delta}\frac{R_\delta}{R_{dc}} = \frac{R_{eff}}{R_\delta}\Delta \tag{6.64}$$

and:

$$\frac{R_{\text{eff}}}{R_\delta} = \frac{\dfrac{R_{\text{eff}}}{R_{\text{dc}}}}{\Delta} \tag{6.65}$$

Example 6.1

Calculate the ratio of effective resistance to DC resistance $R_{\text{eff}}/R_{\text{dc}}$ in a copper foil of thickness d carrying the pulsed current waveform shown in Figure 6.9. This waveform is often encountered in switch mode power supplies.

The Fourier series of $i(t)$ is:

$$i(t) = I_{\text{dc}} + \hat{I}_1 \sin \omega t + \hat{I}_3 \sin 3\omega t + \ldots = \frac{I_o}{2} + \frac{2I_o}{\pi}\left[\sin \omega t + \frac{1}{3}\sin 3\omega t + \ldots\right]$$

The rms value of the current is $I_{\text{rms}} = \dfrac{I_o}{\sqrt{2}}$.

The proximity effect factor due to $i(t)$ is found from Equation 6.62:

$$\begin{aligned}\frac{R_{\text{eff}}}{R_{\text{dc}}} &= \frac{I_{\text{dc}}^2 + \sum\limits_{n=1,\ \text{odd}}^{\infty} k_{p_n} I_n^2}{I_{\text{rms}}^2} \\ &= \frac{\left[\dfrac{I_o}{2}\right]^2 + \left[\dfrac{\sqrt{2}I_o}{\pi}\right]^2 \left[1^2 k_{p_1} + \dfrac{1}{3^2}k_{p_3} + \ldots\right]}{\left[\dfrac{I_o}{\sqrt{2}}\right]^2} \\ &= \frac{1}{2} + \frac{4}{\pi^2}\sum_{n=1,\ \text{odd}}^{\infty} \frac{k_{p_n}}{n^2}\end{aligned}$$

k_{p_n} is the proximity effect factor due to the nth harmonic found from Equation 6.47.

The ratio R_{eff}/R_δ is found from Equation 6.65:

$$\frac{R_{\text{eff}}}{R_\delta} = \frac{\dfrac{1}{2} + \dfrac{4}{\pi^2}\sum\limits_{n=1,\ \text{odd}}^{\infty}\dfrac{k_{p_n}}{n^2}}{\Delta} = \frac{0.5}{\Delta} + \frac{4}{\pi^2}\sum_{n=1,\ \text{odd}}^{\infty}\frac{k_{p_n}}{n^2\Delta}$$

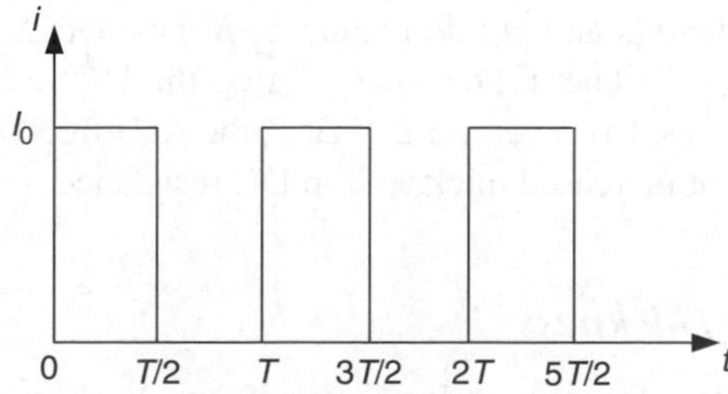

Figure 6.9 Pulsed current waveform.

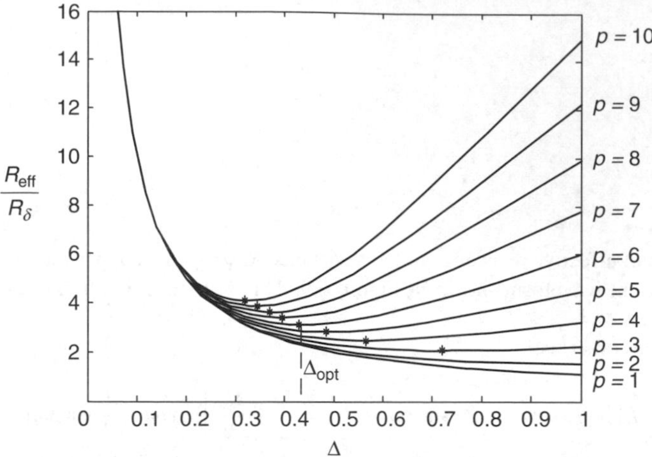

Figure 6.10 Plot of R_{eff}/R_δ versus Δ for various numbers of layers.

Taking Equation 6.47 for k_{p_n}:

$$\frac{R_{eff}}{R_\delta} = \frac{0.5}{\Delta} + \frac{4}{\pi^2} \sum_{n=1,\ \mathrm{odd}}^{N} \frac{1}{n^{\frac{3}{2}}} \left[\frac{\sinh 2\Delta_n + \sin 2\Delta_n}{\cosh 2\Delta_n - \cos 2\Delta_n} + \frac{2(p^2-1)}{3} \frac{\sinh \Delta_n - \sin \Delta_n}{\cosh \Delta_n + \cos \Delta_n} \right]$$

The expression is plotted in Figure 6.10 for $N = 13$ harmonics.

The plots in Figure 6.10 may be interpreted for a number of situations. Consider a case where the frequency of the waveform is fixed. Both R_δ and δ_0 are fixed and the plots may be interpreted as the effective resistance as a function of the layer thickness. Beginning with small values as the layer thickness increases, there is a reduction in resistance. However, as the thickness approaches values comparable to the skin depth at the fundamental frequency, proximity effects become dominant and the resistance increases accordingly. For values of $\Delta > 3$, the current distribution at the centre approaches zero and the resistance levels off at a value based on the current being confined to region at each side of the layer and with a thickness δ_0. In a practical design, values of $\Delta < 2$ would normally be considered.

The horizontal axis in Figure 6.10 is related to foil thickness. The plots show that, for any given number of layers, there is an optimum value of Δ, labelled Δ_{opt}, which minimizes the loss as a function of layer thickness. For $0 < \Delta < \Delta_{opt}$, the DC resistance decreases as the thickness of the foil increases. However, for $\Delta > \Delta_{opt}$, the AC effects on resistance are greater than the mitigating effect of increased thickness on DC resistance.

6.3.1 *The Optimum Thickness*

The plot in Figure 6.10, for the pulsed waveform in Figure 6.9, is redrawn in Figure 6.11 in 3-D and the locus of the minimum value of AC resistance is shown.

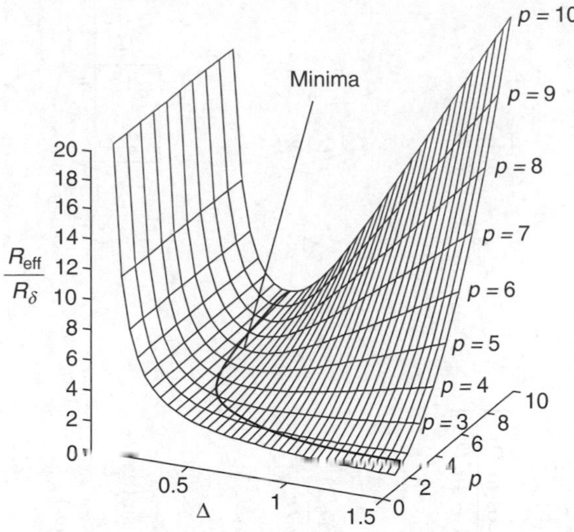

Figure 6.11 Plot of $R_{\mathrm{eff}}/R_{\delta}$ versus Δ for various numbers of layers.

The plot in Figure 6.9 is for the pulsed waveform in Figure 6.8. This type of plot can be generated for any waveform, once its Fourier coefficients are known, by following the procedure in Example 6.1. The Fourier series for various waveforms encountered in power electronics are given in Table 6.1.

For each value of p in Figure 6.11, there is an optimum value of Δ where the AC resistance of the winding is minimum. These optimum points lie on the line marked 'minima' in Figure 6.11, and the corresponding value of the optimum layer thickness is:

$$d_{\mathrm{opt}} = \Delta_{\mathrm{opt}}\delta_0 \tag{6.66}$$

Recognizing that we would have to construct a plot similar to Figure 6.11 for each waveform, a more direct approach is desirable.

The approximation given by Equation 6.54 determines the AC resistance factor at the nth harmonic frequency and may be stated as:

$$k_{p_n} = 1 + \frac{\Psi}{3}n^2\Delta^4. \tag{6.67}$$

on the basis that the skin depth at the nth harmonic is $\delta_n = \dfrac{\delta_0}{\sqrt{n}}$.

Substituting Equation 6.67 into Equation 6.62 yields:

$$\frac{R_{\mathrm{eff}}}{R_{\mathrm{dc}}} = \frac{I_{\mathrm{dc}}^2 + \sum_{n=1}^{\infty} I_n^2 + \frac{\Psi}{3}\Delta^4 \sum_{n=1}^{\infty} n^2 I_n^2}{I_{\mathrm{rms}}^2} \tag{6.68}$$

Table 6.1 Formulae for the optimum thickness of a winding for various waveforms

Current Waveform	I_{rms} and I'_{rms}	Fourier series, $i(t)$	Δ_{opt}
1. 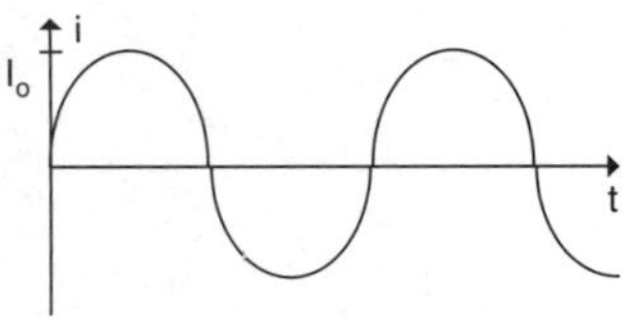	$I_{rms} = \frac{I_o}{\sqrt{2}}$ $I'_{rms} = \frac{2\pi}{T\sqrt{2}} I_o$	$I_o \sin(\omega t)$	$\Delta_{opt} = \sqrt[4]{\frac{1}{\Psi}}$
2. 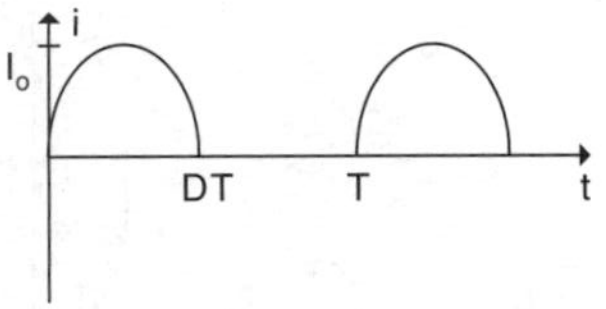	$I_{rms} = I_o \sqrt{\frac{D}{2}}$ $I'_{rms} = I_o \frac{\pi}{DT} \sqrt{\frac{D}{2}}$	$\frac{2DI_o}{\pi} + \sum_{n=1}^{\infty} \frac{4DI_o}{\pi} \left\{ \frac{\cos(n\pi D)}{(1 - 4n^2D^2)} \right\}^a$ $\times \cos(n\omega t)$	$\Delta_{opt} = \sqrt[4]{\frac{4D^2}{\Psi}}$
3. 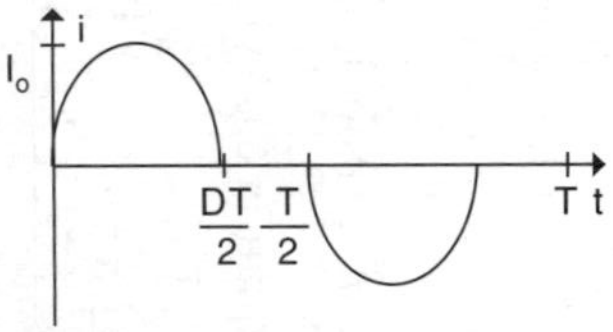	$I_{rms} = I_o \sqrt{\frac{D}{2}}$ $I'_{rms} = I_o \frac{2\pi}{DT} \sqrt{\frac{D}{2}}$	$\sum_{n=1,\ \mathrm{odd}}^{\infty} \frac{4DI_o}{\pi} \left\{ \frac{\cos(n\pi D/2)}{(1 - n^2D^2)} \right\}^a$ $\times \cos(n\omega t)$	$\Delta_{opt} = \sqrt[4]{\frac{D^2}{\Psi}}$
4. 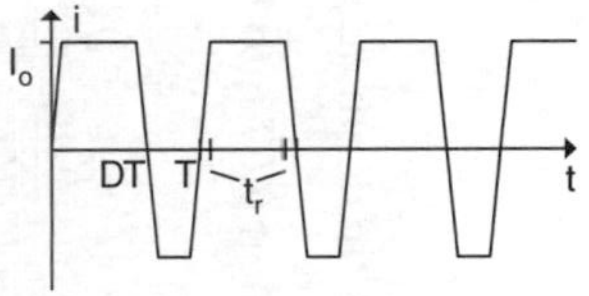	$I_{rms} = I_o \sqrt{1 - \frac{8t_r}{3T}}$ $I'_{rms} = I_o \sqrt{\frac{4}{t_r T}}$	$I_o(2D - 1) + \sum_{n=1}^{\infty} \frac{4I_o}{n\pi} \sin(n\pi D)$ $\times \mathrm{sinc}\left(n\pi \frac{2t_r}{T}\right) \cos(n\omega t)$	$\Delta_{opt} = \sqrt[4]{\frac{\left[1 - \frac{8t_r}{3T}\right] \pi^2 \frac{t_r}{T}}{\Psi}}$
5. 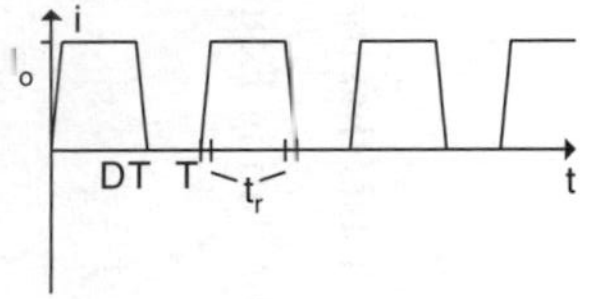	$I_{rms} = I_o \sqrt{D - \frac{4t_r}{3T}}$ $I'_{rms} = I_o \sqrt{\frac{2}{t_r T}}$	$I_o\left(D - \frac{t_r}{T}\right) + \sum_{n=1}^{\infty} \frac{2I_o}{n\pi} \sin\left[n\pi\left(D - \frac{t_r}{T}\right)\right]$ $\times \mathrm{sinc}\left(n\pi \frac{t_r}{T}\right) \cos(n\omega t)$	$\Delta_{opt} = \sqrt[4]{\frac{\left[D - \frac{4t_r}{3T}\right] 2\pi^2 \frac{t_r}{T}}{\Psi}}$

6.	$I_{\text{rms}} = I_o\sqrt{D - \frac{8t_r}{3T}}$ $I'_{\text{rms}} = I_o\sqrt{\frac{4}{t_r T}}$	$\sum_{n=1,\text{ odd}}^{\infty} \frac{4I_o}{n\pi}\sin\left[n\pi\left(D - \frac{t_r}{T}\right)\right] \times \text{sinc}\left(n\pi\frac{t_r}{T}\right)\cos(n\omega t)$	$\Delta_{\text{opt}} = \sqrt[4]{\frac{\left[D - \frac{8t_r}{3T}\right]\pi^2\frac{t_r}{T}}{\Psi}}$
7.	$I_{\text{rms}} = I_o\sqrt{\frac{1}{3}}$ $I'_{\text{rms}} = \frac{2I_o}{T\sqrt{D(1-D)}}$	$\sum_{n=1}^{\infty} \frac{2I_o \text{sinc}(n\pi D)}{\pi n(1-D)}\sin(n\omega t)$	$\Delta_{\text{opt}} = \sqrt[4]{\frac{\pi^2 D(1-D)}{3\Psi}}$
8.	$I_{\text{rms}} = I_o\sqrt{\frac{D}{3}}$ $I'_{\text{rms}} = \frac{2I_o}{\sqrt{D}T}$	$\frac{I_o D}{2} + \sum_{n=1}^{\infty} I_o \sin c^2\left(\frac{n\pi D}{2}\right)\cos(n\omega t)$	$\Delta_{\text{opt}} = \sqrt[4]{\frac{\pi^2 D}{3\Psi}}$
9.	$I_{\text{rms}} = I_o\sqrt{\frac{D}{3}}$ $I'_{\text{rms}} = \frac{4I_o}{\sqrt{D}T}$	$\sum_{n=1,\text{ odd}}^{\infty} I_o \sin c^2\left(\frac{n\pi D}{4}\right)\cos(n\omega t)$	$\Delta_{\text{opt}} = \sqrt[4]{\frac{\pi^2 D}{12\Psi}}$

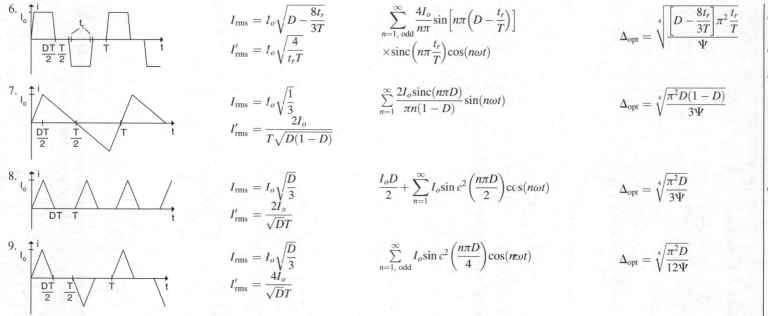

[a]In waveform 2 for $n = k = 1/2D \in$ N (the set of natural numbers), and in waveform 3 for $n = k = 1/D \in$ N, the {expression in curly brackets} is replaced by $\pi/4$.
$\Psi = (5p^2 - 1)/15$, p = No. of layers, sinc $(x) = \sin(x)/x$.

The rms value of the current in terms of its harmonics is:

$$I_{\text{rms}}^2 = I_{\text{dc}}^2 + \sum_{n=1}^{\infty} I_n^2 \tag{6.69}$$

The derivative of $i(t)$ in Equation 6.57 is:

$$\frac{di}{dt} = -\omega \sum_{n=1}^{\infty} nc_n \sin(n\omega t + \varphi_n) \tag{6.70}$$

and the rms value of the derivative of the current is:

$$I_{\text{rms}}'^2 = \omega^2 \sum_{n=1}^{\infty} \frac{n^2 c_n^2}{2} = \omega^2 \sum_{n=1}^{\infty} n^2 I_n^2 \tag{6.71}$$

which, upon substitution into Equation 6.68 using Equation 6.69, yields:

$$\frac{R_{\text{eff}}}{R_{\text{dc}}} = 1 + \frac{\Psi}{3}\Delta^4 \left[\frac{I'_{\text{rms}}}{\omega I_{\text{rms}}}\right]^2 \tag{6.72}$$

This is a straightforward expression for the effective resistance of a winding with an arbitrary current waveform, and it may be evaluated without knowledge of the Fourier coefficients of the waveform.

Taking Equation 6.72 with Equation 6.64:

$$\frac{R_{\text{eff}}}{R_\delta} = \frac{1}{\Delta} + \frac{\Psi}{3}\Delta^3 \left[\frac{I'_{\text{rms}}}{\omega I_{\text{rms}}}\right]^2 \tag{6.73}$$

Setting the derivative to zero yields the optimum value of Δ:

$$\frac{d}{d\Delta}\left(\frac{R_{\text{eff}}}{R_\delta}\right) = -\frac{1}{\Delta^2} + \Psi\Delta^2 \left[\frac{I'_{\text{rms}}}{\omega I_{\text{rms}}}\right]^2 = 0 \tag{6.74}$$

The optimum value of Δ is then:

$$\Delta_{\text{opt}} = \frac{1}{\sqrt[4]{\Psi}} \sqrt{\frac{\omega I_{\text{rms}}}{I'_{\text{rms}}}} \tag{6.75}$$

Substituting this result back into Equation 6.72 produces a very simple expression for the optimum value of the effective AC resistance with an arbitrary periodic current waveform:

$$\left(\frac{R_{\text{eff}}}{R_{\text{dc}}}\right)_{\text{opt}} = \frac{4}{3} \tag{6.76}$$

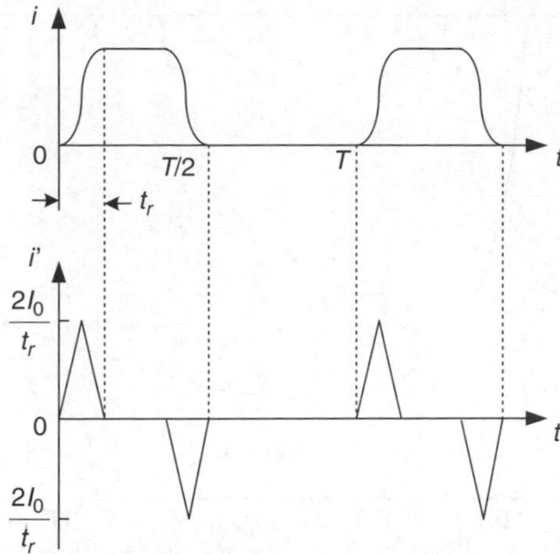

Figure 6.12 Pulsed current waveform and its derivative.

Snelling [4] has already established this result for sinusoidal excitation. The corresponding value for wire conductors with sinusoidal excitation is 3/2 [4].

We may also write Equation 6.72 in terms of Δ_{opt}:

$$\frac{R_{\text{eff}}}{R_{\text{dc}}} = 1 + \frac{1}{3}\left(\frac{\Delta}{\Delta_{\text{opt}}}\right)^4 \tag{6.77}$$

We now have a set of simple formulae with which to find the optimum value of the foil or layer thickness of a winding and its effective AC resistance, where these formulae are based on the rms value of the current waveform and the rms value of its derivative. The formulae have been applied to each of the waveforms in Table 6.1 and the results are presented therein.

Example 6.2

Calculate the ratio of effective resistance to DC resistance $R_{\text{eff}}/R_{\text{dc}}$ in a copper foil of thickness d carrying the pulsed current waveform shown in Figure 6.12.

This waveform is a more realistic version of the waveform in Example 6.1, with a rise and fall time along with a slower characteristic at the start and end of the rising and falling fronts. The derivative is also shown. This waveform is typical of the input in a forward converter. Find the optimum value of Δ for six layers and $t_r/T = 4\%$.

This waveform has a Fourier series:

$$i(t) = I_o\left(\frac{1}{2} - \frac{t_r}{T}\right) + \sum_{n=1}^{\infty} \frac{4I_o}{n^3\pi^3\left(\frac{t_r}{T}\right)^2}\left[1 - \cos\left(\frac{n\pi t_r}{T}\right)\right]\sin\left(n\pi\left(\frac{1}{2} - \frac{t_r}{T}\right)\right)\cos\left(n\left(\omega t - \frac{\pi}{2}\right)\right)$$

The rms value of $i(t)$ is:

$$I_{\text{rms}} = I_o\sqrt{0.5 - \frac{37t_r}{30T}}$$

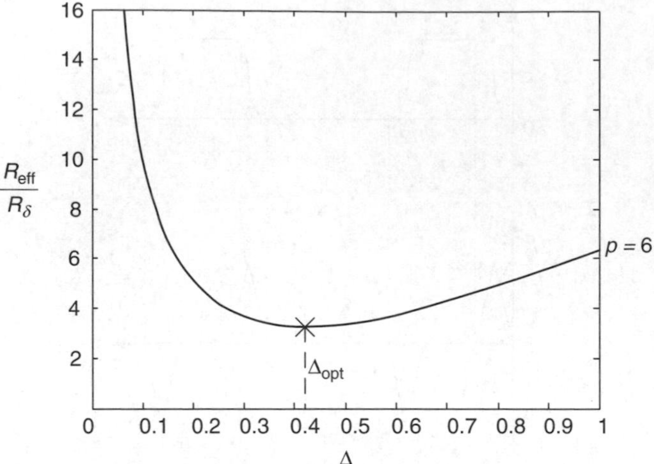

Figure 6.13 Optimum value of foil thickness.

Following the procedure outlined in Example 6.1, $R_{\text{eff}}/R_{\text{dc}}$ is plotted in Figure 6.13 for $p=6$ and $t_r/T=4\%$. The optimum value of Δ is 0.418.

The MATLAB program for this example is listed at the end of the chapter.

Example 6.3

Repeat Example 6.2 using the approximation in Equation 6.75 to find the optimum value of Δ.

The rms value of $i(t)$ is:

$$I_{\text{rms}} = I_o\sqrt{0.5 - \frac{37t_r}{30T}} = I_o\sqrt{0.5 - \frac{37}{30}0.04} = 0.6713I_o$$

The rms value of the derivative of $i(t)$ is:

$$\frac{I'_{\text{rms}}}{\omega} = \frac{I_o}{\pi}\sqrt{\frac{2T}{3t_r}} = \frac{I_o}{\pi}\sqrt{\frac{2}{(3)(0.04)}} = 1.2995I_o$$

The optimum value of Δ is:

$$\Delta_{\text{opt}} = \frac{1}{\sqrt[4]{\Psi}}\sqrt{\frac{\omega I_{\text{rms}}}{I'_{\text{rms}}}} = \frac{1}{\sqrt[4]{\frac{(5)(6)^2-1}{15}}}\sqrt{\frac{0.6713}{1.2995}} = 0.387$$

The value obtained by Fourier analysis is 0.418, and this represents an error of 7.4%. However, the Fourier series of the waveform in Figure 6.12 is not readily available. If waveform 5 in Table 6.1 were used to approximate it, the optimum value of Δ would be:

$$\Delta_{\text{opt}} = \sqrt[4]{\frac{\left[D - \frac{4t_r}{3T}\right] 2\pi^2 \frac{t_r}{T}}{\Psi}} = \sqrt[4]{\frac{\left[0.5 - \frac{(4)(0.04)}{3}\right] 2\pi^2 (0.04)}{\frac{(5)(6)^2 - 1}{15}}} = 0.414$$

On the other hand, the optimum value of Δ obtained by Fourier analysis of waveform 5 is 0.448, which would represent an error of 7.2% when compared to the Fourier analysis of the waveform in Figure 6.12. Waveforms with known Fourier series are often approximations to the actual waveform and can give rise to errors which are of the same order as the approximation, but the approximation is simpler to evaluate.

Example 6.4 Push-pull Converter

Calculate the optimum thickness for the winding in the push-pull converter in Example 5.3.

In Example 5.3, there are six layers, $p = 6$, the duty cycle is $D = 0.67$, the rise time is $t_r = 2.5\%$ and the frequency is $f = 50$ kHz.

The closest waveform to the input voltage shown in Figure 6.12 is number 6 in Table 6.1, and the optimum layer thickness is:

$$\Delta_{\text{opt}} = \sqrt[4]{\frac{\left[D - \frac{8t_r}{3T}\right] \pi^2 \frac{t_r}{T}}{(5p^2 - 1)15}} = \sqrt{\frac{\left[0.67 - \frac{(8)(0.025)}{3}\right] \pi^2 (0.025)}{[(5)(6)^2 - 1]/15}} = 0.3342$$

For copper at 50 kHz, the skin depth is:

$$\delta_0 = \frac{66}{\sqrt{f}} = \frac{66}{\sqrt{(50 \times 10^3)}} = 0.295 \text{ mm}$$

The optimum foil thickness is equal to:

$$d_{\text{opt}} = \Delta_{\text{opt}} \delta_0 = (0.3342)(0.295) = 0.1 \text{ mm}$$

and the AC resistance at the optimum layer thickness is:

$$R_{\text{eff}} = \frac{4}{3} R_{\text{dc}}$$

The DC resistance of the primary and secondary windings made of 0.1 mm × 30 mm foil is 3.29 mΩ, and therefore the AC effective resistance is 4.40 mΩ.

A 2 mm diameter bare copper wire has the same copper area as a in Figure 6.14. The skin effect factor of the round conductor, as given by Equation 6.17, is:

$$k_s = 0.25 + (0.5)\left(\frac{r_o}{\delta_0}\right) = 0.25 + (0.5)\left(\frac{1.0}{0.295}\right) = 1.945$$

The AC resistance of this round conductor due to skin effect is now 3.3 × 1.945 = 6.42 mΩ.

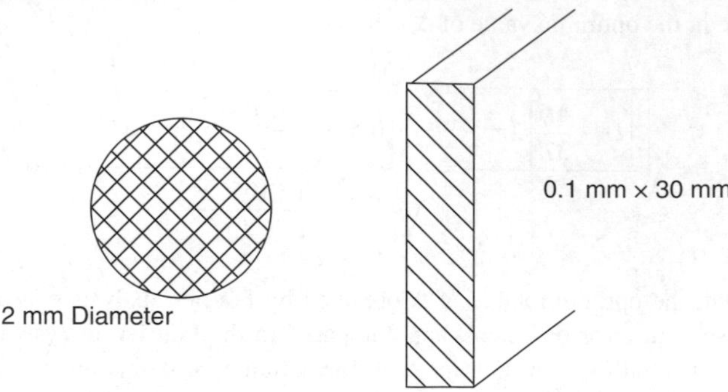

Figure 6.14 Round versus foil conductor.

This is for the fundamental frequency only. Adding the loss for the higher frequencies would increase this value further, compared with the proximity effect factor of $4/3 = 1.33$. Clearly, in this case, the choice of a foil conductor is vastly superior.

6.4 Reducing Proximity Effects by Interleaving the Windings

Consider a transformer with a turns ratio of 2:3, as shown in Figure 6.15. This is the same transformer that was discussed in Section 6.2. The mmf distribution given by Equations 6.19 and 6.21 is also shown.

The current density at the surface of the layers in Figure 6.15 is related to the electric field intensity E ($J = \sigma E$) which, in turn, is related to the magnitude of the magnetic field intensity H. The current density in the windings before interleaving is illustrated in Figure 6.16.

The loss is proportional to J^2 and, taking the sum of the squares of the current density in each layer as an indication of the total loss, the loss is 14.4, taking 1 as the base unit

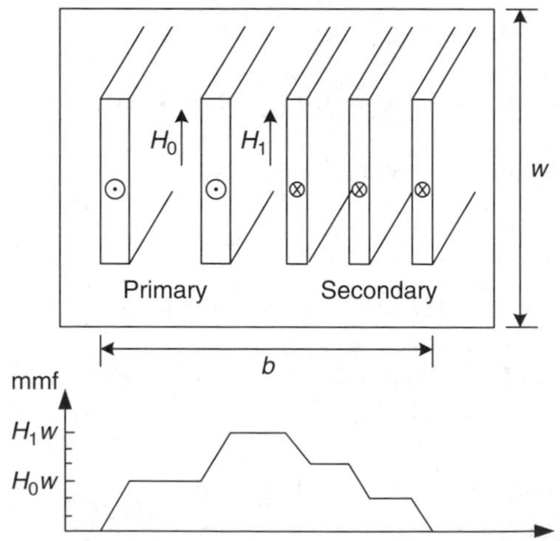

Figure 6.15 Transformer windings before interleaving.

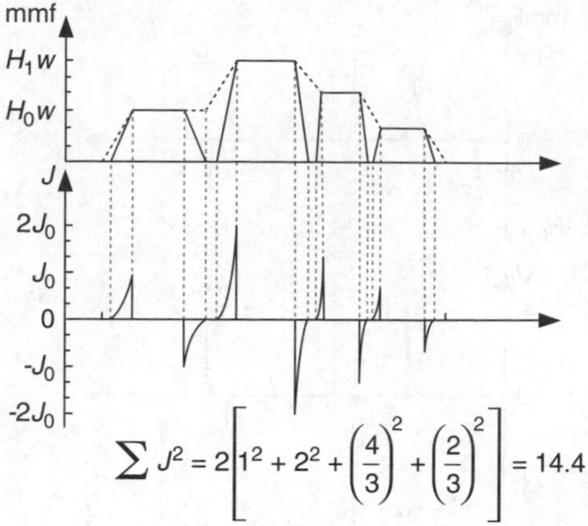

Figure 6.16 Current density distribution before interleaving.

corresponding to J_0. It makes sense, therefore, that the surface value of H should be kept to a minimum. The current density distribution determines the shape of the mmf distribution, and this is shown. The dotted line represents the low frequency distribution. We will return to this point later.

The proximity effect loss may be reduced by interleaving the windings, alternating the primary and secondary layers. This is shown in Figure 6.17 with the current density shown in Figure 6.18.

Adding up the contributions of J^2 in this case yields a total of 4.4, showing an indicative reduction by a factor of more than 3 in the proximity effect loss.

Interleaving also reduces the leakage inductance and interwinding capacitance.

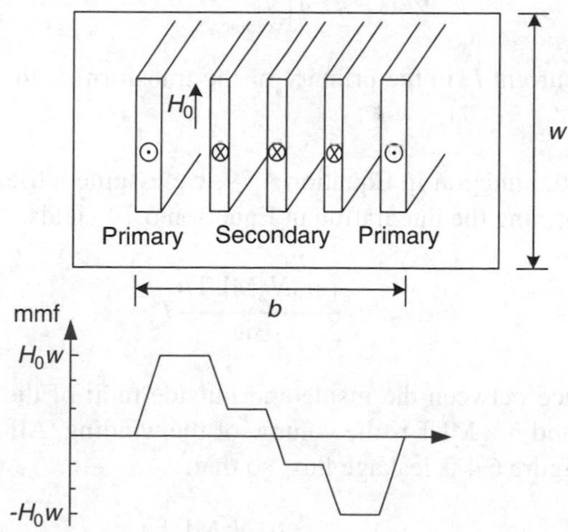

Figure 6.17 Transformer windings after interleaving.

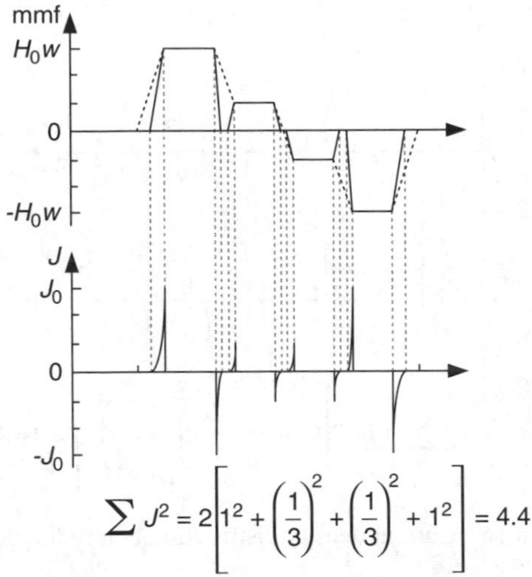

Figure 6.18 Current density distribution after interleaving.

6.5 Leakage Inductance in Transformer Windings

Consider the windings shown in Figure 6.4, the magnetic field is directed along the plane of the layers. The energy per unit volume associated with the H field in Figure 6.4 is given by Equation 2.31.

$$dW_m = \frac{1}{2}\mu_0 H^2 \tag{6.78}$$

The energy associated with the winding in Figure 6.4 is:

$$W_m = \frac{1}{2}\mu_0 \int_{\text{volume}} H^2 dV \tag{6.79}$$

For N_p turns carrying current I_p in the primary of the transformer, the maximum value of H would be $\frac{N_p I_p}{w}$.

In order to evaluate the integral in Equation 6.79, we assume a triangular shape for the H field distribution. Performing the integration in Equation 6.79 yields:

$$W_m = \frac{1}{2}\frac{\mu_0 N_p^2 \text{MLT} b}{3w} I_p^2 \tag{6.80}$$

where b is the difference between the inside and outside radii of the windings, MLT is the mean length of a turn and $b \times$ MLT is the volume of the winding. All the flux in the window of the transformer in Figure 6.4 is leakage flux, so that:

$$L_{l1} + L_{l2} = \frac{\mu_0 N_p^2 \text{MLT} b}{3w} \tag{6.81}$$

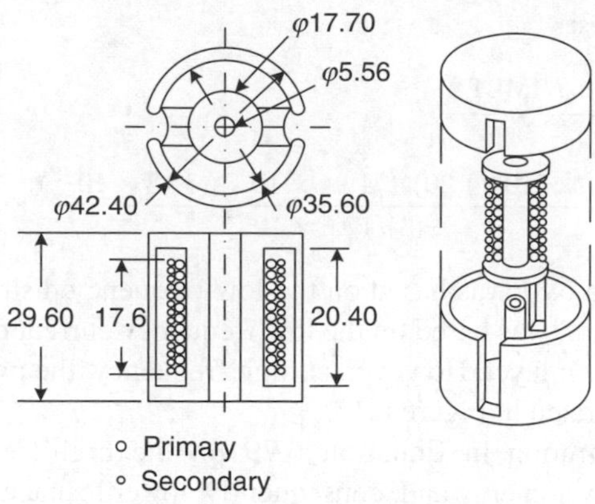

Figure 6.19 Pot core, all dimensions in mm.

L_{l1} and L_{l2} are the components of the leakage flux associated with the primary and secondary windings, as described in Chapter 4.

Evidently, the total leakage is directly related to the total volume occupied by the windings, and the greater spread of windings along a core (increased w) reduces leakage effects. Equation 6.81 shows that the leakage inductance may be reduced by using fewer turns; the most direct approach is to interleave the windings. The ratio b/w suggests that the winding should be placed in a long, narrow window.

This is the ideal case and it takes no account of fringing effects or magnetic energy stored in the windings themselves; however, it does give a reasonable estimate of the leakage inductance in a typical transformer. A more accurate estimate may be obtained by finite element analysis.

Example 6.5

Calculate the leakage inductance for the windings in the pot core shown in Figure 6.19 with the dimensions given in Table 6.2.

The MLT is based on the average radius of the coils:

$$\mathrm{MLT} = 2 \times \pi \times (9.86 + 13.04)/2 = 71.94\,\mathrm{mm}$$

The breadth of the winding is $b = 13.04 - 9.86 = 3.18\,\mathrm{mm}$

Table 6.2 Coil dimensions

Coil 1:	Number of turns	20
	Inside radius	9.86 mm
	Outside radius	11.51 mm
	Width	17.6 mm
Coil 2:	Number of turns	20
	Inside radius	11.39 mm
	Outside radius	13.04 mm
	Width	17.6 mm

The leakage inductance is:

$$\begin{aligned} L_{l1} + L_{l2} &= \frac{\mu_0 N_p^2 \text{MLT}\, b}{3w} \\ &= \frac{(4 \times 10^{-7})(20)^2(71.94 \times 10^{-3})(3.18 \times 10^{-3})}{(3)(17.6 \times 10^{-3})} \times 10^6 = 2.2\ \mu\text{H} \end{aligned}$$

The result in Equation 6.81 was based on the low frequency distribution of the H field in Figure 6.4 and that, in turn, was based on the low frequency current distribution, i.e. constant current density across each layer. However, at high frequency, the proximity effect redistributes the current, as illustrated in Figure 6.15.

Performing the integration in Equation 6.79 on the high-frequency distribution on Figure 6.15 results in lower energy and, consequently, lower leakage inductance. Remember, the real part of the expression in Equation 6.47 represents the loss as predicted by Dowell's formula. The imaginary part of the expression represents the leakage reactance, and this can be easily deduced:

$$\begin{aligned} \frac{L_{l\,ac}}{L_l} &= \frac{3}{2p^2\Delta^2} \text{Im}\left(\Delta(1+j) \left[\coth\left(\Delta(1+j)\right) + \frac{2(p^2-1)}{3} \tanh\left(\frac{\Delta}{2}(1+j)\right) \right] \right) \\ &= \frac{3}{2p^2\Delta} \left[\frac{\sinh 2\Delta - \sin 2\Delta}{\cosh 2\Delta - \cos 2\Delta} + \frac{2(p^2-1)}{3} \frac{\sinh \Delta + \sin \Delta}{\cosh \Delta + \cos \Delta} \right] \end{aligned} \tag{6.82}$$

This is plotted in Figure 6.20. There is a fall off in leakage inductance for $\Delta > 1$.

It is common practice to depend on leakage flux to provide the resonant inductor in resonant power supplies. The dramatic fall-off when the skin depth is comparable to the layer thickness must be considered in the design.

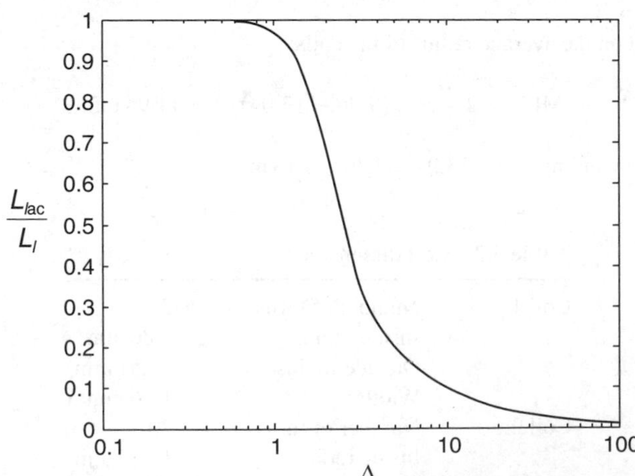

Figure 6.20 Leakage inductance at high frequency.

6.6 Problems

6.1 Derive the formula for the optimum value of Δ for waveform 2 in Table 6.1.

6.2 Verify that the Fourier series for the waveform 3 in Table 6.1 is correct.

6.3 Verify that the Fourier series for the waveform 4 in Table 6.1 is correct.

6.4 Compare the results for the optimum value of Δ for waveform 5 in Table 6.1 obtained by Fourier analysis, with the result given by Δ_{opt} for $p=6$. $D=0.5$ and $t_r/T=0.4\%$.

6.5 Derive the formula for the optimum value of Δ for waveform 6 in Table 6.1.

6.6 Compare the results for the optimum value of Δ for waveform 7 in Table 6.1 obtained by Fourier analysis with the result given by Δ_{opt} for $p=6$. $D=0.4$ and $t_r/T=0.4\%$.

6.7 Write a MATLAB program to generate the 3-D plot of Figure 6.11 for waveform 8 in Table 6.1.

6.8 Derive Equation 6.82 from Dowell's formula.

6.9 Calculate the leakage inductance in the push-pull transformer in Example 5.3.

MATLAB Program for Figure 6.2

```
% Figure 6.2 Current distribution in a circular conductor

for f = [1e3,10e3,100e3,1000e3]; %Hz
r0 = 1.25e-3; %m
mu0 = 4*pi*10^-7;
sigma = 1/(1.72*10^-8);
delta0=1/sqrt(pi*f*mu0*sigma); %skin depth
w = 2*pi*f;
m = sqrt(1j*w*mu0*sigma);

% besseli (Modified Bessel function of first kind)
r = [-1.25e-3:0.001e-3:1.25e-3];
I1 = besseli(0,m*r);
I0 = besseli(0,m*r0);
I = abs(I1/I0);

% plot
plot(r,I,'k','LineWidth',2)
title('High Frequency Effects in the Windings')
xlabel('Wire radius')
ylabel('|J(r)/J(r0)|')
axis([-1.25e-3 1.25e-3 0 1])
grid off
hold on
end
```

MATLAB Program for Figure 6.3

```
% Figure 6.3 Rac/Rdc due to skin effect

n = 0;
for delta = [0:0.01:10];
mr0 = (1+1j)*delta;

I0 = besseli(0,mr0);
I1 = besseli(1,mr0);

R = real((mr0.*(I0/(2*I1))));
n = n+1;
V(n) = R;
D(n) = delta;
end
plot(D,V,'k','LineWidth',2)
title('Rac/Rdc due to skin effect')
xlabel('Ratio of Radius to Depht of Penetration')
ylabel('Rac/Rdc')
axis([0 10 0 6])
grid off
```

MATLAB Program for Figure 6.7

```
% Figure 6.7 Proximity effect factor for sinusoidal excitation

for p = [1:10]
n = 0;
    for delta = [0.1:0.01:10];

    R1 = real(delta*(1+1j)*(coth(delta*(1+1j))+((2*(p^2-1))/3)
*tanh((delta/2)*(1+1j))));

    R2 = delta*[(sinh(2*delta)+sin(2*delta))/(cosh(2*delta)-cos
(2*delta))+((2*(p^2-1))/3)*((sinh(delta)-sin(delta))/(cosh(delta)+
cos(delta)))];

    n = n+1;
    V(n) = R2;
    D(n) = delta;
    end

loglog(D,V,'k','LineWidth',2)
```

```
title('Figure 6.7 Proximity effect factor for sinusoidal excitation')
xlabel('Delta')
ylabel('kp')
axis([0.1 10 1 1000])
grid off
hold on
end
```

MATLAB Program for Figure 6.10

```
% Figure 6.10 Plot of Reff/Rdelta versus Delta for various numbers of
% layers

for p = [1:10]
u=0;

    for delta = [0:0.001:1];
       sum = 0;
       y = 0;
       for n = [1,3,5,7,9,11,13]

          y =
(1/n^(3/2))*((sinh(2*sqrt(n)*delta)+sin(2*sqrt(n)*delta))/(cosh
(2*sqrt(n)*delta)-cos(2*sqrt(n)*delta))+((2*(p^2-1))/3)*((sinh
(sqrt(n)*delta)-sin(sqrt(n)*delta))/(cosh(sqrt(n)*delta)+cos(sqrt
(n)*delta))));
          sum = sum+y;

       end
       R = (0.5/delta)+(4/pi^2)*sum;
       u = u+1;
       V(u) = R;
       D(u) = delta;
    end

plot(D,V,'k','LineWidth',1)

title('Figure 6.10 Plot of Reff/Rdelta versus Delta for various numbers
of layers')
xlabel('Delta')
ylabel('kr')
axis([0 1 0 16])
grid off
hold on
end
```

MATLAB Program for Figure 6.11

```
% Figure 6.11 Plot of Reff/Rdelta versus Delta for various numbers of
% layers

for p = [1:10]
u=1;

    for delta = [0.01:0.04:1.5];
        sum = 0;
        y = 0;
        for n = [1,3,5,7,9,11,13]

            y =
(1/n^(3/2))*((sinh(2*sqrt(n)*delta)+sin(2*sqrt(n)*delta))/(cosh
(2*sqrt(n)*delta)-cos(2*sqrt(n)*delta))+((2*(p^2-1))/3)*((sinh
(sqrt(n)*delta)-sin(sqrt(n)*delta))/(cosh(sqrt(n)*delta)+cos(sqrt
(n)*delta))));
            sum = sum+y;

        end
        R = (0.5/delta)+(4/pi^2)*sum;
        V(p,u) = R;
        D(u) = delta;
        u=u+1;

    end

end

mesh(D,1:10,V)
title('Figure 6.11 Plot of Reff/Rdelta versus Delta for various numbers
of layers')
xlabel('D')
ylabel('p')
zlabel('V')
axis([0 1.5 1 10 0 20])
grid off
hold on

for p = [0.1:0.1:10]
u = 1;

    for delta = [0.01:0.01:1.5];
        sum = 0;
        y = 0;
```

```
        for n = [1,3,5,7,9,11,13]
            y =
(1/n^(3/2))*((sinh(2*sqrt(n)*delta)+sin(2*sqrt(n)*delta))/(cosh
(2*sqrt(n)*delta)-cos(2*sqrt(n)*delta))+((2*(p^2-1))/3)*((sinh
(sqrt(n)*delta)-sin(sqrt(n)*delta))/(cosh(sqrt(n)*delta)+cos(sqrt
(n)*delta))));
            sum = sum+y;

        end
        R = (0.5/delta)+(4/pi^2)*sum;
        V(round(p*10),u) = R;
        D(u) = delta;
        u=u+1;

    end
end

for p=0.1:0.1:10
[krmin,delopt] = min(V(round(p*10),:));
A(round(p*10)) = delopt/100;
B(round(p*10)) = p;
C(round(p*10)) = krmin+0.1;
end

plot3(A,B,C,'k','LineWidth',2)
axis([0 1.5 0 10 0 20])
hold on
```

MATLAB Program for Figure 6.13

```
% Figure 6.13 Optimum value of foil thickness

p = 6;
a = 0.04; % a = tr/T
I0 = 1;
u = 0;

     for delta = [0:0.001:1];
         sum = 0;
         y = 0;
             Irms = I0*sqrt(0.5-37*a/30);
             Idc = I0*(0.5-a);

             for n = [1:19]
             deltan = sqrt(n)*delta;
             In = ((2*sqrt(2)*I0)/(n^3*pi^3*a^2))*[1-cos(n*pi*a)]*sin
(n*pi*(0.5-a));
```

```
            Kpn = deltan*((sinh(2*deltan)+sin(2*deltan))/...
                (cosh(2*deltan)-cos(2*deltan))+...
                ((2*(p^2-1))/3)*((sinh(deltan)-sin(deltan))/...
                (cosh(deltan)+cos(deltan))));
            y = Kpn*In^2;
            sum = sum+y;
            end

        R = ((Idc^2+sum)/Irms^2)/delta;
        R = R/1;
        u = u+1;
        V(u) = R;
        D(u) = delta;
    end

plot(D,V,'k','LineWidth',2)

title('Figure 6.13 Optimum value of foil thickness')
xlabel('Delta')
ylabel('kr')
axis([0 1 0 16])
grid off
hold on
```

MATLAB Program for Figure 6.20

```
% Figure 6.20 Leakage inductance at high frequency

p = 6;
n=0;
    for delta = [0:0.1:100];

    R1 =
(3/(2*p^2*delta^2))*imag(delta*(1+1j)*(coth(delta*(1+1j))+
((2*(p^2-1))/3)*tanh((delta/2)*(1+1j))));
    R2 = 1-(1/30)*((p^2-1)/p^2)-((2^8)/factorial(7))*delta^4;

    n = n+1;
    V(n) = R1;
    W(n) = R2;
    D(n) = delta;
    end

semilogx(D,V,'b','LineWidth',2)
title('Leakage inductance')
```

```
xlabel('Delta')
ylabel('FL')
axis([0 100 0 1])
grid off
hold on
```

References

1. Dowell, P.L. (1966) Effects of eddy currents in transformer windings. *Proceedings of the Institution of Electrical Engineers* **113** (8), 1387–1394.
2. Hurley, W.G., Gath, E., and Breslin, J.G. (2000) Optimizing the AC resistance of multilayer transformer windings with arbitrary current waveforms. *IEEE Transactions on Power Electronics* **15** (2), 369–376.
3. Ferreira, J.A. (1994) Improved analytical modeling of conductive losses in magnetic components. *IEEE Transactions on Power Electronics* **9** (1), 127–131.
4. Snelling, E.C. (1988) *Soft Ferrites: Properties and Applications*, 2nd edn, Butterworths, London.

Further Reading

1. Bartoli, M., Reatti, A., and Kazimierczuk, M.K. (1994) High-frequency models of ferrite core inductors. *Proceedings of the IEEE Industrial Electronics, Control and Instrumentation, IECON*, pp. 1670–1675.
2. Bartoli, M., Reatti, A., and Kazimierczuk, M.K. (1994) Modelling iron-powder inductors at high frequencies. *Proceedings of the IEEE Industry Applications Conference, IAS*, pp. 1225–1232.
3. Bennett, E. and Larson, S.C. (1940) Effective resistance to alternating currents of multilayer windings. *Transactions of the American Institute of Electrical Engineers* **59** (12), 1010–1017.
4. Carsten, B. (1986) High frequency conductor losses in switchmode magnetics. *Proceedings of the High Frequency Power Converter Conference*, pp. 155–176.
5. Cheng, K.W.E. and Evans, P.D. (1994) Calculation of winding losses in high-frequency toroidal inductors using single strand conductors. *IEE Proceedings-Electric Power Applications B* **141** (2), 52–62.
6. Cheng, K.W.E. and Evans, P.D. (1995) Calculation of winding losses in high frequency toroidal inductors using multistrand conductors. *IEE Proceedings-Electric Power Applications B* **142** (5), 313–322.
7. Dale, M.E. and Sullivan, C.R. (2006) Comparison of single-layer and multi-layer windings with physical constraints or strong harmonics. *Proceedings of the IEEE International Symposium on Industrial Electronics*, pp. 1467–1473.
8. Dauhajre, A. and Middlebrook, R.D. (1986) Modelling and estimation of leakage phenomenon in magnetic circuits. *Proceedings of the IEEE Power Electronics Specialists Conference, PESC*, pp. 213–216.
9. Erickson, R.W. (2001) *Fundamentals of Power Electronics*, 2nd edn, Springer, Norwell, MA.
10. Evans, P.D. and Chew, W.M. (1991) Reduction of proximity losses in coupled inductors. *IEE Proceedings-Electric Power Applications B* **138** (2), 51–58.
11. Ferreira, J.A. (2010) *Electromagnetic Modelling of Power Electronic Converters (Power Electronics and Power Systems)*, 1st edn, Springer, Norwell, MA.
12. Hanselman, D.C. and Peake, W.H. (1995) Eddy-current effects in slot-bound conductors. *IEE Proceedings-Electric Power Applications B* **142** (2), 131–136.
13. Hui, S.Y.R., Zhu, J.G., and Ramsden, V.S. (1996) A generalized dynamic circuit model of magnetic cores for low- and high-frequency applications. II. Circuit model formulation and implementation. *IEEE Transactions on Power Electronics* **11** (2), 251–259.
14. Hurley, W.G. and Wilcox, D.J. (1994) Calculation of leakage inductance in transformer windings. *IEEE Transactions on Power Electronics* **9** (1), 121–126.
15. Hurley, W.G., Wilcox, D.J., and McNamara, P.S. (1991) Calculation of short circuit impedance and leakage impedance in transformer windings. *Proceedings of the IEEE Power Electronics Specialists Conference, PESC*, pp. 651–658.

16. Hurley, W.G., Wolfle, W.H., and Breslin, J.G. (1998) Optimized transformer design: inclusive of high-frequency effects. *IEEE Transactions on Power Electronics* **13** (4), 651–659.
17. Jiankun, H. and Sullivan, C.R. (2001) AC resistance of planar power inductors and the quasidistributed gap technique. *IEEE Transactions on Power Electronics* **16** (4), 558–567.
18. Judd, F. and Kressler, D. (1977) Design optimization of small low-frequency power transformers. *IEEE Transactions on Magnetics* **13** (4), 1058–1069.
19. Kassakian, J.G. and Schlecht, M.F. (1988) High-frequency high-density converters for distributed power supply systems. *Proceedings of the IEEE* **76** (4), 362–376.
20. Kazimierczuk, M.K. (2009) *High-Frequency Magnetic Components*, John Wiley & Sons, Chichester.
21. McAdams, W.H. (1954) *Heat Transmission*, 3rd edn, McGraw-Hill, New York.
22. McLyman, C.W.T. (2004) *Transformer and Inductor Design Handbook*, 3rd edn, Marcel Dekker Inc., New York.
23. Muldoon, W.J. (1978) Analytical design optimization of electronic power transformers. *Proceedings of Power Electronics Specialists Conference, PESC*, pp. 216–225.
24. Pentz, D.C. and Hofsajer, I.W. (2008) Improved AC-resistance of multiple foil windings by varying foil thickness of successive layers. *The International Journal for Computation and Mathematics in Electrical and Electronic Engineering* **27** (1), 181–195.
25. Perry, M.P. (1979) Multiple layer series connected winding design for minimum losses. *IEEE Transactions on Power Apparatus and Systems* **PAS-98** (1), 116–123.
26. Petkov, R. (1996) Optimum design of a high-power, high-frequency transformer. *IEEE Transactions on Power Electronics* **11** (1), 33–42.
27. Ramo, S., Whinnery, J.R., and Van Duzer, T. (1984) *Fields and Waves in Communication Electronics*, 2nd edn, John Wiley & Sons, New York.
28. Sagneri, A.D., Anderson, D.I., and Perreault, D.J. (2010) Transformer synthesis for VHF converters. *Proceedings of the International Power Electronics Conference, IPEC*, pp. 2347–2353.
29. Schaef, C. and Sullivan, C.R. (2012) Inductor design for low loss with complex waveforms. *Proceedings of the IEEE Applied Power Electronics Conference and Exposition, APEC*, pp. 1010–1016.
30. Sullivan, C.R. (1999) Optimal choice for number of strands in a litz-wire transformer winding. *IEEE Transactions on Power Electronics* **14** (2), 283–291.
31. Sullivan, C.R. (2001) Cost-constrained selection of strand diameter and number in a litz-wire transformer winding. *IEEE Transactions on Power Electronics* **16** (2), 281–288.
32. Sullivan, C.R. (2007) Aluminum windings and other strategies for high-frequency magnetics design in an era of high copper and energy costs. *Proceedings of the IEEE Applied Power Electronics Conference, APEC*, pp. 78–84.
33. Sullivan, C.R., McCurdy, J.D., and Jensen, R.A. (2001) Analysis of minimum cost in shape-optimized Litz-wire inductor windings. *Proceedings of the IEEE Power Electronics Specialists Conference, PESC*, pp. 1473–1478.
34. Sullivan, C.R. and Sanders, S.R. (1996) Design of microfabricated transformers and inductors for high-frequency power conversion. *IEEE Transactions on Power Electronics* **11** (2), 228–238.
35. Urling, A.M., Niemela, V.A., Skutt, G.R., and Wilson, T.G. (1989) Characterizing high-frequency effects in transformer windings-a guide to several significant articles. *Proceedings of the IEEE Applied Power Electronics Conference and Exposition, APEC*, pp. 373–385.
36. Van den Bossche, A. (2005) *Inductors and Transformers for Power Electronics*, 1st edn, CRC Press, New York.
37. Vandelac, J.P. and Ziogas, P.D. (1988) A novel approach for minimizing high-frequency transformer copper losses. *IEEE Transactions on Power Electronics* **3** (3), 266–277.
38. Venkatraman, P.S. (1984) Winding eddy current losses in switch mode power transformers due to rectangular wave currents. *Proceedings of the 11th National Solid-State Power Conversion Conference, Powercon 11*, pp. A1.1–A1.11.
39. Williams, R., Grant, D.A., and Gowar, J. (1993) Multielement transformers for switched-mode power supplies: toroidal designs. *IEE Proceedings-Electric Power Applications B* **140** (2), 152–160.
40. Xi, N. and Sullivan, C.R. (2004) Simplified high-accuracy calculation of eddy-current loss in round-wire windings. *Proceedings of the IEEE Power Electronics Specialists Conference, PESC*, pp. 873–879.
41. Xi, N. and Sullivan, C.R. (2009) An equivalent complex permeability model for litz-wire windings. *IEEE Transactions on Industry Applications* **45** (2), 854–860.

42. Xu, T. and Sullivan, C.R. (2004) Optimization of stranded-wire windings and comparison with litz wire on the basis of cost and loss. *Proceedings of the IEEE Power Electronics Specialists Conference, PESC*, pp. 854–860.
43. Zhu, J.G., Hui, S.Y.R., and Ramsden, V.S. (1996) A generalized dynamic circuit model of magnetic cores for low- and high-frequency applications. I. Theoretical calculation of the equivalent core loss resistance. *IEEE Transactions on Power Electronics* **11** (2), 246–250.
44. Zimmanck, D.R. and Sullivan, C.R. (2010) Efficient calculation of winding-loss resistance matrices for magnetic components. *Proceedings of the IEEE Workshop on Control and Modeling for Power Electronics, COMPEL*, pp. 1–5.
45. Ziwei, O., Thomsen, O.C., and Andersen, M. (2009) The analysis and comparison of leakage inductance in different winding arrangements for planar transformer. *Proceedings of the IEEE Power Electronics and Drive Systems, PEDS*, pp. 1143–1148.

7

High Frequency Effects in the Core[1]

We saw in Chapter 6 that skin and proximity effects have a major influence on the operation of a transformer or inductor at high frequencies due to the circulation of eddy currents, the main effect being to increase the winding loss. Eddy currents can also flow in the core material, since it has finite resistivity. In the past, laminations were used at power frequencies to overcome the eddy current effects of the low resistivity of silicon steel. In power electronics, the operating frequencies are constantly being increased, with new materials; the high but finite resistivity of materials such as ferrites means that upper limits on operating frequency apply. In this chapter, analysis is provided to establish the frequency limitations due to eddy currents in the core material.

7.1 Eddy Current Loss in Toroidal Cores

Eddy current loss arises in the same way that skin effect loss arises in conductors, namely due to alternating flux in a conducting medium. Consider the toroidal core in Figure 7.1.

When an alternating current flows in the coil, eddy currents flow as shown in accordance with the principle of Lenz's law. The flux level is reduced at the centre of the core and bunched to the surface. This has two effects: the net flux is reduced, with a consequential reduction in inductance, and eddy current loss appears. Both of these effects can be represented as impedance in the coil, with the real part representing eddy current loss and the imaginary part representing the inductance. These quantities are clearly frequency-dependent.

Consider a coil fully wound on the core of Figure 7.1. For the purposes of calculating self inductance and eddy current loss, it is reasonable to neglect leakage effects. The winding

[1] Part of this chapter is reproduced with permission from [1] Hurley, W.G. Wilcox, D.J. and McNamara, P.S. (1991) Calculation of short circuit impedance and leakage impedance in transformer windings. *Proceedings of the IEEE Power Electronics Specialists Conference, PESC*, pp. 651–658.

Transformers and Inductors for Power Electronics: Theory, Design and Applications, First Edition.
W. G. Hurley and W. H. Wölfle.

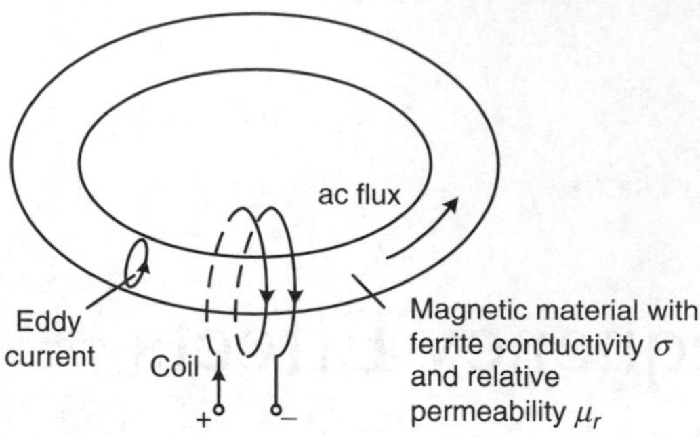

Figure 7.1 Eddy current loss in a toroidal core.

may be replaced by a current sheet, as shown in Figure 7.2, with a surface current density K_f A/m.

In the absence of leakage, the magnetic field intensity H is z directed only and the electric field E has a φ component only.

Taking the general form of Maxwell's equations in a linear homogeneous isotropic medium (Equation 6.3):

$$\nabla \times \mathbf{H} = \sigma E = J$$

so that:

$$-\frac{\partial H_z}{\partial r} = \sigma E_\phi \tag{7.1}$$

and using Equation 6.2:

$$\nabla \times \mathbf{E} = -\mu_0 \frac{\partial \mathbf{H}}{\partial t}$$

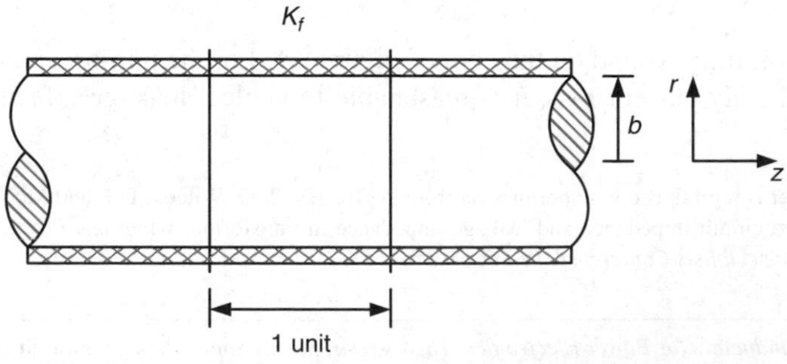

Figure 7.2 Current sheet.

yielding:

$$\frac{1}{r}\frac{\partial(rE\phi)}{\partial r} = j\omega\mu_r\mu_0 H_z \tag{7.2}$$

Eliminating H gives an expression for E:

$$\frac{d^2E}{dr^2} + \frac{1}{r}\frac{dE}{dr} - \frac{E}{r^2} + j\omega\mu_r\mu_0\sigma E = 0 \tag{7.3}$$

This Bessel's equation has a solution of the form:

$$E(r) = AI_1(\Gamma_0 r) + BK_1(\Gamma_0 r) \tag{7.4}$$

where:

$$\Gamma_0 = \sqrt{j\omega\mu_r\mu_0\sigma} = (1+j)\delta \tag{7.5}$$

and δ is the skin depth in the core.

$$\delta = \frac{1}{\sqrt{\pi f \mu_r\mu_0\sigma}} \tag{7.6}$$

I_1 and K_1 are modified Bessel functions of the first and second kind respectively. The constant $B = 0$ in Equation 7.4 since $E(r) = 0$ at $r = 0$.

The constant A is found by applying Ampere's law to a contour just above and below the current sheet in Figure 7.2, noting that $H = 0$ outside the current sheet means that:

$$H(b) = K_f \tag{7.7}$$

From Equations 7.4 and 7.7:

$$H(b) = -\frac{\Gamma_0}{j\omega\mu_r\mu_0}AI_0(\Gamma_0 b) = K_f \tag{7.8}$$

and A is then:

$$A = -j\omega\mu_r\mu_0\frac{K_f}{\Gamma_0 I_0(\Gamma_0 b)} \tag{7.9}$$

The electric field is found from Equation 7.4:

$$E(r) = -j\omega\mu_r\mu_0\frac{I_1(\Gamma_0 r)}{\Gamma_0 I_0(\Gamma_0 b)} \tag{7.10}$$

At the surface of the core at $r = b$, the electric field is:

$$E(b) = -j\omega\mu_r\mu_0 b\frac{I_1(\Gamma_0 b)}{\Gamma_0 b I_0(\Gamma_0 b)}K_f \tag{7.11}$$

The induced voltage at the surface at $r = b$ due to the current in a segment of length ℓ is:

$$V = -\int_0^{2\pi} E(b)bd\varphi = -2\pi r E(r)|_{r=b} \tag{7.12}$$

and the impedance is:

$$Z = \frac{V}{I} = \left.\frac{-2\pi r E(r)}{K_f \ell}\right|_{r=b} \tag{7.13}$$

Substituting for $E(b)$, given by Equation 7.11, gives the impedance for a one turn coil:

$$Z = \frac{j\omega\mu_r\mu_0 2\pi b^2}{K_f \ell} K_f \frac{I_1(\Gamma_0 b)}{\Gamma_0 b I_0(\Gamma_0 b)} = j\omega\mu_r\mu_0 \frac{A_c}{\ell_c} \frac{2I_1(\Gamma_0 b)}{\Gamma_0 b I_0(\Gamma_0 b)} \tag{7.14}$$

where $A_c = \pi b^2$ is the cross-sectional area of the core and l_c is the length of the core. For N turns, the self impedance of the coil is:

$$Z = j\omega\mu_r\mu_0 N^2 \frac{A_c}{\ell_c} \frac{2I_1(\Gamma_0 b)}{\Gamma_0 b I_0(\Gamma_0 b)} \tag{7.15}$$

This result may be written down in terms of the low frequency asymptotic value of the inductance of the toroid L_0:

$$L_0 = \frac{\mu_r\mu_0 N^2 A_c}{\ell_c} \tag{7.16}$$

Thus:

$$Z = j\omega L_0 \frac{2I_1(\Gamma_0 b)}{\Gamma_0 b I_0(\Gamma_0 b)} \tag{7.17}$$

The self impedance of the coil given by Equation 7.15 has two components:

$$Z = R_s + j\omega L_s \tag{7.18}$$

where R_s represents the eddy current loss in the core and L_s is the self inductance. Both R_s and L_s are frequency dependent. This dependence is best illustrated by numerical approximations to the quantities given by Equation 7.17.

7.1.1 Numerical Approximations

The term $\Gamma_0 b$ may be rewritten in terms of the skin depth δ:

$$\Gamma_0 b = \sqrt{j\omega\mu_r\mu_0\sigma}\, b = (1+j)\frac{b}{\delta} = \sqrt{2}\sqrt{j}\Delta \tag{7.19}$$

Δ is a dimensionless parameter that is the ratio of the radius of the toroidal core cross-section to the skin depth.

The function $\dfrac{2I_1(\Gamma_0 b)}{\Gamma_0 b I_0(\Gamma_0 b)}$ may be approximated as follows by asymptotic expansions of the Bessel functions:

$$\mathrm{Re}\left\{\frac{2I_1(\sqrt{2}\sqrt{j}\Delta)}{\sqrt{2}\sqrt{j}\Delta I_0(\sqrt{2}\sqrt{j}\Delta)}\right\} = \begin{cases} 1 - \dfrac{\Delta^4}{12 + 1.43\Delta^4} & \Delta < 2.1 \\ \dfrac{1}{\Delta} + \dfrac{1}{16\Delta^3} + \dfrac{1}{16\Delta^4} & \Delta > 2.1 \end{cases} \tag{7.20}$$

$$\mathrm{Im}\left\{\frac{2I_1(\sqrt{2}\sqrt{j}\Delta)}{\sqrt{2}\sqrt{j}\Delta I_0(\sqrt{2}\sqrt{j}\Delta)}\right\} = \begin{cases} -\dfrac{\Delta^2}{4} + \dfrac{\Delta^6}{\dfrac{384}{11} + 4.16\Delta^4} & \Delta < 2.8 \\ -\dfrac{1}{\Delta} + \dfrac{1}{2\Delta^2} + \dfrac{1}{16\Delta^3} & \Delta > 2.8 \end{cases} \tag{7.21}$$

The resistance and inductance terms of the core impedance may be evaluated by these approximations:

$$R_s = \begin{cases} \omega L_0\left(\dfrac{\Delta^2}{4} - \dfrac{\Delta^6}{\dfrac{384}{11} + 4.16\Delta^4}\right) & \Delta < 2.8 \\ \omega L_0\left(\dfrac{1}{\Delta} - \dfrac{1}{2\Delta^2} - \dfrac{1}{16\Delta^3}\right) & \Delta > 2.8 \end{cases} \tag{7.22}$$

$$L_s = \begin{cases} L_0\left(1 - \dfrac{\Delta^4}{12 + 1.43\Delta^4}\right) & \Delta < 2.1 \\ L_0\left(\dfrac{1}{\Delta} + \dfrac{1}{16\Delta^3} + \dfrac{1}{16\Delta^4}\right) & \Delta > 2.1 \end{cases} \tag{7.23}$$

7.1.2 Equivalent Core Inductance

L_0 in Equation 7.16 is recognized as the classical expression for the inductance of a toroid at low frequencies. L_0 was derived in Example 2.7; the derivation was based on the assumption of uniform flux density in the core, i.e. no eddy currents. L_s from Equation 7.18 is plotted in Figure 7.3 as a function of Δ. The program is listed at the end of the chapter.

The inductance is reduced to 50% of its low frequency value at $\Delta \cong 2$; that is, when the diameter of the core is approximately four skin depths.

Example 7.1

Calculate the maximum operating frequency of an inductor, made of a toroidal ferrite core with a circular cross-sectional diameter of 1.0 cm, that will ensure that the inductance of a toroidal inductor does

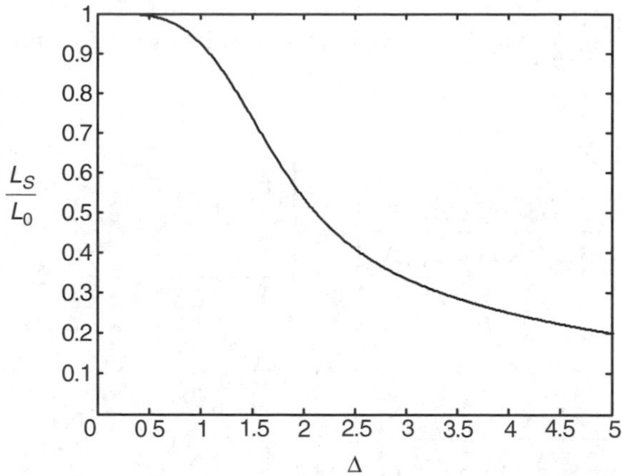

Figure 7.3 Self inductance in a toroidal core as a function of frequency.

not drop below 80% of its low frequency value. The ferrite material is Mn-Zn with relative permeability $\mu_r = 2500$ and electrical resistivity of 1 Ωm.

The approximation in Equation 7.23 means that $L_s/L_o = 0.8$ and:

$$1 - \frac{\Delta^4}{12 + 1.43\Delta^4} = 0.8$$

Solve to give $\Delta = 1.35$.

The diameter of the core is $2b$ and this means that $2b = 2 \times \Delta \times \delta = 1.7\delta = 1.0$ cm.

The maximum skin depth is then $1.0/1.7 = 0.588$ cm.

The conductivity is $\sigma = 1(\Omega\text{m})^{-1}$.

The skin depth is:

$$\delta = \frac{1}{\sqrt{(\pi)f_{\max}(2500)(4\pi \times 10^{-7})(1)}} \times 10^2 = \frac{1006}{\sqrt{f_{\max}}} = 0.588 \text{ cm}$$

Taking this value of skin depth means that the maximum frequency is 2.93 MHz in this example. This is well above the recommended operating frequency for a core of this material and cross-section.

7.1.3 Equivalent Core Resistance

The approximation given by Equation 7.22 for R_s gives us further insight into the nature of the core loss in terms of operating frequency and maximum flux density.

Invoking the approximation in Equation 7.22 for $\Delta < 2.8$ and taking the first term yields:

$$R_s = \omega L_0 \frac{\Delta^2}{4} = \pi f^2 \sigma \left(\frac{\mu_r \mu_0 N}{l_c} \right)^2 \frac{l_c \pi b^2}{2} \tag{7.24}$$

Noting that the volume of the core is simply $l_c\pi b^2 = l_c Ac$, the instantaneous power loss is i^2R_s and the instantaneous power loss per unit volume is:

$$p = \frac{\pi f^2 \sigma}{2}\left(\frac{\mu_r \mu_0 Ni}{l_c}\right)^2 \tag{7.25}$$

Applying Ampere's law, using Equation 7.7 with the current $Ni = K_f l_c$, means that the term inside the bracket in Equation 7.25 is the magnetic flux density. We have assumed that the outside diameter of the toroid is much greater than the diameter of the core cross-section to take the length of the winding as the mean core length. Assuming the magnetic flux density $B = B_{\max}\sin\omega t$ at the surface of the core, the average value of $(B_{\max}\sin\omega t)^2$ is $B_{\max}/2$ and the average power loss in the core due to eddy currents is:

$$p = \frac{\pi f^2 \sigma B_{\max}^2 \pi b^2}{4} \tag{7.26}$$

This result shows that the core loss may be reduced by increasing the electrical resistivity (reducing the electrical conductivity) of the core material. The total loss due to eddy currents is proportional to the square of the radius of the cross-section. The use of a smaller core cross-section to reduce eddy current loss suggests the use of laminated layers of core material for high-frequency operation.

Example 7.2

Calculate the equivalent series resistance and the average eddy current loss in the core of Example 7.1 that operates at a maximum flux density of 0.4 T at 1 MHz with 100 turns. The outside diameter of the toroid is 8 cm.

The inside diameter of the toroid is (8 – 2) cm = 6 cm and the average diameter is 7 cm. The mean length of the magnetic path is $\pi \times 7$ cm.

$$L_0 = \frac{(2500)(4\pi \times 10^{-7})(100^2)(\pi)(0.5 \times 10^{-2})^2}{\pi(7 \times 10^{-2})} = 0.0112\,\text{H}$$

$$\delta = \frac{1006}{\sqrt{(1 \times 10^6)}} = 1.006\,\text{cm}$$

$$\Delta = \frac{0.5}{1.006} = 0.497$$

$$R_s = (2\pi)(1 \times 10^6)(0.0112)\left(\frac{0.497^2}{4} - \frac{0.497^6}{384/11 + (4.16)(0.497^4)}\right) = 4315.45\,\Omega$$

$$p = \frac{\pi(1 \times 10^6)^2(1)(0.4)^2\pi(0.5 \times 10^{-2})^2}{4} = 9.87\,\text{W/cm}^3$$

Example 7.3

Estimate the total core loss for the core in Example 7.2 based on the manufacturer's data for N87 Mn-Zn.

$$P_{\mathrm{fe}} = K_c f^{\alpha} B_{\mathrm{max}}^{\beta}$$

For N87 Mn-Zn the parameters of the Steinmetz equation may be taken from Table 1.1, the total core loss per unit volume is

$$p = (16.9)(1 \times 10^6)^{1.25}(0.4)^{2.35} = 62.05\ \mathrm{W/cm^3}$$

This suggests that the hysteresis loss is 62.05−9.87 = 52.18 W/cm^3.

7.2 Core Loss

The celebrated general Steinmetz equation [2] for core loss is commonly used to describe the total core loss under sinusoidal excitation (Equation 1.29):

$$P_{\mathrm{fe}} = K_c f^{\alpha} B_{\mathrm{max}}^{\beta} \tag{1.29}$$

P_{fe} is the time-average core loss per unit volume; B_{max} is the peak value of the flux density with sinusoidal excitation at the frequency f; K_c, α and β are constants that may be found from manufacturers' data.

Non-sinusoidal excitation is normal for typical power electronics applications. The non-linear nature of ferromagnetic materials means that it is not a simple case of adding the individual frequency components of a Fourier series. Simply using the peak value of the flux density would underestimate the total core loss. The improved General Steinmetz Equation (iGSE) [3] addresses this issue by modifying the original equations while retaining the original coefficients, K_c, α and β.

The time-average power loss using the iGSE is:

$$\begin{aligned} P_v &= \frac{1}{T}\int_0^T k_i \left|\frac{dB(t)}{dt}\right|^{\alpha} |\Delta B|^{\beta-\alpha} dt \\ &= k_i |\Delta B|^{\beta-\alpha} \frac{1}{T}\int_0^T k_i \left|\frac{dB(t)}{dt}\right|^{\alpha} dt \\ &= \quad k_i |\Delta B|^{\beta-\alpha} \left|\frac{dB(t)}{dt}\right|^{\alpha} \end{aligned} \tag{7.27}$$

Here, ΔB is the peak-to-peak flux density of the excitation. k_i is given by:

$$k_i = \frac{K_c}{2^{\beta-1}\pi^{\alpha-1}\displaystyle\int_{2\pi}^{0} |\cos(\theta)|^{\alpha} d\theta} \tag{7.28}$$

A useful approximation for k_i is:

$$k_i = \frac{K_c}{2^{\beta-1}\pi^{\alpha-1}\left(1.1044 + \dfrac{6.8244}{\alpha + 1.354}\right)} \tag{7.29}$$

Example 7.4

Calculate the core loss in the forward converter of Example 5.2.

Recall that the frequency is 25 kHz and the duty cycle is 0.75. The core material specifications are given in Table 5.5 and the core specifications are given in Table 5.6. These tables are repeated below for convenience:

The flux waveform is shown in Figure 7.4.

Normally, iGSE should be used together with a piecewise linear model (PWL) applied to the flux waveform in Figure 7.4, and the power loss per unit volume is:

$$P_v = k_i|\Delta B|^{\beta-\alpha}\frac{1}{T}\left[\int_0^{DT}\left|\frac{\Delta B}{DT}\right|^{\alpha}dt + \int_{DT}^{T}\left|\frac{\Delta B}{(1-D)T}\right|^{\alpha}dt\right]$$

Rearranging:

$$P_v = k_i|\Delta B|^{\beta-\alpha}\frac{1}{T}\left[|\Delta B|^{\alpha}(DT)^{1-\alpha} + |\Delta B|^{\alpha}[(1-D)T]^{1-\alpha}\right]$$

Table 5.5 Material specifications

K_c	37.2
α	1.13
β	2.07
B_{sat}	0.4T

Table 5.6 Core and winding specifications

A_c	$1.25\ \text{cm}^2$
W_a	$1.78\ \text{cm}^2$
A_p	$2.225\ \text{cm}^4$
V_c	$11.5\ \text{cm}^3$
k_f	1.0
k_u	0.4
MLT	6.9 cm
ρ_{20}	$1.72\ \mu\Omega$-cm
α_{20}	0.00393

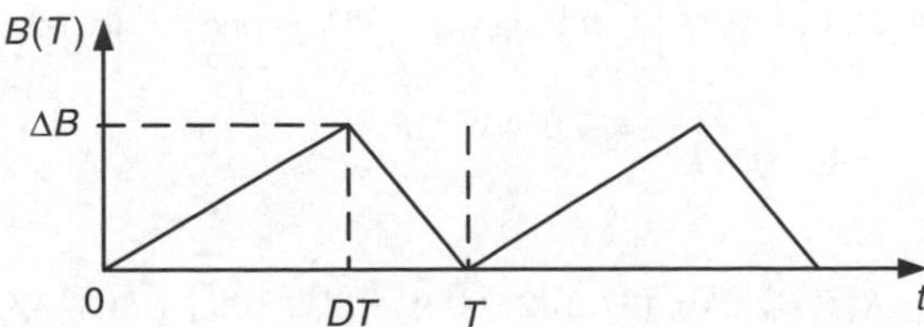

Figure 7.4 Flux waveform in the transformer of a forward converter.

With further simplification:

$$P_v = k_i|\Delta B|^\beta \frac{1}{T^\alpha}\left[D^{1-\alpha} + (1-D)^{1-\alpha}\right]$$

k_i can be calculated by using Equation 7.29:

$$k_i = \frac{K_c}{2^{\beta-1}\pi^{\alpha-1}\left(1.1044 + \dfrac{6.8244}{\alpha + 1.354}\right)}$$

$$= \frac{37.2}{2^{2.07-1}\pi^{1.13-1}\left(1.1044 + \dfrac{6.8244}{1.13 + 1.354}\right)} = 3.964$$

ΔB can be calculated by using Faraday's law:

$$\Delta B = \frac{V_P DT}{N_P A_c} = \frac{(12)(0.75)}{(9)(2.125 \times 10^{-4})(25\,000)} = 0.32\ \text{T}$$

The core loss per unit volume is:

$$P_v = k_i|\Delta B|^\beta \frac{1}{T^\alpha}\left[D^{1-\alpha} + (1-D)^{1-\alpha}\right] = 5.56(0.198)^{2.07}(25\,000)^{1.13}\left[0.75^{1-1.13} + (1-0.75)^{1-1.13}\right]$$

$$= 7.815 \times 10^4\ \text{W/m}^3$$

The total core loss is then $11.5 \times 10^{-6} \times 7.815 \times 10^4 = 0.899$ W.
The result using GSE was 0.898 W.

Example 7.5

The ripple on the output inductor of the forward converter of Example 5.3 is 0.1 T (peak to peak) and the DC current is 7.5 A. Calculate the core loss per unit volume; assume the same material characteristics as given in Table 5.5.

The voltage waveform on the output inductor is shown in Figure 7.5; the core loss is mainly due to the AC current ripple. The effect of DC bias on the core loss is not taken into account; this effect is described in [4].

The iGSE, together with a PWL model, may be used to calculate the core loss. Since the waveform shape and the corresponding time periods are the same as the transformer in Example 7.4, the same expressions apply.

$$P_v = k_i|\Delta B|^\beta \frac{1}{T^\alpha}\left[D^{1-\alpha} + (1-D)^{1-\alpha}\right] = (3.964)(0.1)^{2.07}(25\,000)^{1.13}\left[0.75^{1-1.13} + (1-0.75)^{1-1.13}\right]$$

$$= 7\,034\ \text{W/m}^3$$

With GSE

$$P_{fe} = K_c f^\alpha B_{\max}{}^\beta = (37.2)(25\,000)^{1.3}(0.1/2)^{2.1} = 7\,032\ \text{W/m}^3$$

The error is less than 1%.

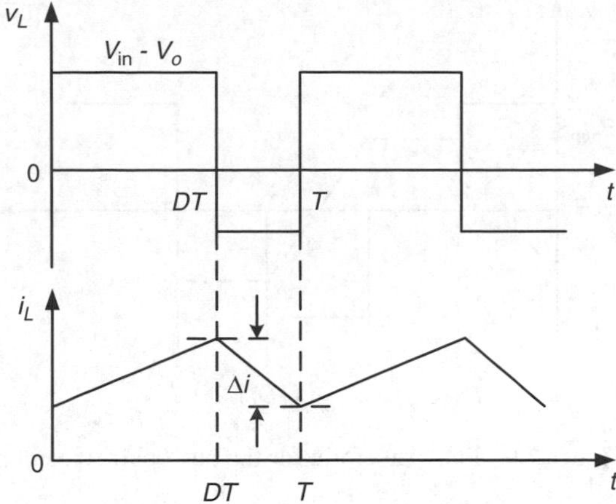

Figure 7.5 Flux waveform for the output inductor in a forward converter.

Example 7.6

Calculate the core loss in a push-pull converter of Example 5.3.

Recall that the frequency is 50 kHz and the duty cycle is 0.67 with input voltage of 36 V. The voltage waveform factor is $K_v = 4.88$. The core material specifications are given in Table 5.8 and the core specifications are given in Table 5.9. These tables are repeated below for convenience:

The peak value of the flux density in the core is:

$$B_{\max} = \frac{\sqrt{D}V_{dc}}{K_v f N_p A_c} = \frac{\sqrt{0.67}(36)}{(4.88)(50\,000)(6)(1.73 \times 10^{-4})} = 0.116\,\mathrm{T}$$

The flux waveform is shown in Figure 7.6.

Table 5.8 Material specifications

K_c	9.12
α	1.24
β	2.0
B_{sat}	0.4 T

Table 5.9 Core and winding specifications

A_c	1.73 cm^2
W_a	2.78 cm^2
A_p	4.81 cm^4
V_c	17.70 cm^3
k_f	1.0
k_u	0.4
MLT	7.77 cm
ρ_{20}	1.72 μΩ-cm
α_{20}	0.00393

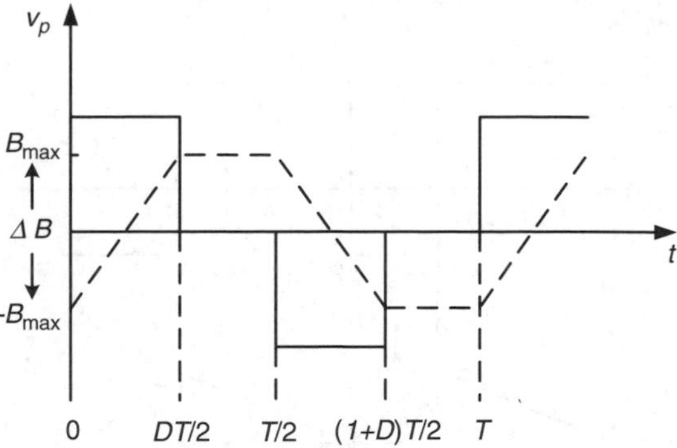

Figure 7.6 Flux waveform for the push-pull converter.

Taking iGSE with a piecewise linear model (PWL) applied to the flux waveform in Figure 7.6, the power loss per unit volume is:

$$P_v = k_i|\Delta B|^{\beta-\alpha}\frac{1}{T}\left[\int_0^{DT/2}\left|\frac{\Delta B}{DT/2}\right|^{\alpha}dt + \int_{T/2}^{(1+D)T/2}\left|\frac{\Delta B}{DT/2}\right|^{\alpha}dt\right]$$

Rearranging and simplifying:

$$P_v = k_i|\Delta B|^{\beta-\alpha}\frac{1}{T}\left[|2\Delta B|^{\alpha}(DT)^{1-\alpha}\right]$$

k_i is found from Equation 7.29:

$$k_i = \frac{K_c}{2^{\beta-1}\pi^{\alpha-1}\left(1.1044 + \dfrac{6.8244}{\alpha+1.354}\right)} = \frac{9.12}{2^{2.0-1}\pi^{1.24-1}\left(1.1044 + \dfrac{6.8244}{1.24+1.354}\right)} = 0.9275$$

The core loss per unit volume is:

$$\begin{aligned} P_v &= k_i|\Delta B|^{\beta-\alpha}\frac{1}{T}\left[|2\Delta B|^{\alpha}(DT)^{1-\alpha}\right] \\ &= (0.9275)(0.232)^{2.0-1.24}(50\,000)\left[(2\times 0.232)^{1.24}\times(0.67/(50\,000))^{1-1.24}\right] \\ &= 0.871\times 10^5\ \mathrm{W/m^3} \end{aligned}$$

The total core loss is then $(17.71\times 10^{-6})(0.871\times 10^5) = 1.543$ W.
Using GSE, taking equation :

$$P_{\mathrm{fe}} = V_c K_c f^{\alpha} B_{\max}^{\beta} = (17.70\times 10^{-6})(9.12)(50\,000)^{1.24}(0.116)^{2.0} = 1.458\ \mathrm{W}$$

The difference is approximately 6% when compared to iGSE.

7.3 Complex Permeability

Returning to Equation 7.15, the equivalent series impedance may be written in terms of this equation, with the approximations in Equations 7.20 and 7.21 for $\Delta < 2$:

$$\begin{aligned} Z &= R_s + j\omega L_s \\ &= \omega\mu_r \frac{\Delta^2}{4} L_1 + j\omega\mu_r \left(1 - \frac{\Delta^4}{12 - 1.43\Delta^4}\right) L_1 \end{aligned} \tag{7.30}$$

Or, in terms of complex permeability:

$$Z = j\omega L_1 (\mu'_{rs} - j\mu''_{rs}) \tag{7.31}$$

Here, L_1 is the low frequency inductance of the toroid with a relative permeability of 1 (air):

$$L_1 = \frac{\mu_0 N^2 A_c}{\ell_c} \tag{7.32}$$

and the series complex relative permeability is defined as:

$$\mu_{rs} = \mu'_{rs} - j\mu''_{rs} \tag{7.33}$$

Based on the analysis provided above, we can find the components of the relative permeability as:

$$\mu'_{rs} = \mu_r \, \mathrm{Re}\left(\frac{2I_1(\Gamma_0 b)}{\Gamma_0 b I_0(\Gamma_0 b)}\right) = \mu_r \left(1 - \frac{\Delta^4}{12 - 1.43\Delta^4}\right) \tag{7.34}$$

$$\mu''_{rs} = -\mu_r \, \mathrm{Im}\left(\frac{2I_1(\Gamma_0 b)}{\Gamma_0 b I_0(\Gamma_0 b)}\right) = \mu_r \frac{\Delta^2}{4} \tag{7.35}$$

μ_r is simply the relative permeability of the core material as before.

The loss tangent is defined in terms of the loss angle δ:

$$\tan\delta = \frac{\mu''_{rs}}{\mu'_{rs}} = \frac{R_s}{\omega L_s} \tag{7.36}$$

The Q factor is:

$$Q = \frac{\omega L_s}{R_s} = \frac{\mu'_{rs}}{\mu''_{rs}} \tag{7.37}$$

Manufacturers provide plots of μ_r vs. frequency so that the inductance is calculated by using the appropriate value of μ_r in the low frequency toroidal equation (Equation 7.16). This achieves the same results as Figure 7.3. However, Δ is a function of the core diameter and measurements on one core are not entirely appropriate for a different-sized core of the same material.

Equation 7.16 is based on a homogeneous isotropic core. In particular, it is assumed that the relative permeability and electrical resistivity are constant. It turns out that the electrical resistivity of a ferrite is frequency-dependent. Snelling [5] gives typical values for Mn-Zn ferrites as 1 Ωm at low frequencies and 0.001 Ωm at high frequencies.

There are two factors which contribute to resistivity in polycrystalline ferrites such as Mn-Zn. There is a granular structure, in which the resistance across the grain boundaries is on the order of one million times greater than that of the ferrite material inside the grains. At low frequencies, the grain boundaries dominate the resistivity, while at high frequencies, capacitive effects shunt the grain boundaries and the ferrite material within the grains' resistivity dominates the overall resistivity.

The resistivity was measured on a sample of P type Mn-Zn ferrite and the variation with frequency is illustrated in Figure 7.7.

Manufacturers present the imaginary part of the complex relative permeability μ''_{rs} to allow the designer to find the inductance by inserting the frequency-dependent value of relative permeability in the classical toroidal inductance formula given by Equation 7.16. In effect, the manufacturer generates the complex permeability by taking measurements on a toroidal core of known dimensions, and deduces the complex permeability from the measured impedance as given by Equation 7.30.

The ratio of L_s/L_0 is given by:

$$\frac{L_s}{L_0} = \begin{cases} 1 - \dfrac{\Delta^4}{12 + 1.43\Delta^4} & \Delta < 2.1 \\ \dfrac{1}{\Delta} + \dfrac{1}{16\Delta^3} + \dfrac{1}{16\Delta^4} & \Delta > 2.1 \end{cases} \tag{7.38}$$

Recall that L_0, given by Equation 7.16, is $L_0 = \mu_r L_1$, and L_1 is given by Equation 7.32.

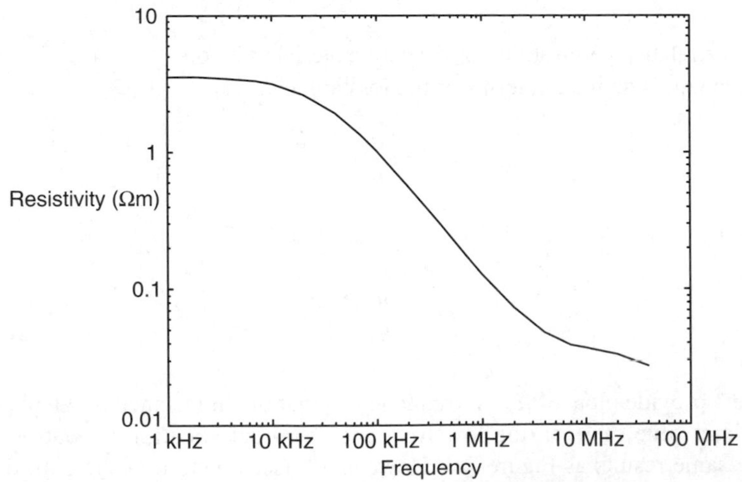

Figure 7.7 Resistivity of P type ferrite.

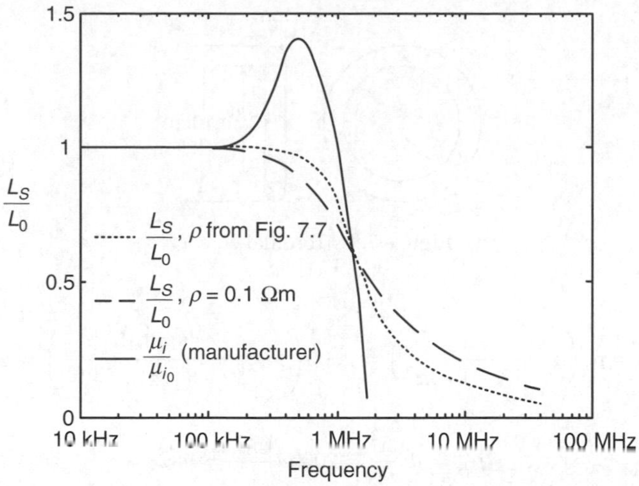

Figure 7.8 Initial permeability of P type ferrite.

Equation 7.38 is plotted in Figure 7.8 for a 10 mm diameter P type ferrite core ($b = 5$ mm). One calculation was performed using a single low-frequency value of resistivity, 0.1 Ωm, and another calculation was performed with values taken from Figure 7.7. The manufacturer supplied initial permeability as a function of frequency for P type Mn-Zn ferrite, and this is normalized to the low-frequency value ($\mu_{i0} = 2500$). Clearly, Equation 7.38 is a very good fit for the manufacturer's measurements, even in the case where a single value of resistivity is used.

The peak in the manufacturer's data at 600 kHz is due to a complex capacitive effect at domain walls in the ferrite. It is important to stress that only one value of relative permeability was used in the calculations in Figure 7.8. The permeability is constant, and the reduction in self inductance at high frequency is due to the electrical resistivity of the core material giving rise to eddy currents. The manufacturer accounted for the reduction in inductance at high frequency by supplying a frequency-dependent permeability for use in the classical toroidal formula. In fact, the variation is entirely predictable and is calculable with Equation 7.38, knowing the electrical resistivity and the relative permeability of the core material and the physical dimensions of the core.

Example 7.7

Calculate the series impedance of a toroidal inductor wound with 100 turns on a toroid with the dimensions shown in Figure 7.9 at 1 MHz. Take $\rho = 0.1$ Ωm and $\mu_r = 2000$.

$$\delta = \frac{1}{\sqrt{\pi(10^6)(2000)(4\pi \times 10^{-7})(10)}} \times 10^2 = 0.356 \text{ cm}$$

$$\Delta = \frac{0.5}{0.356} = 1.40$$

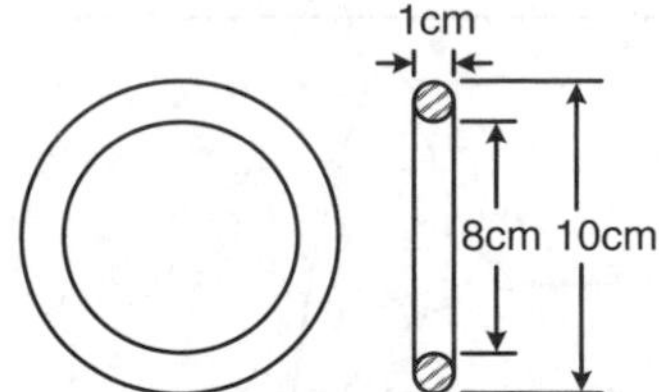

Figure 7.9 Toroidal core.

$$\mu'_{rs} = \mu_r\left(1 - \frac{\Delta^4}{12 - 1.43\Delta^4}\right) = 2000\left(1 - \frac{(1.40)^4}{12 - (1.43)(1.40)^4}\right) = 819$$

$$\mu''_{rs} = \mu_r\frac{\Delta^2}{4} = 2000\frac{(1.4)^2}{4} = 980$$

$$L_1 = \frac{\mu_0 N^2 A_c}{\ell_c} = \frac{(4\pi \times 10^{-7})(100)^2\pi(0.5 \times 10^{-2})^2}{\pi(9 \times 10^{-2})} = 3.49 \times 10^{-6}\ \text{H}$$

$$\begin{aligned} Z &= j\omega L_1(\mu'_{rs} - j\mu''_{rs}) = j(2\pi \times 10^6)(3.49 \times 10^{-6})(819 - j980) \\ &= 21{,}489 + j17{,}962\ \Omega \end{aligned}$$

7.4 Laminations

The factors which influence the choice of core lamination material and thickness may be deduced by considering a laminated plate in a uniform magnetic field, as shown in Figure 7.10.

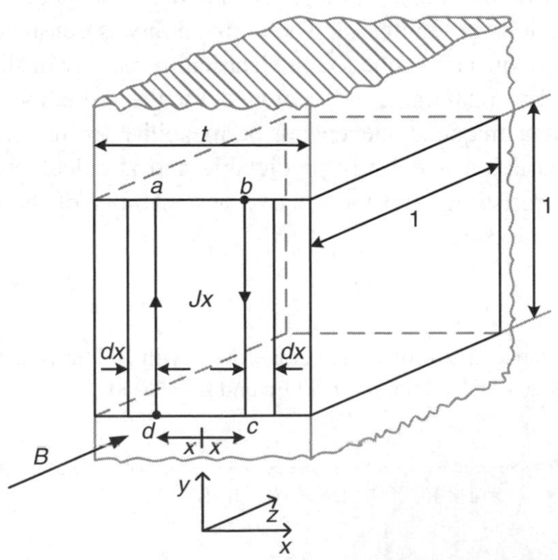

Figure 7.10 Lamination in a uniform magnetic field.

Consider a volume of magnetic lamination material of unit height and depth, with a uniform sinusoidal flux density $B = B_{max}\sin\omega t$ at right angles to the face of the lamination. The alternating flux induces eddy currents, as represented by the current density J_x. Applying Faraday's law in the form of Equation 1.14 around the loop *abcd*, with contributions from the sides *ab* and *cd* at a distance x from the centre line, yields an expression for the induced emf E_x:

$$2E_x = -2\frac{dB}{dt} \tag{7.39}$$

where E_x has the same direction as current density J_x and is related to it by the microscopic form of Ohm's law:

$$J_x = \sigma E_x \tag{7.40}$$

where σ is the conductivity of the material in the lamination. The instantaneous power loss per unit volume due to J_x is:

$$p(t) = \frac{1}{\sigma}J_x^2 = \sigma E_x^2 = \sigma x^2\left(\frac{dB}{dt}\right)^2 \tag{7.41}$$

The instantaneous power loss in a tranche of thickness *dx* with unit height and depth is:

$$p(t) = \sigma x^2\left(\frac{dB}{dt}\right)^2 dx \tag{7.42}$$

Integrating from $x = 0$ *to* $x = t/2$ and taking both sides of the centre plane into account gives the instantaneous power loss in a volume of unit height and depth, in a lamination of thickness t, as:

$$p(t) = 2\sigma\left(\frac{dB}{dt}\right)^2\int_0^{\frac{t}{2}} x^2 dx = \frac{t^3}{12}\sigma\left(\frac{dB}{dt}\right)^2 \tag{7.43}$$

In a transformer core, there are $1/t$ laminations in a unit thickness so that, in a unit cube, the instantaneous eddy current loss per unit volume is:

$$p(t) = \frac{t^2}{12}\sigma\left(\frac{dB}{dt}\right)^2 \tag{7.44}$$

The average value of $\left(\frac{dB}{dt}\right)^2$ is the average value of $(\omega B_{max}\cos\omega t)^2$ which is $\omega^2 B_{max}^2/2$, so that the average power loss per unit volume of the core is:

$$\langle p\rangle = \frac{t^2}{12}\sigma\omega^2 B_{max}^2\frac{1}{2} = \frac{\pi^2 f^2\sigma t^2 B_{max}^2}{6} \tag{7.45}$$

Example 7.8

A power transformer, operating at 50 Hz would have laminations of thickness 0.3 mm, with $B_{max} = 1.5$ T and $\sigma = 2\times10^6\ (\Omega\text{m})^{-1}$, giving:

$$\langle p\rangle = \frac{\pi^2(50)^2(2\times10^6)(0.3\times10^{-3})^2(1.5)^2}{6} = 1.66\ \text{kW/m}^3$$

With a density of 7650 kg/m^3, this corresponds to 0.22 W/kg. The core loss for silicon steel in Table 1.1 is 5.66 kW/m^3, which includes the hysteresis loss.

This simplified analysis assumes that the presence of the eddy currents does not disturb the magnetic field, and a suitable choice of lamination would ensure this. The skin depth in the material used in the above example would be 0.5 mm for $\mu_r = 10\,000$. In general, the thickness of the lamination should be less than the skin depth given by Equation 1.21 at the operating frequency in order to ensure a uniform distribution of the flux density.

The result in Equation 7.45 shows that the eddy current loss under AC operating conditions is proportional to the square of frequency and the square of the maximum flux density. While this is somewhat simplified, it shows that eddy current loss is an important design consideration for iron cores.

For ferrites, the conductivity σ would be in the order of 2 $(\Omega\text{-m})^{-1}$ and, therefore, eddy current loss would not arise until frequencies into the MHz range are encountered.

7.5 Problems

7.1 Write a MATLAB program to compare the exact value of Z given by Equation 7.15, with the approximations given by Equations 7.22 and 7.23.

7.2 Repeat Example 7.1 for Ni-Zn ferrite with a relative permeability of 400 and electrical resistivity of 10 000 Ωm.

7.3 Repeat Example 7.2 for Ni-Zn ferrite with a relative permeability of 400 and electrical resistivity of 10 000 Ωm.

7.4 Evaluate the integral in Equation 7.28 for $\alpha = 1$ and $\alpha = 2$.

7.5 Calculate the core loss in Example 3.3 using iGSE.

7.6 Calculate the eddy current loss per unit volume for amorphous alloy tape at 25 kHz. The tape thickness is 40 μm, the maximum flux density is 0.2 T and the resistivity of the material is 1.3 μΩm.

MATLAB Program for Figure 7.3

```
% Figure 7.3 : inductance in a toroidal core as a function of frequency
n=0;
Lo=1;

for delta = [0:0.01:10];
I0 = besseli(0,sqrt(2)*sqrt(j)*delta);
I1 = besseli(1,sqrt(2)*sqrt(j)*delta);

Ls = Lo*real(2*I1/(sqrt(2)*sqrt(j)*delta*I0));

n = n+1;
V(n) = Ls/Lo;
D(n) = delta;
end

plot(D,V,'k','LineWidth',2)
titlc('Self')
```

```
xlabel('Delta')
ylabel('Ls/Lo')
axis([0 5 0 1])
grid off
```

MATLAB Program for Figure 7.8

```
% figure 7.8 : Initial permeability of P type ferrite

n = 0;
Lo = 1;
b = 0.005; %m
ro = 0.1; %omega m
mur = 2500;
mu0 = 4*pi*10^-7;

for f1 = [10000:100000:100000000];
delta = b*sqrt(pi*f1*mu0*mur/ro);

I0 = besseli(0,sqrt(2)*sqrt(1j)*delta);
I1 = besseli(1,sqrt(2)*sqrt(1j)*delta);
Ls = real(2*I1/(sqrt(2)*sqrt(1j)*delta*I0));

n = n+1;
V(n) = Ls;
D(n) = f1;
end

semilogx(D,V,'k','LineWidth',2)
title('Self')
xlabel('frequency')
ylabel('Ls/Lo')
axis([10000 100000000 0 1.5])
grid off
hold on

n = 0;
f2
=[100,200,400,700,1000,2000,4000,7000,10000,20000,40000,70000,
100000,200000,400000,700000,1000000,2000000,4000000,7000000,
10000000,20000000,40000000];
for delta =
[0.00263,0.00374,0.00530,0.00701,0.00838,0.01186,0.01704,
0.02291,0.02812,0.04340,0.07188,0.11353,0.15553,0.29685,0.57167,
```

```
0.98783,1.38301,2.58237,4.53450,6.65482,8.16618,12.22863,
19.11912];
I0 = besseli(0,sqrt(2)*sqrt(1j)*delta);
I1 = besseli(1,sqrt(2)*sqrt(1j)*delta);
L = real(2*I1/(sqrt(2)*sqrt(1j)*delta*I0));
n = n+1;
U(n) = L;
end
semilogx(f2,U,'b','LineWidth',2)

f3 = [10000,100000,150000,200000,250000,300000,350000,400000,
450000,500000,550000,600000,700000,800000,900000,1000000,1500000,
1750000];
mu = [1,1,1,1.04,1.10,1.18,1.28,1.35,1.38,1.40,1.38,1.36,1.30,1.20,
1.10,1,0.40,0];
semilogx(f3,mu,'r','LineWidth',2)
```

References

1. Hurley, W.G., Wilcox, D.J., and McNamara, P.S. (1991) Calculation of short circuit impedance and leakage impedance in transformer windings. *Proceedings of the IEEE Power Electronics Specialists Conference, PESC* pp. 651–658.
2. Steinmetz, C.P. (1984) On the law of hysteresis. *P. Proceedings of the IEEE*, **72** (2), 197–221.
3. Venkatachalam, K., Sullivan, C.R., Abdallah, T., and Tacca, H. (2002) Accurate prediction of ferrite core loss with nonsinusoidal waveforms using only Steinmetz parameters. *Proceedings of IEEE Workshop on Computers in Power Electronics, COMPEL*, pp. 36–41.
4. Muhlethaler, J., Biela, J., Kolar, J.W., and Ecklebe, A. (2012) Core losses under the DC bias condition based on Steinmetz parameters. *IEEE Transactions on Power Electronics* **27** (2), 953–963.
5. Snelling, E.C. (1988) *Soft Ferrites: Properties and Applications*, 2nd edn, Butterworths, London.

Further Reading

1. Del Vecchio, R.M., Poulin, B., Feghali, P.T. *et al.* (2001) *Transformer Design Principles: With Applications to Core-Form Power Transformers*, 1st edn, CRC Press, Boca Raton, FL.
2. Di, Y. and Sullivan, C.R. (2009) Effect of capacitance on eddy-current loss in multi-layer magnetic films for MHz magnetic components. *Proceedings of the IEEE Energy Conversion Congress and Exposition, ECCE*, pp. 1025–1031.
3. Dowell, P.L. (1966) Effects of eddy currents in transformer windings. *Proceedings of the Institution of Electrical Engineers* **113** (8), 1387–1394.
4. Dwight, H.B. (1961) *Tables of Integrals and Other Mathematical Data*, 4th edn, Macmillan, London.
5. Erickson, R.W. (2001) *Fundamentals of Power Electronics*, 2nd edn, Springer, Norwell, MA.
6. Goldberg, A.F., Kassakian, J.G., and Schlecht, M.F. (1989) Issues related to 1–10-MHz transformer design. *IEEE Transactions on Power Electronics* **4** (1), 113–123.
7. Han, Y., Cheung, G., Li, A. *et al.* (2012) Evaluation of magnetic materials for very high frequency power applications. *IEEE Transactions on Power Electronics* **27** (1), 425–435.
8. Hurley, W.G. and Wilcox, D.J. (1994) Calculation of leakage inductance in transformer windings. *IEEE Transactions on Power Electronics* **9** (1), 121–126.
9. McLachlan, N.W. (1955) *Bessel Functions for Engineers*, 2nd edn, Clarendon Press, Oxford.

10. McLyman, C.W.T. (1997) *Magnetic Core Selection for Transformers and Inductors*, 2nd edn, Marcel Dekker Inc., New York.
11. McLyman, C.W.T. (2002) *High Reliability Magnetic Devices*, 1st edn, Marcel Dekker Inc., New York.
12. McLyman, C.W.T. (2004) *Transformer and Inductor Design Handbook*, 3rd edn, Marcel Dekker Inc., New York.
13. E.S. MIT (1943) *Magnetic Circuits and Transformers (MIT Electrical Engineering and Computer Science)*, The MIT Press, Cambridge, MA.
14. Muhlethaler, J., Biela, J., Kolar, J.W., and Ecklebe, A. (2012) Improved core-loss calculation for magnetic components employed in power electronic systems. *IEEE Transactions on Power Electronics* **27** (2), 964–973.

Section Three
Advanced Topics

8

Measurements

Traditionally, measurements for transformers and inductors have been focused on power frequency operation. These approaches remain relevant today, but they must be modified or replaced to take account of the high frequencies encountered in power electronics. Measurement of inductance provides several challenges, since it is not always a single value, particularly when saturation is involved. Two methods for inductance measurement are treated – one involving DC current and another involving AC signals.

Traditionally, the measurements of losses in transformers have been carried out by the established open-circuit and short-circuit tests. However, these tests must be adapted to the field of power electronics in a manner that allows us to measure the core loss over a wide frequency range and to present the data in a format compatible with manufacturers' data sheets.

The measurement of the *B-H* loop is described for core materials. Knowledge of winding capacitance is important, particularly when there are resonant frequencies involved. Winding capacitance also plays a role in the dynamic response when a step change in voltage is shorted by the winding capacitance. A suitable measurement of capacitance in transformer windings is presented, along with some basic formulas to estimate the capacitance in typical winding configurations.

8.1 Measurement of Inductance

The measurement of inductance has traditionally been carried out using the Anderson bridge, named after Alexander Anderson, who was president and professor of physics at the National University of Ireland, Galway (formerly Queen's College Galway) in the late nineteenth century. The Anderson Bridge operated in much the same way as the Wheatstone bridge. However, the advent of modern network analyzers has consigned the Anderson bridge to the history museum.

Inductance can be highly variable as saturation takes effect. We may divide inductance measurement into two categories: the quiescent or DC value; and the incremental or AC value. The DC measurement of inductance is easily obtained from the response of an applied step function of voltage. The incremental value may be found from the application of an AC signal.

Transformers and Inductors for Power Electronics: Theory, Design and Applications, First Edition.
W. G. Hurley and W. H. Wölfle.

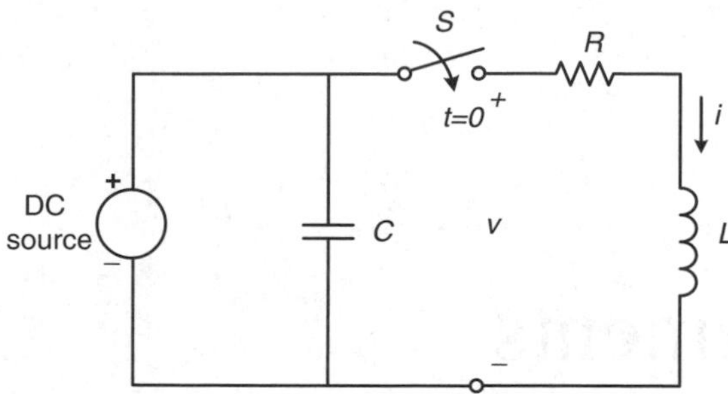

Figure 8.1 Inductance measurement by step voltage method.

8.1.1 *Step Voltage Method*

The inductor may be modelled by a series combination of resistance and inductance, as shown in Figure 8.1.

The simplest way to apply the step voltage is to charge a large capacitor – an ultracapacitor is ideal. The capacitor is initially charged from a power supply and the switch S is closed at $t = 0$. The capacitor provides a constant voltage, and the current rises in the inductor and may be measured by a small non-inductive precision sampling resistor.

The general solution for the setup in Figure 8.1 is:

$$V = Ri + \frac{d\lambda}{dt} \tag{8.1}$$

In discretized form, this becomes, for constant V:

$$V = Ri(k) + \frac{\lambda(k) - \lambda(k-1)}{\Delta t} \qquad k = 1, 2, 3, \ldots \tag{8.2}$$

Samples of current are taken at intervals of time Δt, so Equation 8.2 may be rewritten:

$$\lambda(k) = V\Delta t - Ri(k)\Delta t + \lambda(k-1) \tag{8.3}$$

$\lambda(k)$ is the flux linkage at time $t_k = k\,\Delta t$ that is after k time steps and the corresponding value of current is $i(k)$.

This is best illustrated by an example.

Example 8.1

Table 8.1 shows the data for the current in an inductor following a step change of voltage. $V = 70$ V, the time step is 0.24 ms and $R = 2.5\ \Omega$. Calculate the value of the inductance and estimate the value of current at the onset of saturation.

Inserting the values given above into Equation 8.3 yields:

$$\lambda(t_k) = (70)(0.24 \times 10^{-3}) - (2.5)(0.24 \times 10^{-3})i(t_k) + \lambda(t_{k-1})$$

Table 8.1 Inductance measurements

k	t_k(ms)	$i(k)$ (A)	$\lambda(k)$ (Wb-T)	$L(i_k)$ (mH)
1	0.24	0.32	$16.608*10^{-3}$	51.9
2	0.48	0.60	$33.048*10^{-3}$	55.5
3	0.72	0.92	$49.296*10^{-3}$	53.5
4	0.96	1.20	$65.376*10^{-3}$	54.5
5	1.20	1.44	$81.312*10^{-3}$	56.5
6	1.44	1.84	$92.008*10^{-3}$	52.3
7	1.68	2.40	$112.368*10^{-3}$	46.8
8	1.92	3.40	$127.128*10^{-3}$	32.4
9	2.16	4.44	$141.264*10^{-3}$	31.8

or

$$\lambda(t_k) = (16.8 \times 10^{-3}) - (0.60 \times 10^{-3})i(t_k) + \lambda(t_{k-1})$$

The inductance at each value of current is:

$$L(i_k) = \frac{\lambda(k)}{i(k)}$$

$L(i_k)$ is shown in Figure 8.2. Above 2 A, the inductance falls off to indicate that the onset of saturation was reached in the core at that point.

8.1.2 *Incremental Impedance Method*

This method measures the AC or incremental inductance as a function of its DC bias. For this purpose a DC current is fed through the inductor, while an AC voltage source is applied to

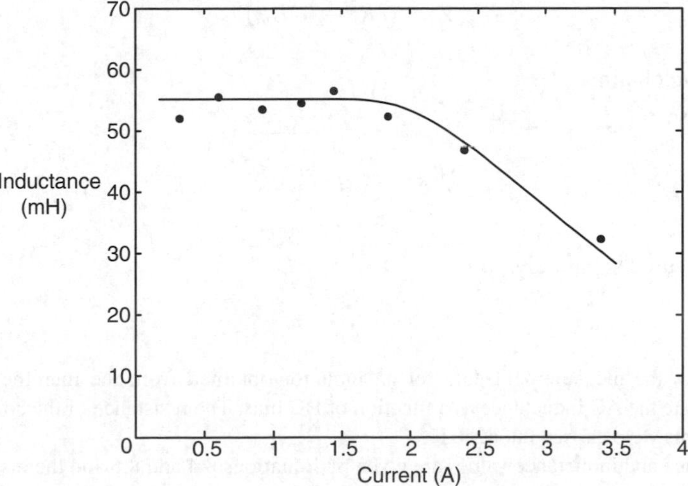

Figure 8.2 Inductance and incremental inductance measured by the step voltage method.

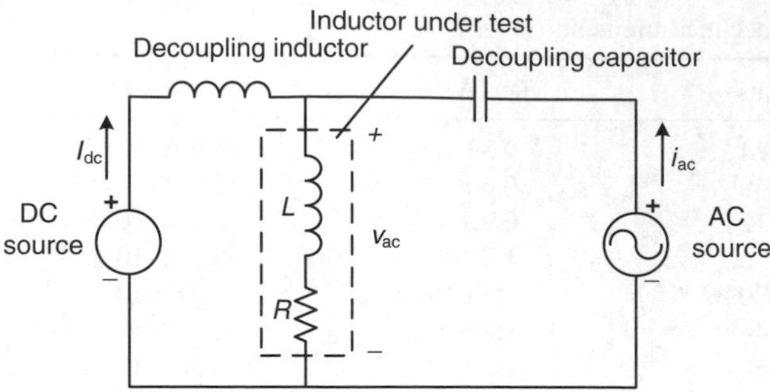

Figure 8.3 Inductance measurement by incremental impedance.

the inductor driving a small AC current through the inductor as shown in Figure 8.3. The incremental inductance may be calculated from the AC voltage and current.

In this circuit, the AC and the DC circuits must be totally decoupled from each other, otherwise the DC supply would short-circuit the AC voltage on the device under test with its large internal capacitance. The DC current through the inductor can be set independently from the AC values. The DC supply is decoupled from the AC circuit with a large decoupling inductor L_c (typically 1 H). The AC source is decoupled from the DC circuit by a large capacitor C_c (typically 1500 μF). The impedance can be calculated from the measured AC voltage v_{ac} and current i_{ac} on the inductor under test.

$$Z = \frac{v_{ac}}{i_{ac}} \tag{8.4}$$

The impedance of the inductor is given by the effective inductance and the resistance of the coil.

$$Z = \sqrt{R^2 + (\omega L_{ac})^2} \tag{8.5}$$

From which we obtain:

$$L_{ac} = \frac{1}{\omega}\sqrt{\frac{v_{ac}^2}{i_{ac}^2} - R^2} \tag{8.6}$$

R may be measured separately.

Example 8.2

Table 8.2 shows the measurement data for an inductor obtained from the incremental impedance method. Calculate the AC inductance as a function of DC bias. The resistance of the coil was 4.0 Ω and the measurements were carried out at 50 Hz.

The impedance and inductance values are given by Equations 8.4 and 8.6 and the results are summarized in Table 8.2. The inductance is plotted in Figure 8.4.

Table 8.2 Inductance measurements

I_{dc}(A)	v_{ac}(V)	i_{ac}(mA)	$Z(\Omega)$	L_{ac}(mH)
0.000	1.105	52.70	20.97	65.5
0.250	1.103	51.40	21.46	67.1
0.500	1.112	53.90	20.63	64.5
0.750	1.115	58.10	19.19	59.8
1.000	1.128	67.50	16.71	51.7
1.250	1.115	83.70	13.32	40.5
1.500	1.184	109.00	10.86	32.2
1.750	1.193	131.00	9.11	26.1
2.000	1.182	161.00	7.34	19.6
2.250	1.147	179.00	6.41	15.9
2.500	1.105	187.00	5.91	13.9
2.750	1.080	191.00	5.65	12.7
3.000	1.040	195.00	5.33	11.2
3.250	1.010	199.00	5.08	9.9
3.500	0.984	197.00	4.99	9.5
3.750	0.966	201.00	4.81	8.5
4.000	0.944	210.00	4.50	6.5

8.2 Measurement of the *B-H* Loop

The *B-H* loop is of interest because we need knowledge of the magnetic parameters such as B_{sat}, the saturation flux density, the coercive force H_c and the residual flux B_r. These terms are explained in Chapter 2.

The simplest set up is a coil of N turns on a toroidal core, as illustrated in Figure 8.5.

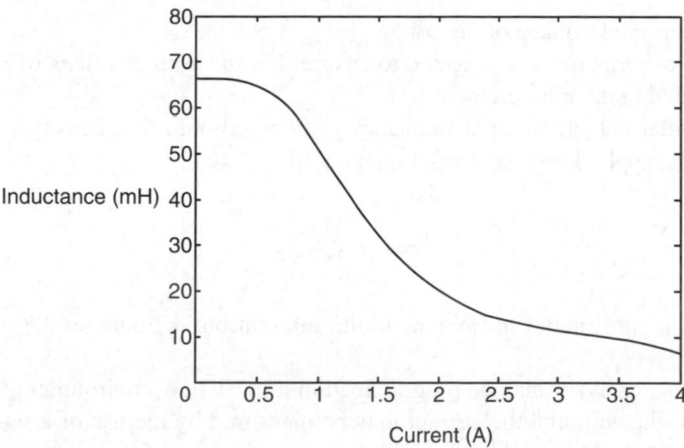

Figure 8.4 AC inductance measured by the incremental impedance method.

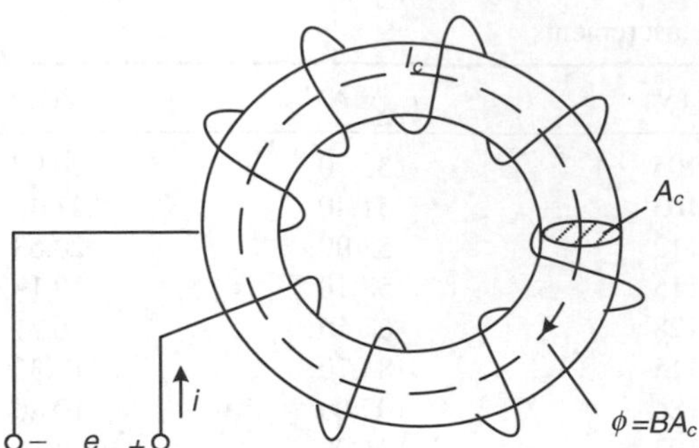

Figure 8.5 Toroidal core for *B-H* measurement.

Recall from Chapter 2, Equation 2.1:

$$\oint_C = \mathbf{H} \cdot d\mathbf{l} = Ni$$

The magnetic field intensity may be obtained directly from the current measurement.

$$H_c = \frac{N}{l_c} i \tag{8.7}$$

where l_c is the mean length of the magnetic path in the test specimen.

The flux density in the coil is found from Faraday's law, assuming negligible winding resistance:

$$B = \frac{1}{NA_c} \int v dt \tag{8.8}$$

where v is the terminal voltage of the coil.

The number of turns must be selected to ensure that the correct values of B_{sat} and H_c are correctly included in the measurement.

For a sinusoidal voltage input at frequency *f*, the maximum flux density is related to the peak value of the applied voltage, from Equation 8.8:

$$B_{\max} = \frac{V_{\text{peak}}}{2\pi f N A_c} \tag{8.9}$$

It is a straightforward matter to implement the integration in Equation 8.8 with the set-up shown in Figure 8.6.

The gain-phase analyzer may be programmed in a LabVIEW environment to integrate the input voltage to the coil, and the current may be measured by the use of a sampling resistor R_{ref}. The signal generator sets the frequency for the power amplifier.

The normal magnetization curve may be obtained by taking the *B*/*H* ratio at the tips of the hysteresis loops for different values of $B_{\max}$.

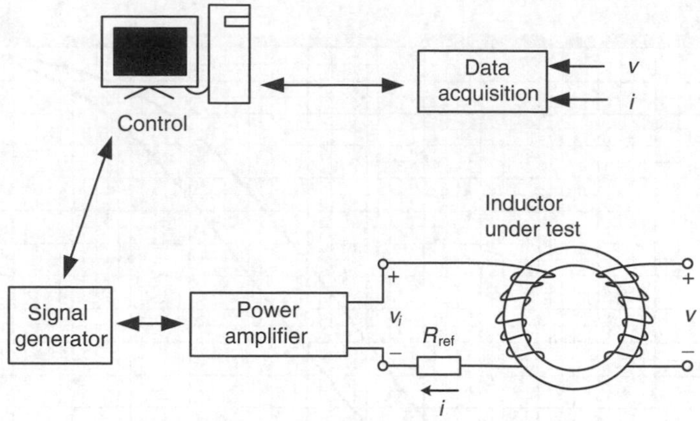

Figure 8.6 Test set up for measurement of the *B-H* loop.

Example 8.3

Select the number of turns for a closed core made of Mn-Zn ferrite, with mean magnetic path length of 75.5 mm and a cross-sectional area of 81.4 mm^2, that yields a value of maximum flux density B_{max} of 400 mT and coercive force H_c of 155 A/m at a current of 1.15 A. Calculate the rms value of the applied voltage for 50 Hz excitation.

Using Equation 8.7, the number of turns is:

$$N = \frac{(155)(75.5 \times 10^{-3})}{1.15} = 10 \text{ turns}$$

The rms value of the applied voltage at 400 mT and 50 Hz excitation is:

$$V_{rms} = 4.44fNB_{max}A_c = (4.44)(50)(10)(0.4)(81.4 \times 10^{-6}) = 72.3 \text{ mV}$$

$$V_{max} = \sqrt{2}(72.3) = 102.2 \text{ mV}$$

So the amplitude of v is in the order of 100 mV for $B_{sat} = 0.4$ T at 50 Hz excitation.

The *B-H* loop was generated for the PC40 material by winding ten turns on a core with the dimensions given above. The *B-H* loop is shown in Figure 8.7.

We may also obtain the hysteresis loss per unit volume by integrating the *B-H* characteristic to find the area enclosed by the *B-H* loop.

8.3 Measurement of Losses in a Transformer

The main parameters in a transformer may be measured by two simple tests, namely the short-circuit test that forces rated current through the windings at a low voltage, and the open-circuit test that is carried out at rated voltage to include the magnetizing current.

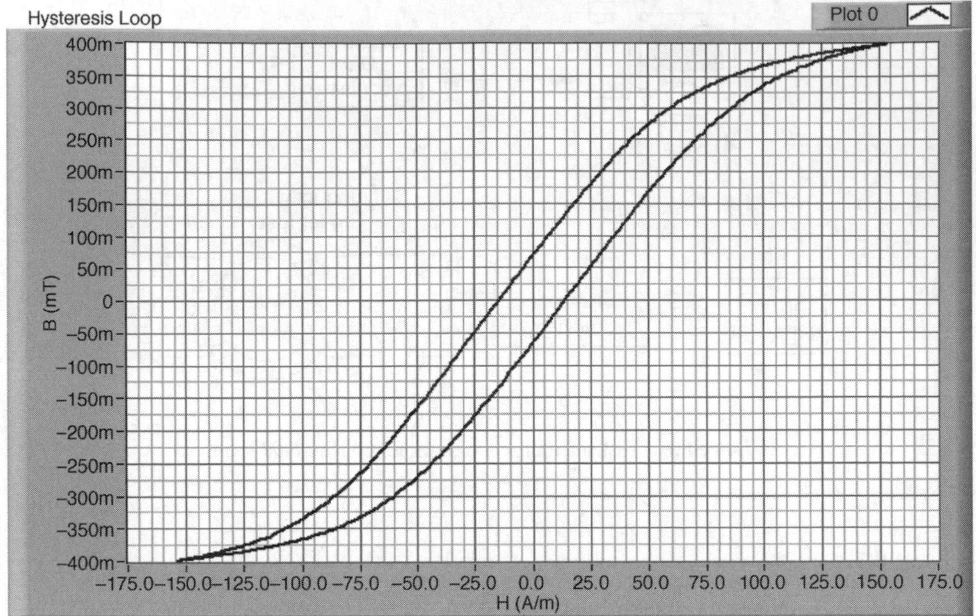

Figure 8.7 Hysteresis loop for PC40 material at 25 °C.

The tests may be carried out on either the low voltage or high voltage side of the transformer. These tests are traditionally associated with 50 Hz or 60 Hz power transformers, with the test power readily supplied at line frequency. However, in power electronics applications, the transformers operate at hundreds of kHz. In this case, we can measure the core loss at high frequency by using a power amplifier and a gain-phase analyzer. It is also possible to apply a DC step voltage to the transformer, with one winding short-circuited to infer the leakage reactance from the step response.

8.3.1 Short-Circuit Test (Winding/Copper Loss)

With one winding short-circuited, typically 10% of rated voltage on the other winding is sufficient to establish rated full load current. For convenience, we will short the secondary winding, take measurements in the primary winding and refer the secondary quantities as appropriate.

Measure:

- V_{sc} short-circuit primary voltage
- I_{sc} short-circuit primary (rated) current
- P_{sc} short-circuit power (measured with a wattmeter).

The core loss is negligible, since the input voltage is very low. For that reason, the core circuit parameters are shown dotted in Figure 8.8.

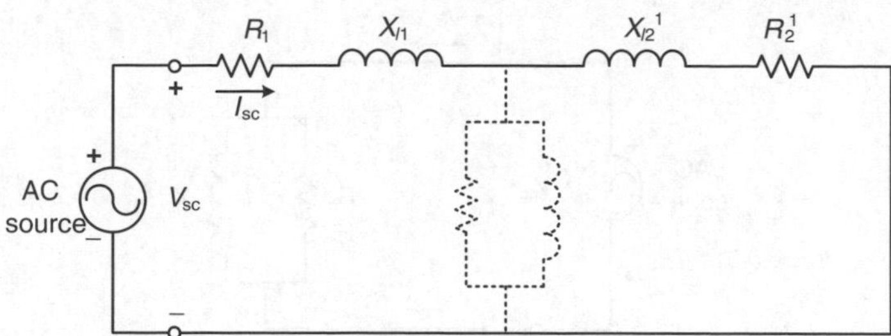

Figure 8.8 Transformer short-circuit test.

Invoking the transformer analysis of Chapter 4 for the equivalent circuit of Figure 8.8, the equivalent impedance Z_{eq} looking into the terminals of the transformer is given by the short-circuit impedance Z_{sc}:

$$Z_{eq} = Z_{sc} = \frac{V_{sc}}{I_{sc}} \tag{8.10}$$

By examination of Figure 8.8, the real part of Z_{eq} is the equivalent resistance R_{eq} of the windings. This is made up of the resistance of the primary winding and the resistance of the secondary winding reflected into the primary. The primary to secondary turns ratio is a:

$$R_{eq} = R_{sc} = \frac{P_{sc}}{I_{sc}^2} \tag{8.11}$$

$$R_{eq} = R_1 + a^2 R_2 \tag{8.12}$$

The imaginary part of Z_{eq} is the equivalent leakage reactance of the windings. This consists of the leakage reactance of the primary winding and the leakage reactance of the secondary winding reflected into the primary winding:

$$X_{eq} = X_{sc} = \sqrt{(Z_{sc}^2 - R_{sc}^2)} \tag{8.13}$$

$$X_{eq} = X_{l1} + a^2 X_{l2} \tag{8.14}$$

As a first approximation, it is reasonable to assume that $R_1 = a^2 R_2$, $X_{l1} = a^2 X_{l2}$ in a well-designed transformer. A more realistic approach is to take the ratios of $\frac{R_1}{R_2}$ and $\frac{X_{l1}}{X_{l2}}$ as the ratio of the DC resistance of the individual windings, which may be easily measured.

8.3.2 Open-Circuit Test (Core/Iron Loss)

With rated voltage on the primary winding and with the secondary winding open-circuited, the magnetizing current flows in the primary winding. The voltage drops in R_{eq} and X_{eq} are very small due to small magnetizing current and the power input is very nearly equal to the core loss.

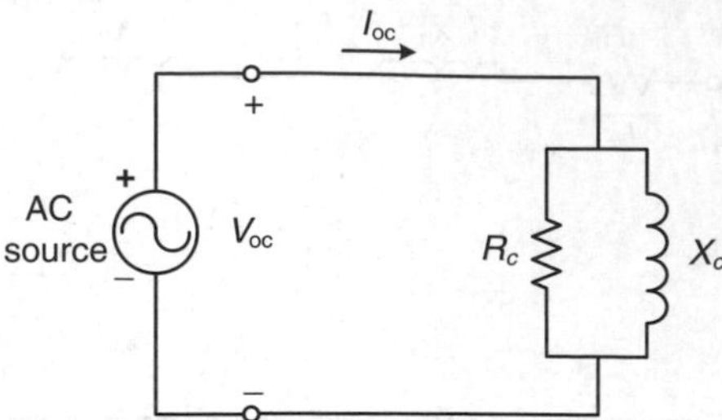

Figure 8.9 Transformer open-circuit test.

Measure:

- V_{oc} open-circuit primary voltage
- I_{oc} open-circuit primary current
- P_{oc} open-circuit power (use a wattmeter)

The equivalent circuit for these measurements is shown in Figure 8.9. Again, by invoking the analysis of Chapter 4, we may deduce the core parameters.

The core reactance is given by:

$$\frac{1}{Z_\phi} = \frac{1}{R_c} + \frac{1}{jX_c} \tag{8.15}$$

where R_c represents the core loss and X_c represents the magnetizing reactance of the core.

$$\frac{1}{Z_\phi} = \frac{I_{oc}}{V_{oc}} \tag{8.16}$$

The core resistance is:

$$R_c = \frac{V_{oc}^2}{P_{oc}} \tag{8.17}$$

and the core reactance is:

$$X_c = \frac{1}{\sqrt{\dfrac{1}{Z_\phi^2} - \dfrac{1}{R_c^2}}} \tag{8.18}$$

These two tests provide sufficient data to characterize the transformer in terms of its equivalent circuit.

Example 8.4

In a 1000 VA, 220: 110 V transformer, the following readings were obtained on the low voltage side of the transformer:

$$\begin{aligned}\text{Short-circuit test}: V_{\text{sc}} &= 8.18\text{ V}\\ I_{sc} &= 9.1\text{ A}\\ P_{sc} &= 27.4\text{ W}\\ \text{Open-circuit test}: V_{\text{oc}} &= 110\text{ V}\\ I_{\infty} &= 0.53\text{ A}\\ P_1 &= 18.5\text{ W}\end{aligned}$$

Determine the equivalent resistance and reactance of the transformer.
The equivalent impedance is:

$$Z_{\text{eq}} = \frac{8.18}{9.1} = 0.9\,\Omega$$

The equivalent resistance is:

$$R_{\text{eq}} = P_{\text{sc}}/I_{\text{sc}}{}^2 = 27.4/(9.1)^2 = 0.33\,\Omega$$

and the equivalent leakage reactance is:

$$X_{\text{eq}} = \sqrt{(Z_{\text{sc}}^2 - R_{\text{sc}}^2)} = \sqrt{(0.9^2 - 0.33^2)} = 0.84\,\Omega$$

The measurements were taken on the low voltage side, so we must convert these to the high voltage side by multiplying by the square of the turns ratio:
Turns ratio = 220/110 = 2
On the high voltage side:

$$\begin{aligned}R_{eq}^H &= (2)^2(0.33) = 1.32\,\Omega\\ X_{eq}^H &= (2)^2(0.84) = 3.36\,\Omega\end{aligned}$$

Taking the results of the open-circuit test, the core impedance is:

$$Z_\phi = \frac{V_{\text{oc}}}{I_{\text{oc}}} = \frac{110}{0.53} = 207.5\,\Omega$$

and the core equivalent resistance is:

$$R_c = \frac{V_{\text{oc}}^2}{P_{\text{oc}}} = \frac{110^2}{18.5} = 654\,\Omega$$

Determine the core parameters:

$$X_c = \frac{1}{\sqrt{\left(\frac{1}{Z_\phi^2} - \frac{1}{R_c^2}\right)}} = \frac{1}{\sqrt{\left(\frac{1}{207.5^2} - \frac{1}{654^2}\right)}} = 218.8\,\Omega$$

On the high voltage side:

$$R_c^H = (2)^2(654) = 2616\,\Omega$$
$$X_c^H = (2)^2(218.8) = 875.2\,\Omega$$

The efficiency of the transformer may be obtained from the measurements taken in the open-circuit test.

Example 8.5

Find the efficiency of the transformer in Example 8.4 when it is connected to a load with a power factor of 0.8.

The rated current in the primary (high voltage) winding is:

$$I_1 = \frac{\text{VA}}{V_1} = \frac{1000}{220} = 4.55\text{ A}$$

The total copper loss is:

$$I_1^2 R_{eq}^H = (4.55)^2(1.32) = 27.3\text{ W}$$

The core loss is the power loss measured in the open-circuit test = 18.5 W.

The total losses are 27.3 W + 18.5 W = 45.80 W.

The load is connected to the low voltage side and the output current is 4.55 A × 2 = 9.1 A at 110 V with a power factor of 0.8. The output power is:

$$VI\cos\varphi = (110)(9.1)(0.8) = 801\text{ W}$$

The input supplies this power as well as the losses. The input power is 801 W + 45.8 W = 846.8 W.

The efficiency is:

$$\text{Efficiency} = \frac{\text{Power Out}}{\text{Power In}} = \left(\frac{801}{846.8}\right)(100) = 94.6\%$$

8.3.3 *Core Loss at High Frequencies*

Normally, a wattmeter is used for the transformer tests at line frequency but, at the frequencies encountered in power electronics, a gain-phase meter must replace the wattmeter. The gain-phase analyzer can be readily programmed to measure voltage, current and phase angle and the set-up is shown in Figure 8.10. The core is made from the material under test (the shape of the core is not important). The core should not contain air gaps; recall from Chapter 2 that the air gap determines the level of magnetic field intensity in the core.

It is generally desirable to measure the core loss over a wide frequency range while maintaining a constant flux level in the core. Recall from Chapter 4, for sine wave excitation,

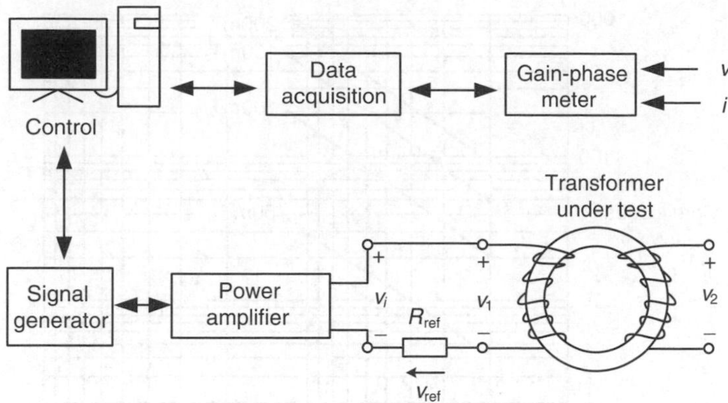

Figure 8.10 Test set-up for measurement of core loss at high frequency.

the transformer equation is:

$$E_{rms} = 4.44 f N B_{max} A_c \tag{8.19}$$

Constant flux can be maintained over the desired range of frequencies by keeping the ratio E_{rms}/f constant. In the case of core loss, the open-circuit test informs our approach.

The signal generator in Figure 8.10 sets the frequency for the power amplifier, which is programmed to give the correct output open-circuit voltage for each frequency, based on Equation 8.19. For the purpose of this calculation, E_{rms} in Equation 8.19 is interpreted as the correct open-circuit voltage for each frequency. The input voltage adjustment for frequency is easily implemented in the LabVIEW environment. The voltage is measured in the open-circuit winding to avoid errors associated with voltage drops due to current flow in the test winding. The gain-phase analyzer measures the input current by sampling the voltage v_{ref} across the non-inductive resistor R_{ref}, the output voltage v_2 and the corresponding phase θ between v_{ref} and v_2.

The core loss is now:

$$P_{fe} = a v_2 \frac{v_{ref}}{R_{ref}} \cos\theta \tag{8.20}$$

The primary to secondary turns ratio a is included because the voltage is measured in the secondary winding and the current is measured in the primary winding.

The LabVIEW controller may be programmed to present the data in the format shown in Figure 8.11.

The core loss is usually expressed in watts per unit volume (kW/m^3), as shown in Figure 8.11 for PC40 Mn-Zn ferrite core material at three different flux levels. Since the core heats up as a result of the test, it is important to place the transformer in a controlled oven to stabilize its temperature, and also to ensure that the subsequent test is carried out very quickly. This is easily achieved in a programmable test set-up.

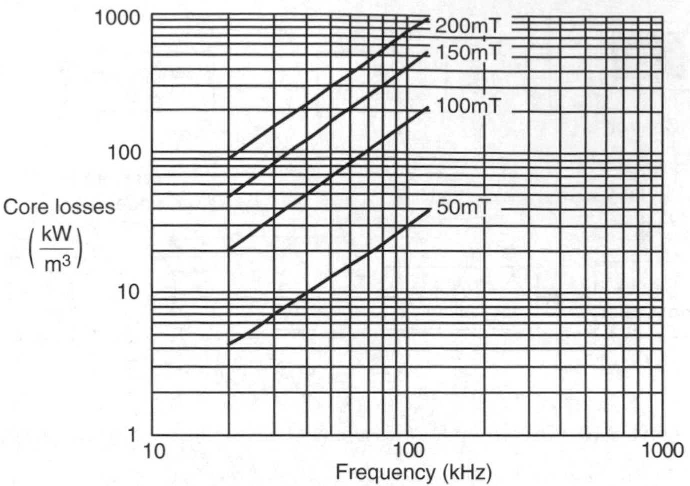

Figure 8.11 Core loss as a function of frequency at 25 °C.

The core loss in Figure 8.11 may be expressed in the empirical Steinmetz equation, as described in Chapter 4 using Equation 1.29:

$$P_{\text{fe}} = K_c f^{\alpha} B_{\text{max}}^{\beta}$$

Example 8.6

Determine the parameters α, β and K_c for the material in whose core loss data is shown in Figure 8.11.

Essentially we have three unknowns, so we need three data points. Pick two points at 20 kHz, corresponding to maximum flux densities of 50 mT and at 200 mT respectively, and pick the third point at 100 kHz and 200 mT:

- At point A: $f = 20$ kHz, $B_{\text{max}} = 50$ mT and $P_{\text{fe}} = 4.5$ kW/m^3.
- At point B: $f = 20$ kHz, $B_{\text{max}} = 200$ mT and $P_{\text{fe}} = 90$ kW/m^3.
- At point C: $f = 100$ kHz, $B_{\text{max}} = 200$ mT and $P_{\text{fe}} = 700$ kW/m^3.

Taking logarithmic values of ratios given by Equation , the following identities apply:

$$\alpha = \frac{\ln\left(\frac{700}{90}\right)}{\ln\left(\frac{100}{20}\right)} = 1.275$$

$$\beta = \frac{\ln\left(\frac{90}{4.5}\right)}{\ln\left(\frac{200}{50}\right)} = 2.161$$

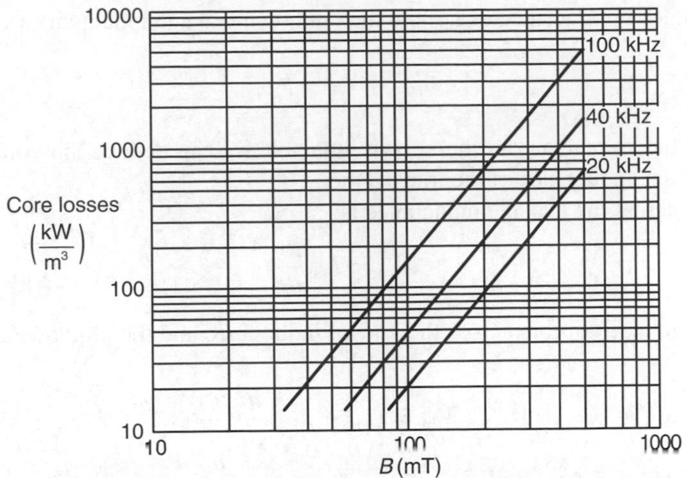

Figure 8.12 Core loss as a function of induction level at 25 °C.

and:

$$K_c = \frac{700 \times 10^3}{(100 \times 10^3)^{1.275}(0.2)^{2.161}} = 9.563$$

Core loss may also be expressed as a function of induction level for different frequencies. We can generate the core loss versus induction level as shown in Figure 8.12, using the parameters we have deduced above.

8.3.4 Leakage Impedance at High Frequencies

The leakage reactance in the short-circuit test can be found by applying a step voltage to the short-circuited transformer through a known inductance L_m (normally a toroid), as illustrated in Figure 8.13.

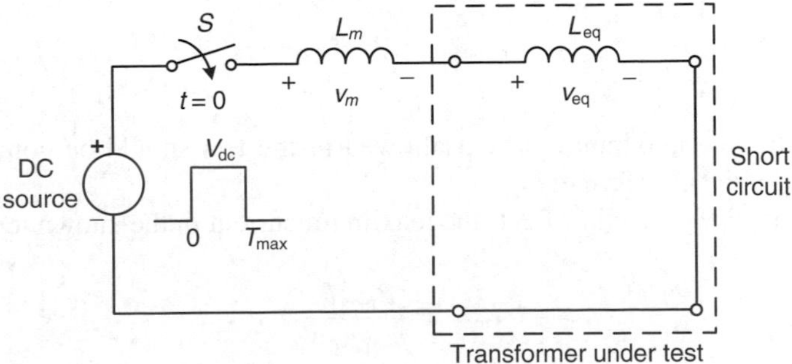

Figure 8.13 Transformer short-circuit test at high frequency.

Taking the known inductance as L_m and the unknown leakage inductance as L_{eq}, then:

$$V_{dc} = V_m + V_{eq} \tag{8.21}$$

By measuring the voltage across the known inductor, we can deduce the voltage across the equivalent leakage inductance of the transformer.

The voltage across the leakage inductance is:

$$V_{eq} = L_{eq}\frac{di}{dt} \tag{8.22}$$

The input DC voltage appears across the known inductance and the unknown leakage inductance:

$$V_{in} = (L_m + L_{eq})\frac{di}{dt} \tag{8.23}$$

We have neglected the effect of the winding resistance because, at high frequencies, the leakage inductance will dominate the voltage, in which case the ratio of the inductances is directly related to the ratio of the voltages, since *di/dt* is common. In that case, the leakage inductance is simply:

$$\frac{L_{eq}}{L_m} = \frac{V_{eq}}{V_m} = \frac{V_{dc} - V_m}{V_m} = \frac{V_{dc}}{V_m} - 1 \tag{8.24}$$

We must ensure that the core of the known toroidal inductor is not saturated. From the analysis of Section 2.4.2, in a toroid of N turns and cross-sectional area A_c:

$$V_m = NA_c\frac{dB}{dt} \tag{8.25}$$

The maximum flux density is reached at T_{max}. Therefore, the voltage V_{in} should be applied for a maximum pulse width of T_{max} that satisfies:

$$B_{max} = \frac{V_m}{NA_c}T_{max} \tag{8.26}$$

and:

$$T_{max} = \frac{NA_c}{V_m}B_{max} \tag{8.27}$$

As a general rule, the maximum current allowed in the test should be comparable to the nominal current of the transformer.

From Ampere's law of Section 1.2.1, the maximum current in the known toroid is:

$$I_{max} = \frac{B_{max}l_c}{\mu_{eff}\mu_0 N} \tag{8.28}$$

where μ_{eff} is the effective permeability of the core.

Example 8.7

The following readings were obtained from a leakage test on a 50 kHz transformer.

The input voltage of 5 V was applied to a known inductance of 3 mH, connected in series with a transformer with a shorted secondary. The known inductor was built with 500 turns on a powder iron core ($B_{max} = 0.3$ T, $\mu_{eff} = 75$) with a cross-sectional area of 1 cm^2 and a magnetic path length of 9.8 cm. The voltage measured across the known inductor was 2.3 V. Calculate the width of the applied voltage pulse and the leakage inductance of the transformer.

The maximum pulse width is:

$$T_{max} = \frac{(500)(1.0 \times 10^{-4})}{2.3}(0.3) \times 10^3 = 6.5 \text{ ms}$$

The leakage inductance is:

$$L_{eq} = 3\left(\frac{5}{2.3} - 1\right) = 3.52 \text{ mH}$$

As a general rule, the maximum current allowed in the test should be comparable to the nominal current of the transformer.

From Ampere's law in Section 1.2.1, the maximum current is:

$$I_{max} = \frac{B_{max} l_c}{\mu_{eff} \mu_0 N} = \frac{(0.3)(9.8 \times 10^{-2})}{(75)(4\pi \times 10^{-7})(500)} = 0.62 \text{ A}$$

8.4 Capacitance in Transformer Windings

The capacitance of a transformer or inductor winding is of interest in power electronic switching circuits, because the capacitance may provide a short circuit to a step change in voltage, causing over-current circuitry to trip. Flyback circuits are particularly prone to this problem. The second issue relates to resonances in the windings, so the designer needs to know the effective capacitance of the windings.

Figure 8.14 shows the classical lumped-parameter representation of a transformer winding. The mutual impedance between winding sections i and j is represented by Z_{ij}.

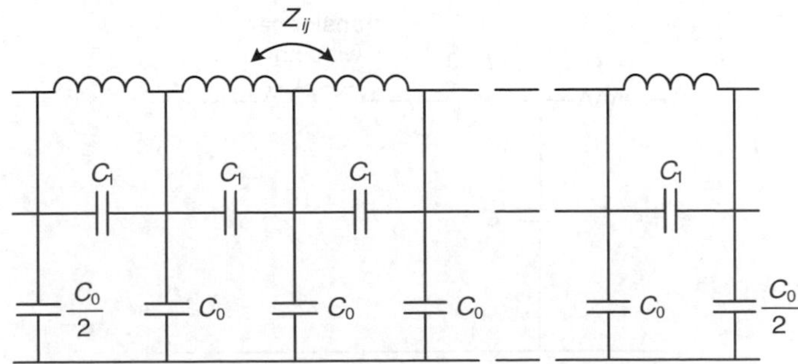

Figure 8.14 Lumped parameter transformer model.

C_0 represents the distributed shunt capacitance between each section and the core, and C_1 represents the series capacitance between sections of the winding. The winding is continuous, but it is normal to discretize the winding for analysis. 10–20 sections are normally sufficient to represent the winding accurately, i.e. as regards natural resonant frequencies. The model is similar to that of a transmission line, except that mutual coupling exists between sections and there is also series capacitance. The presence of mutual coupling makes transformer analysis much more complex than transmission line analysis.

8.4.1 Transformer Effective Capacitance

For n sections of winding, the total shunt capacitance is:

$$C_g = nC_0 \tag{8.29}$$

and the total series capacitance is:

$$C_s = \frac{C_1}{n} \tag{8.30}$$

The network of capacitances C_0 and C_1 in Figure 8.14 has an equivalent capacitance C_{eff} [1]. This is the equivalent capacitance placed across the transformer terminals that would draw the same charging current as the capacitance distribution in the transformer winding shown in Figure 8.14. The effective capacitance is given by:

$$C_{\text{eff}} = \sqrt{C_g C_s} \tag{8.31}$$

C_{eff} cannot be measured directly due to the presence of the coil inductance. However, it can be measured by suddenly discharging a known capacitance C_k into the winding and recording the instantaneous voltage drop, as shown in Figure 8.15.

By conservation of charge:

$$C_{\text{eff}} = C_k \frac{\Delta E}{E_o - \Delta E} \tag{8.32}$$

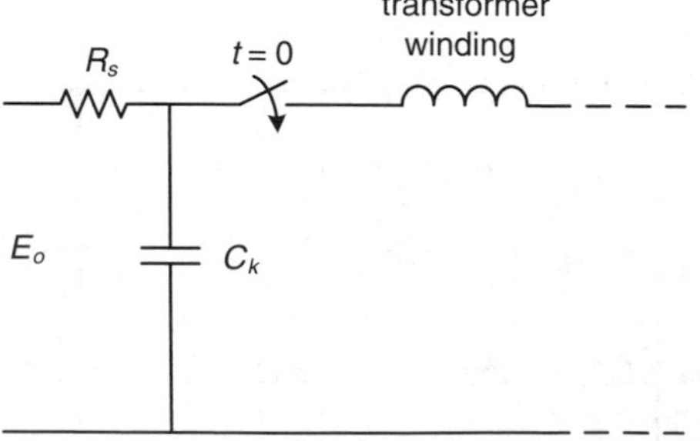

Figure 8.15 Circuit to measure effective capacitance.

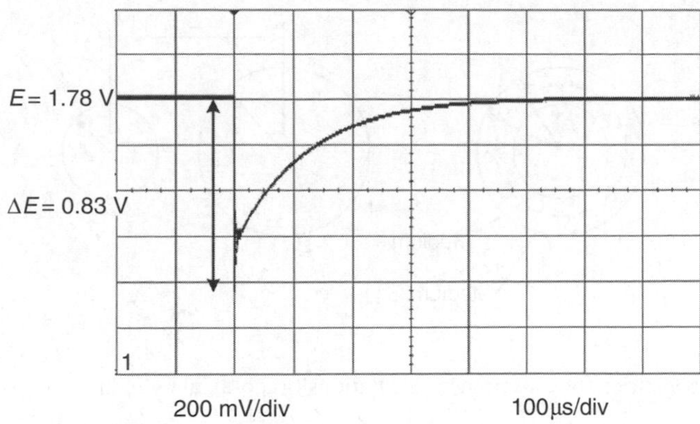

Figure 8.16 Measurement of effective capacitance.

where E_o is the initial voltage on C_k and ΔE is the drop in voltage at the instant of discharge [1]. R_s is chosen so that the time constant for charging C_k is much longer than the transient time in the winding after the switch is closed. Due to the inductance of the winding, the circuit will oscillate after the initial discharge and, eventually, the capacitances will recharge to E_o through R_s. C_k is normally chosen to be of the same order of magnitude as C_{eff}.

Example 8.8

Calculate the effective capacitance of the transformer for which the test results of Figure 8.16 were obtained. The value of the known capacitance is 500 pF.

$$C_{\text{eff}} = C_k \frac{\Delta E}{E_o - \Delta E} = (500)\left(\frac{0.83}{1.78 - 0.83}\right) = 437 \text{ pF}$$

8.4.2 Admittance in the Transformer Model

The admittance terms in the transformer model of Figure 8.14 are made up of the shunt capacitance C_0 and the series capacitance C_1. The individual capacitances that make up the shunt capacitance C_0 and the series capacitance C_1 may be comprised of many different combinations of sections of windings. However, the capacitance values in most configurations can be calculated using three formulae for the geometries shown in Figure 8.17.

Co-Axial Cylinders

This configuration consists of a cylinder of radius r_1 inside a cylinder of radius r_2, and the radius of the boundary between the two dielectric mediums is R. The capacitance between the cylinders is:

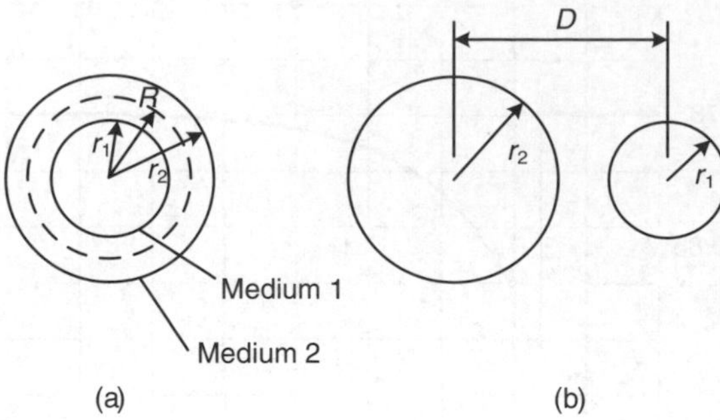

Figure 8.17 Geometries for capacitance calculations (a) co-axial cylinders, (b) external cylinders.

$$C = \frac{2\pi\varepsilon_0 l_c}{\dfrac{\ln\left(\dfrac{R}{r_1}\right)}{\varepsilon_{r1}} + \dfrac{\ln\left(\dfrac{r_2}{R}\right)}{\varepsilon_{r2}}} \tag{8.33}$$

l_c is the cylinder length and ε_{r1} and ε_{r2} are the relative permittivities of the dielectrics in mediums 1 and 2 respectively. Typically, one medium is air and the other medium is insulation.

External Cylinders

This configuration consists of two cylinders of radius r_1 and r_2 respectively, at a distance D apart. The capacitance between the cylinders is:

$$C = \frac{2\pi\varepsilon_r\varepsilon_0 l_c}{\cosh^{-1}\left\{\dfrac{D^2 - r_1^2 - r_2^2}{2r_1 r_2}\right\}} \tag{8.34}$$

Parallel Plates

In many cases, the radius of curvature of a winding is much greater than the separation of the windings under consideration. In this case, it is perfectly reasonable to assume parallel plates. This configuration consists of two parallel plates of area A with a composite dielectric of thicknesses d_1 and d_2. The corresponding dielectric constants are ε_{r1} and ε_{r2} respectively.

$$C = \frac{\varepsilon_0 A}{\dfrac{d_1}{\varepsilon_{r1}} + \dfrac{d_2}{\varepsilon_{r2}}} \tag{8.35}$$

The approach is best illustrated by the following examples.

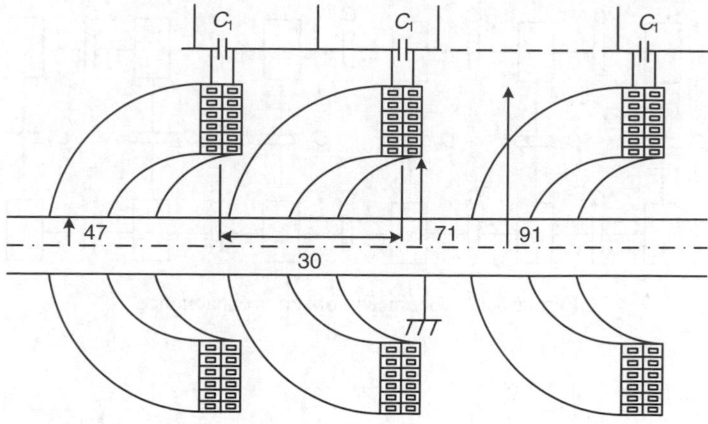

Figure 8.18 Disc winding dimensions.

Example 8.9

A high-voltage disc winding is illustrated in Figure 8.18. The winding consists of 16 discs in series and each disc winding has 12 turns tightly wound in two layers as shown. The disc dimensions are:

- HV winding inside radius = 71.0 mm
- HV winding outside radius = 91.0 mm
- HV winding conductor = 5 × 3.15 mm
- Width of disc winding = 12.06 mm
- Insulation thickness = 0.08 mm
- Dielectric constant insulation = 3.81.

Calculate the series capacitance of the winding.

The individual capacitances that make up the series capacitance C_1 of the disc are shown in Figure 8.19. C_d is the capacitance between each of the adjacent conductors in a disc winding pair and C_t is the capacitance between conductors within a disc. The conductors in Figure 8.19 have rectangular cross-sections; it is equally likely that they may be circular conductors.

In general, both C_d and C_t will depend on the position of the conductor within the disc but, for convenience, it is reasonable to assume average dimensions without a significant loss of accuracy. The shunt capacitance C_1 is then the equivalent capacitance of the network above.

Both capacitances in Figure 8.19 may be found from the parallel plate formula as follows:

The average radius of the disc is 81 mm and the width of the conductors that make up C_t is 5 mm. Therefore, the area of the capacitor plate is:

$$A = 2\pi Rd = (2\pi)(81)(5) = 2545\,\text{mm}^2$$

There are two layers of insulation so the plate separation is 0.16 mm and the capacitance is obtained from Equation 8.35 with one medium:

$$C_t = \frac{(3.81)(8.854 \times 10^{-12})(2545 \times 10^{-6})}{0.16 \times 10^{-3}} \times 10^{12} = 536\,\text{pF}$$

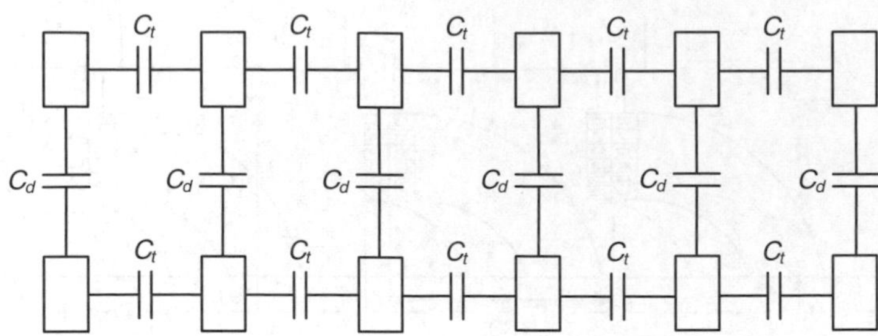

Figure 8.19 Geometry for series capacitance.

The capacitance C_d between the parallel conductors that make up the disc may be modelled as parallel plates. The total width of the disc is 12.06 mm and the conductors and insulation make up 10.16 mm, leaving 0.16 mm of insulation (dielectric constant 3.81) and 1.74 mm of air. The area of the plate is:

$$A = 2\pi Rt = (2\pi)(81)(3.15) = 1603\ \text{mm}^2$$

The capacitance is then given by Equation 8.35:

$$C_d = \frac{(8.854 \times 10^{-12})(1603 \times 10^{-6})}{\dfrac{0.16 \times 10^{-3}}{3.81} + 1.74 \times 10^{-3}} \times 10^{12} = 8.0\ \text{pF}$$

The series capacitance C_1 across the winding in Figure 8.19 is now:

$$C_1 = C_d + \frac{5C_dC_t}{2C_d + C_t} = 8.0 + \frac{(5)(8.0)(536)}{(2)(8.0) + 536} = 46.84\ \text{pF}$$

Example 8.10

Figure 8.20 shows a transformer layout consisting of a primary winding with 39 turns in two layers, and a seven-turn secondary with a reset winding of three turns. The primary winding has two parallel 0.5 mm wires. The screens are made of 0.035 mm copper foil. The main dimensions are shown. Estimate the effective capacitance of the primary winding when measured between the input to the primary winding and the screen with the other end of the winding in open circuit.

The individual capacitances that make up the shunt capacitance C_0 of the primary winding is may be defined as follows:

- C'_g is the capacitance between the primary winding and the screen $W1$;
- C''_g is the capacitance between the primary winding and screen $W3$;
- C_1 is the series capacitance of the primary winding;

The shunt capacitance C_0 of Figure 8.14 is the parallel combination of C'_g and C''_g. C_0 in this case is the same as C_0 in Equation 8.29. The series capacitance C_s in this case is the same as C_1.

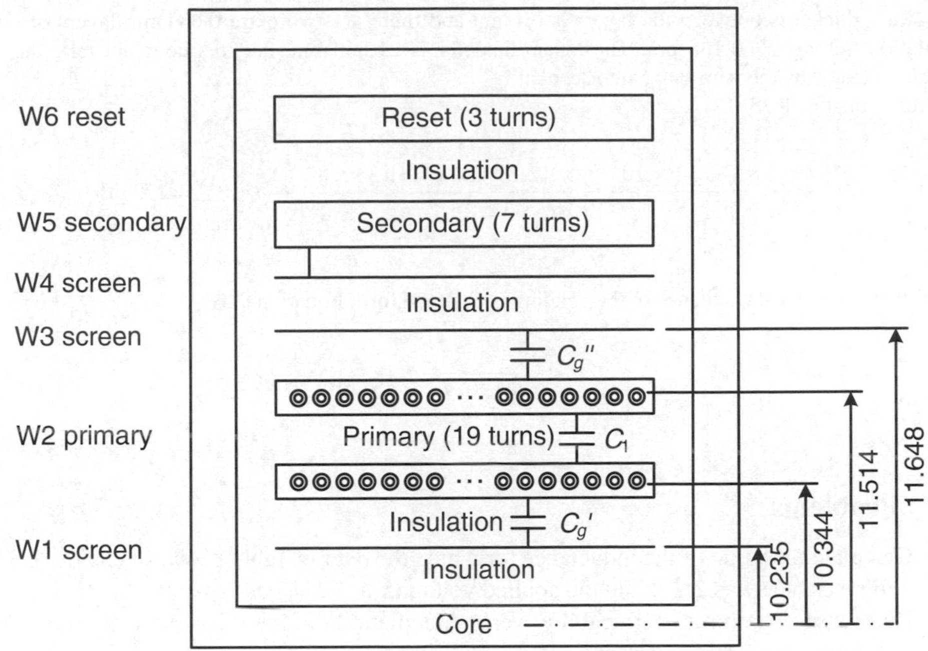

Figure 8.20 Transformer dimensions.

C'_g is made of two concentric cylinders with $r_1 = 10.235$ mm, the insulation thickness is 0.075 mm with a relative permittivity of 3.8, and there is an extra 0.034 mm layer of wire insulation, giving $r_2 = 10.344$ mm. The length of the winding is approximately 20 mm, based on 19 turns of two 0.5 mm wires in parallel.

$$C'_g = \frac{(2\pi)(3.8)(8.854 \times 10^{-12})(20 \times 10^{-3})}{\ln\left(\dfrac{10.344}{10.235}\right)} \times 10^{12} = 399\,\text{pF}$$

C''_g is made of two concentric cylinders with $r_1 = 11.514$ mm, the insulation thickness is 0.100 mm with a relative permittivity of 3.8, and there is an extra 0.034 mm layer of wire insulation, giving $r_2 = 11.648$ mm.

$$C''_g = \frac{(2\pi)(3.8)(8.854 \times 10^{-12})(20 \times 10^{-3})}{\ln\left(\dfrac{11.648}{11.514}\right)} \times 10^{12} = 365\,\text{pF}$$

$$C_g = C'_g + C''_g = 399 + 365 = 764\,\text{pF}$$

The capacitance between the two layers of the primary winding may be considered a parallel plate capacitor, since the separation between the layer is much smaller than the radius of the windings. The insulation thickness between the layers is 0.1 mm and there are two extra 0.034 mm layers of wire insulation, giving $d = 0.168$ mm. The mean diameter is 21.858 mm, but this does not take the air medium created by the wire shape into account.

From Equation 8.35:

$$C_s = \frac{(3.8)(8.854 \times 10^{-12})(\pi)(21.858 \times 10^{-3})(20 \times 10^{-3})}{0.169 \times 10^{-3}} \times 10^{12} = 273\ \text{pF}$$

Finally the effective capacitance of the winding is obtained form Equation 8.31

$$C_{\text{eff}} = \sqrt{764 \times 273} = 457\ \text{pF}$$

8.5 Problems

8.1 Calculate the value of the inductance for which the data in Table 8.3 was generated. The coil resistance was 2.5 Ω and the applied voltage was 72 V.

8.2 Reconstruct the initial permeability versus magnetic field intensity curve for the material in the core tested in Example 8.2.

8.3 A single phase, 100 KVA 1000/100 V transformer gave the following test results with the instruments connected to the high voltage side:

Open-circuit test: 1000 V, 0.6 A, 400 W

Short-circuit test: 50 V, 100 A, 1800 W

(a) Determine the rated voltage and rated current for the high-voltage and low-voltage sides.

(b) Derive the approximate equivalent circuit referred to the HV side.

8.4 Table 8.4 shows the measurement data for an inductor obtained from the incremental impedance method. Calculate the AC inductance as a function of DC bias. The resistance of the coil was 4.0 Ω and the measurements were carried out at 50 Hz.

8.5 Devise a method to estimate the number of turns in a transformer winding based on a turns ratio.

Table 8.3 Inductance measurements

k	t_k(ms)	$i(k)$ (A)
1	1	0.5
2	2	1.6
3	3	6.0
4	4	14.6
5	5	17.3

Table 8.4 Inductance measurements

I_{dc}(A)	v_{ac}(V)	i_{ac}(mA)	$Z(\Omega)$	L_{ac}(mH)
0.000	1.090	42.90	25.41	79.9
0.250	1.086	42.10	25.80	81.2
0.500	1.092	47.00	23.23	72.9
0.750	1.119	60.40	18.53	57.6
1.000	1.143	80.10	14.27	43.6
1.250	1.169	104.00	11.24	33.5
1.500	1.181	126.00	9.37	27.0
1.750	1.176	151.00	7.79	21.3
2.000	1.148	175.00	6.56	16.6
2.250	1.119	190.00	5.89	13.8
2.500	1.085	199.00	5.45	11.8
2.750	1.051	204.00	5.15	10.3
3.000	1.024	208.00	4.92	9.1
3.250	0.998	212.00	4.71	7.9
3.500	0.978	219.00	4.47	6.3
3.750	0.960	222.00	4.32	5.2
4.000	0.934	231.00	4.04	1.9

Reference

1. Heller, B. and Veverka, A. (1968) *Surge Phenomena in Electrical Machines*, Iliffe Books, London.

Further Reading

1. Bi, S., Sutor, A., and Yue, J. (2008) Optimization of a measurement system for the hysteretic characterization of high permeable materials. *Proceedings of the Virtuelle Instrumente in der Praxis VIP 2008*, pp. 22–30.
2. Dalessandro, L., da Silveira Cavalcante, F., and Kolar, J.W. (2007) Self-capacitance of high-voltage transformers. *IEEE Transactions on Power Electronics* **22** (5), 2081–2092.
3. Dauhajre, A. and Middlebrook, R.D. (1986) Modelling and estimation of leakage phenomenon in magnetic circuits. *Proceedings of the IEEE Power Electronics Specialists Conference, PESC*, pp. 213–226.
4. Gradzki, P.M. and Lee, F.C. (1991) High-frequency core loss characterization technique based on impedance measurement. *Proceedings of the Virginia Power Electronic Center Seminar, VPEC*, pp. 1–8.
5. Hai Yan, L., Jian Guo, Z., and Hui, S.Y.R. (2007) Measurement and modeling of thermal effects on magnetic hysteresis of soft ferrites. *IEEE Transactions on Magnetics* **43** (11), 3952–3960.
6. Hui, S.Y.R. and Zhu, J. (1995) Numerical modelling and simulation of hysteresis effects in magnetic cores using transmission-line modelling and the Preisach theory. *IEE Proceedings-Electric Power Applications B* **142** (1), 57–62.
7. Hurley, W.G. and Wilcox, D.J. (1994) Calculation of leakage inductance in transformer windings. *IEEE Transactions on Power Electronics* **9** (1), 121–126.
8. Hurley, W.G., Wilcox, D.J., and McNamara, P.S. (1991) Calculation of short circuit impedance and leakage impedance in transformer windings. *Proceedings of the IEEE Power Electronics Specialists Conference, PESC*, pp. 651–658.
9. Jiles, D.C. and Atherton, D.L. (1984) Theory of ferromagnetic hysteresis (invited). *Journal of Applied Physics* **55** (6), 2115–2120.
10. Kazimierczuk, M.K. (2009) *High-Frequency Magnetic Components*, John Wiley & Sons, Chichester.

11. Kis, P., Kuczmann, M., Füzi, J., and Iványi, A. (2004) Hysteresis measurement in LabView. *Physica B: Condensed Matter*, **343** (1–4), 357–363.
12. Kuczmann, M. and Sarospataki, E. (2006) Realization of the Jiles-Atherton hysteresis model applying the LabVIEW and MATLAB software package. *Journal of Electrical Engineering* **57** (8), 40–43.
13. Kulkarni, S.V. (2004) *Transformer Engineering: Design and Practice*, 1st edn, CRC Press, New York.
14. McLyman, C.W.T. (1997) *Magnetic Core Selection for Transformers and Inductors*, 2nd edn, Marcel Dekker Inc., New York.
15. McLyman, C.W.T. (2002) *High Reliability Magnetic Devices*, 1st edn, Marcel Dekker Inc., New York.
16. McLyman, C.W.T. (2004) *Transformer and Inductor Design Handbook*, 3rd edn, Marcel Dekker Inc., New York.
17. E.S. MIT (1943) *Magnetic Circuits and Transformers (MIT Electrical Engineering and Computer Science)*, The MIT Press, Cambridge, MA.
18. Muhlethaler, J., Biela, J., Kolar, J.W., and Ecklebe, A. (2012) Improved core-loss calculation for magnetic components employed in power electronic systems. *IEEE Transactions on Power Electronics* **27** (2), 964–973.
19. Muhlethaler, J., Biela, J., Kolar, J.W., and Ecklebe, A. (2012) Core losses under the DC bias condition based on steinmetz parameters. *IEEE Transactions on Power Electronics* **27** (2), 953–963.
20. Prabhakaran, S. and Sullivan, C.R. (2002) Impedance-analyzer measurements of high-frequency power passives: techniques for high power and low impedance. *Conference Record of the Industry Applications Conference, IAS*, pp. 1360–1367.
21. Snelling, E.C. (1988) *Soft Ferrites: Properties and Applications*, 2nd edn, Butterworths, London.
22. Steinmetz, C.P. (1984) On the law of hysteresis. *Proceedings of the IEEE* **72** (2), 197–221.
23. Van den Bossche, A. (2005) *Inductors and Transformers for Power Electronics*, 1st edn, CRC Press, New York.
24. Venkatachalam, K., Sullivan, C.R., Abdallah, T., and Tacca, H. (2002) Accurate prediction of ferrite core loss with nonsinusoidal waveforms using only Steinmetz parameters. *Proceedings of IEEE Workshop on Computers in Power Electronics, COMPEL*, pp. 36–41.

9

Planar Magnetics[1]

The relentless drive towards high-density electronic circuits has been a feature of microelectronics for several decades and, typically, improvements in power efficiency have paralleled the shrinkage of electronic circuits. The effects are obvious in IC design; component densities continue to increase, with no sign of letting up. The latest trends in microelectronic techniques and nanotechnology, such as thick film and thin film technologies, are being pushed by the requirements of microelectronics to reduce size and cost and to improve reliability. By extension, low-profile planar magnetic components can be incorporated into the production processes that are already used to fabricate resistors and capacitors, and magnetic materials that are suitable to meet these requirements are constantly under development.

One of the major drawbacks in establishing planar magnetic technology is the lack of accurate analytical models for typical structures. Prototypes are expensive to fabricate and test and, while they offer useful insights into specific designs, they do not always extend to the general case. Analytical models that take account of frequency-dependent losses in the magnetic materials are needed to enhance our understanding of planar magnetics [1–3].

The reduction in the size of magnetic devices is essential for further overall miniaturisation and increased functionality of power conversion systems. Conventional magnetics using bobbins are bulky and the assembly process is difficult to automate. Planar magnetics fabrication and assembly processes have several advantages over conventional magnetics:

- **Low profile:** due to the fabrication process, planar magnetic components have a lower profile than their wire-wound counterparts.

[1] Parts of this chapter are reproduced with permission from [1] Hurley, W.G. and Duffy, M.C. (1995) Calculation of self and mutual impedances in planar magnetic structures. *IEEE Transactions on Magnetics* **31** (4), 2416–2422; [2] Hurley, W.G. and Duffy, M.C. (1997) Calculation of self- and mutual impedances in planar sandwich inductors. *IEEE Transactions on Magnetics* **33** (3), 2282–2290; [3] Hurley, W.G., Duffy, M.C., O'Reilly, S, and O'Mathuna, S.C. (1999) Impedance formulas for planar magnetic structures with spiral windings. *IEEE Transactions on Industrial Electronics* **46** (2), 271–278; [4] Wang, N., O'Donnell, T., Meere, R. *et al.* (2008) Thin-film-integrated power inductor on Si and its performance in an 8-MHz buck converter. *IEEE Transactions on Magnetics* **44** (11), 4096–4099; [5] Mathuna, S.C.O., O'Donnell, T., Wang, N., and Rinne, K. (2005) Magnetics on silicon: an enabling technology for power supply on chip. *IEEE Transactions on Magnetics* **20** (3), 585–592.

Transformers and Inductors for Power Electronics: Theory, Design and Applications, First Edition.
W. G. Hurley and W. H. Wölfle.

- **Automation:** it is difficult to automate the winding of conventional inductors and transformers, whereas the processes used in planar magnetics are based on advanced computer-aided manufacturing techniques.
- **High power densities:** planar inductors and transformers are spread out, which gives them a bigger surface-to-volume ratio than conventional components. The increased surface area enhances the thermal performance and this, in turn, means that the power density is increased. Increased heat convection also means that higher operating frequencies are achievable.
- **Predictable parasitics:** in wire-wound components, it is very difficult to control the winding layout, which can mean more leakage effects and winding capacitance. As a result, significant variations in these electrical parameters appear in devices manufactured at the same time. With planar magnetics, the windings are precise and consistent, yielding magnetic designs with highly controllable and predictable characteristic parameters.

The size of magnetic components may be reduced by operating at high frequency. Planar magnetic components use this principle to reduce component size while taking advantage of microelectronic processing. The magnetic materials used in planar magnetics have finite conductivity and, at sufficiently high frequency, unwanted eddy current loss appears. In general, the number of turns in a planar device tends to be limited by the manufacturing process. The low profile tends to lead to a larger footprint compared with its conventional counterpart. The spreading effect leads to high capacitance between layers and between the windings and the core. In multilayer devices, the interlayer capacitance introduces resonance at high frequencies.

When considering conventional versus planar devices, several issues must be addressed:

- the shape;
- the trade-off between magnetic core area and winding window area;
- the magnetic path length versus the mean length of a turn;
- the surface area.

Planar magnetics have opened up new applications such as coreless transformers for gate drives, radio frequency (rf) inductors and a Power Supply on Chip (PwrSoC). The circuit models have not necessarily changed, but the new layouts of windings and cores require new models for inductance and loss mechanisms.

9.1 Inductance Modelling

The mutual inductance between two filaments in air was treated in Chapter 2. This was extended to formulae for mutual inductance between coils with finite cross-sections by integrating the filamentary formula over the cross-section; the current density was taken as uniform over the section. Accurate formulae emerged, but the lack of computing power meant that results were often presented in look-up tables. Another approach was to develop approximations based on the filament formula, with judicious choice of filament placement.

In the case of planar magnetic components, the current density is not uniform because the aspect ratio of width to height of a section is typically very large. This means that the length of the conducting path is much shorter on the inside of a flat wide coil and, consequently, the current density is higher. The accuracy of the classic formulae is much improved when the proper current density is included in the analysis.

The starting-point for a planar magnetic component is the air-cored spiral inductor. Despite its physical simplicity, this forms the basis for more advanced configurations, such as coils on magnetic substrates [1] and sandwich inductors [2], where the coil is placed between two magnetic substrates. The analysis begins with a formula for the mutual inductance between two planar spiral coils in air that takes full account of the current density distribution across the planar section. The result can be extended to a component with several turns per layer and with several layers.

The next step is to add a magnetic substrate. A ferromagnetic substrate with finite conductivity introduces eddy current loss and hysteresis loss that add to the winding resistance loss. A frequency-dependent mutual impedance formula for this case is derived, which takes the eddy current loss into account. Finally, we will add a second substrate layer in a sandwich configuration.

9.1.1 Spiral Coil in Air

The mutual inductance between two filaments shown in Figure 9.1 is the basis for establishing the general mutual inductance formula for planar structures [1].

The formula has the form:

$$M = \mu_0 \pi a r \int_0^\infty J_1(kr) J_1(ka) e^{-k|z|} dk \tag{9.1}$$

where J_1 is a Bessel function of the first kind, a and r are the filament radius and μ_0 is the permeability of free space.

An alternative to Equation 9.1 can be written in terms of elliptic integrals:

$$M = \mu_0 \sqrt{ar} \frac{2}{f} \left[\left(1 - \frac{f^2}{2} \right) K(f) - E(f) \right] \tag{9.2}$$

where $K(f)$ and $E(f)$ are complete elliptic integrals of the first and second kind, respectively, and where:

$$f = \sqrt{\frac{4ar}{z^2 + (a + r)^2}} \tag{9.3}$$

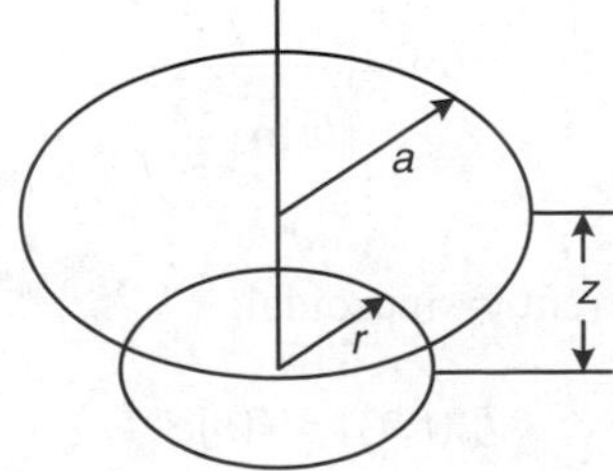

Figure 9.1 Circular concentric filaments in air.

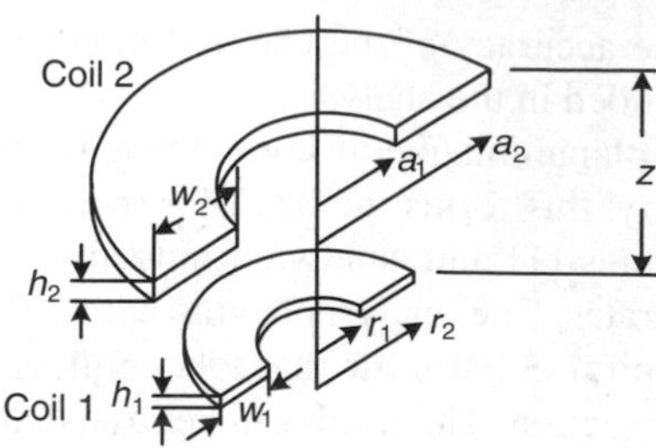

Figure 9.2 Planar coils of rectangular cross-section.

The next step is to consider two circular and concentric planar sections, as depicted in Figure 9.2, with dimensions as shown.

In practice, a spiral coil would connect two sections in series, which can be accurately modelled by the concentric circular coils as illustrated. In Chapter 2, the mutual inductance between sections was found by integrating the filamentary formula (Equation 9.1) over each cross-section, with the assumption that the current density is constant over each section. The approach is adequate when the width and height of the coil section are approximately equal. In a planar structure, however, the ratio of width to height could be as big as 50 : 1. The shorter path on the inside edge of the conducting section means that the resistance to current flow is lower, and therefore the current density is higher on the inside than on the outside.

On the basis of this observation, it is reasonable to assume that there is an inverse relationship between the current density *J*(*r*) and the radius *r*. By the same token, current density in the *z* direction may be taken as constant because of the low profile. Integrating the current density over the cross-section yields the total current *I*:

$$h = \int_{r_1}^{r_2} J(r)dr = I \tag{9.4}$$

The current density has the form:

$$J(r) = \frac{K}{r} \tag{9.5}$$

with *K* a constant.

Solving Equations 9.4 and 9.5 over a cross-section with an inside radius of r_1, outside radius r_2 and height h yields:

$$J(r) = \frac{I}{hr\ln\left(\frac{r_2}{r_1}\right)} \tag{9.6}$$

In the following analysis, the current is sinusoidal:

$$J_\phi(r,t) = J(r)e^{j\omega t} \tag{9.7}$$

where ω is the angular frequency.

The induced voltage in the filament at (r, τ_1) in coil 1 due to the current in an annular section $da \times d\tau_2$ at radius a in coil 2 is:

$$dV = j\omega MJ(a)\, da\, d\tau_2 \tag{9.8}$$

where M is the mutual inductance between the filaments at (r, τ_1) and $(a, z + \tau_2)$.

Integrating Equation 9.8 over the cross-section of coil 2 yields the voltage at (r, τ_1) due to all the current in coil 2:

$$V(r) = j\omega\mu_0\pi r \int_0^\infty \int_{\frac{-h_2}{2}}^{\frac{h_2}{2}} \int_{a_1}^{a_2} aJ(a)J_1(kr)J_1(ka)e^{-k|z+\tau_2-\tau_1|} da\, d\tau_2\, dk \tag{9.9}$$

The power transferred to the annular segment at (r, τ_1) due to coil 2 is:

$$dP = V(r)J(r)dr\, d\tau_1 \tag{9.10}$$

The total power transferred to coil 1 may be obtained by integrating Equation 9.10 over its cross-section:

$$P = j\omega\mu_0\pi \int_0^\infty \int_{\frac{-h_1}{2}}^{\frac{h_1}{2}} \int_{\frac{-h_2}{2}}^{\frac{h_2}{2}} \int_{r_1}^{r_2} \int_{a_1}^{a_2} rJ(r)J_1(kr)aJ(a)J_1(ka)e^{-k|z+\tau_2-\tau_1|} da\, dr\, d\tau_1\, d\tau_2\, dk \tag{9.11}$$

The internal integrals are readily solved, with $J(r)$ given by Equation 9.6, to yield:

$$P = j\omega\mu_0\pi \frac{I_1 I_2}{h_1 \ln\left(\frac{r_2}{r_1}\right) h_2 \ln\left(\frac{a_2}{a_1}\right)} \int_0^\infty S(kr_2, kr_1)S(ka_2, ka_1)Q(kh_1, kh_2)e^{-k|z|}dk \tag{9.12}$$

where:

$$\begin{aligned} Q(kx, ky) &= \frac{2}{k^2}\left[\cosh\left(k\frac{x+y}{2}\right) - \cosh\left(k\frac{x-y}{2}\right)\right] \quad z > \frac{h_1 + h_2}{2} \\ &= \frac{2}{k}\left(h + \frac{e^{-kh} - 1}{k}\right) \quad z = 0, x = y = h \end{aligned} \tag{9.13}$$

and:

$$S(kx, ky) = \frac{J_0(kx) - J_0(ky)}{k} \tag{9.14}$$

but:

$$P = v_2 i_2 = j\omega M_{12} I_1 I_2 \tag{9.15}$$

where M_{12} is the mutual inductance between the two coils. Substituting Equation 9.15 establishes M_{12}:

$$M_{12} = \frac{\mu_0 \pi}{h_1 h_2 \ln\left(\frac{r_2}{r_1}\right) \ln\left(\frac{a_2}{a_1}\right)} \int_0^\infty S(kr_2, kr_1) S(ka_2, ka_1) Q(kh_1, kh_2) e^{-k|z|} dk \tag{9.16}$$

This is the final result and, despite its complex appearance, it is easily solved with numerical integration using MATLAB.

Combined with a simple formula for the DC resistance, the equivalent circuit model includes a resistance in series with an inductance given by Equation 9.16.

Example 9.1

Calculate the self inductance of the four-turn planar coil shown in Figure 9.3. This configuration is commonly used for inductive charging in electric vehicles.

The device consists of four planar coils in series.

We need to calculate the self inductance of coil 1 (same as the self inductance of coil 2) and the self inductance of coil 3 (same as the self inductance of coil 4). Next, we need the mutual inductances M_{12}, M_{13} (same as M_{24}), M_{14} (same as M_{23}) and M_{34}.

The required dimensions for the self inductance calculations are:

$$\text{Coil 1}: r_1 = a_1 = 1.15\,\text{mm},\ r_2 = a_2 = 1.75\,\text{mm},\ h_1 = h_2 = 15\,\mu\text{m},\ z = 0$$

$$\text{Coil 2}: r_1 = a_1 = 1.15\,\text{mm},\ r_2 = a_2 = 1.75\,\text{mm},\ h_1 = h_2 = 15\,\mu\text{m},\ z = 0$$

$$\text{Coil 3}: r_1 = a_1 = 2.00\,\text{mm},\ r_2 = a_2 = 2.60\,\text{mm},\ h_1 = h_2 = 15\,\mu\text{m},\ z = 0$$

$$\text{Coil 4}: r_1 = a_1 = 2.00\,\text{mm},\ r_2 = a_2 = 2.60\,\text{mm},\ h_1 = h_2 = 15\,\mu\text{m},\ z = 0$$

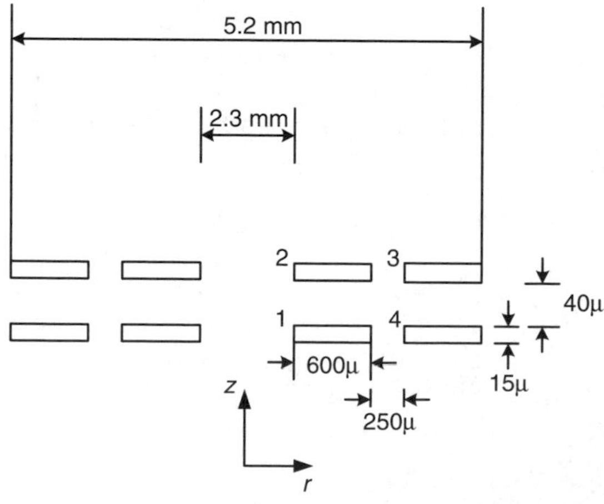

Figure 9.3 Layout of a planar device.

The required dimensions for the mutual inductance calculations are:

$$\text{Coils 1 and 2}: r_1 = a_1 = 1.15\,\text{mm},\ r_2 = a_2 = 1.75\,\text{mm},\ h_1 = h_2 = 15\,\mu\text{m},\ z = 55\,\mu\text{m}$$

$$\text{Coils 1 and 3}: r_1 = 1.15\,\text{mm},\ r_2 = 1.75\,\text{mm},\ h_1 = 15\,\mu\text{m}$$

$$a_1 = 2.00\,\text{mm},\ a_2 = 2.60\,\text{mm},\ h_2 = 15\,\mu\text{m}$$

$$z = 55\,\mu\text{m}$$

$$\text{Coils 1 and 4}: r_1 = 1.15\,\text{mm},\ r_2 = 1.75\,\text{mm},\ h_1 = 15\,\mu\text{m}$$

$$a_1 = 2.00\,\text{mm},\ a_2 = 2.60\,\text{mm},\ h_2 = 15\,\mu\text{m}$$

$$z = 0$$

$$\text{Coils 2 and 3}: \text{same as coils 1 and 4}$$

$$\text{Coils 2 and 4}: \text{same as coils 1 and 3}$$

$$\text{Coils 3 and 4}: r_1 = a_1 = 2.00\,\text{mm},\ r_2 = a_2 = 2.60\,\text{mm},\ h_1 = h_2 = 15\,\mu\text{m},\ z = 55\,\mu\text{m}$$

The total inductance is:

$$L = 2L_{11} + 2L_{33} + 2M_{12} + 4M_{13} + 4M_{14} + 2M_{34}.$$

The MATLAB program to carry out the calculations is listed at the end of this chapter.

The program yielded the following results:

$L_{11} = 4.366$ nH; $L_{33} = 8.320$ nH; $M_{12} = 3.956$ nH; $M_{13} = 2.223$ nH; $M_{14} = 2.229$ nH and $M_{34} = 9.327$ nH. The total inductance of the four-turn coil is 69.75 nH.

9.1.2 Spiral Coil on a Ferromagnetic Substrate

The self inductance of the coil in Figure 9.2 will be enhanced by adding a magnetic substrate. Placing the coil on an ideal magnetic substrate of infinite thickness would double the overall inductance of the coil. Currents flowing in the coil induce eddy currents in a core material with finite conductivity. A general impedance formula that takes frequency-dependent eddy current loss in the substrate into account is required to model the loss for high-frequency operation.

The geometry for the mutual impedance between two filaments placed above a magnetic substrate is shown in Figure 9.4, with appropriate dimensions. The lower filament in Figure 9.4 is at a height d_1 above the substrate of thickness t, so that an insulating layer between a coil and the substrate can be modelled.

The mutual impedance between the two circular concentric filaments in Figure 9.4 is:

$$Z = j\omega M + Z_t^f \tag{9.17}$$

where M is the mutual inductance that would exist in the absence of the substrate, and is the same as Equation 9.1.

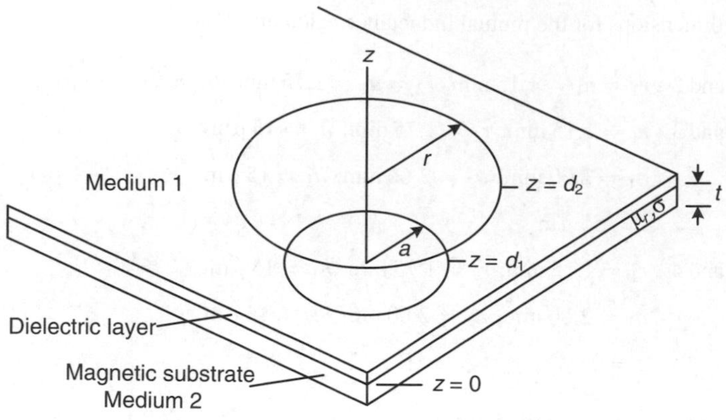

Figure 9.4 Circular concentric filaments on a magnetic substrate.

The additional impedance due to the presence of the substrate may be found by solving Maxwell's equations. For a magnetoquasistatic system, the following forms of Maxwell's equations hold in a linear homogeneous isotropic medium:

$$\nabla \times \mathbf{H} = \mathbf{J}_\phi \tag{9.18}$$

$$\nabla \times \mathbf{E} = -\frac{\partial \mathbf{B}}{\partial t} \tag{9.19}$$

The filamentary turn at $z = d_1$ in Figure 9.4 carries a sinusoidal current $i_\phi(t) = I_\phi\, e^{j\omega t}$. Medium 1 may be air or a dielectric, and medium 2 is the magnetic substrate. The solution is divided into four regions: region 1 ($z \geq d_1$); region 2 ($0 \leq z < d_1$); region 3 ($-t \leq z < 0$); and region 4 ($z < -t$). Regions 1, 2 and 4 corresponds to medium 1 and region 3 corresponds to medium 2.

The following identities apply to the electric field intensity E and the magnetic field intensity H because there is cylindrical symmetry:

$$E_r = 0; E_z = 0; \frac{\partial E_\phi}{\partial \phi} = 0 \tag{9.20}$$

$$H_\phi = 0; \frac{\partial H_r}{\partial \phi} = 0; \frac{\partial H_z}{\partial \phi} = 0 \tag{9.21}$$

Maxwell's equations in each region may be shown to be:

Region 1 ($z \geq d_1$):

$$\frac{\partial H_r}{\partial z} - \frac{\partial H_z}{\partial r} = I_\phi \delta(r - a)\delta(z - d_1) \tag{9.22}$$

$$\frac{\partial E_\phi}{\partial z} = j\omega\mu_0 H_r \tag{9.23}$$

$$\frac{1}{r}\frac{\partial(rE_\phi)}{\partial r} = -j\omega\mu_0 H_z \tag{9.24}$$

Eliminating H gives the following result for E_ϕ:

$$\frac{\partial^2 E_\phi}{\partial z^2} + \frac{\partial^2 E_\phi}{\partial r^2} + \frac{1}{r}\frac{\partial E_\phi}{\partial r} - \frac{E_\phi}{r^2} = j\omega\mu_0 I_\phi \delta(r-a)\delta(z-d_1) \tag{9.25}$$

Region 2 ($0 \le z < d_1$):

There is no current in this region and the electric field is given by:

$$\frac{\partial^2 E}{\partial z^2} + \frac{\partial^2 E}{\partial r^2} + \frac{1}{r}\frac{\partial E}{\partial r} - \frac{E}{r^2} = 0 \tag{9.26}$$

Region 3 ($-t \le z < 0$):

In this region, $J_\phi = \sigma E_\phi$ and the electric field is:

$$\frac{\partial^2 E_\phi}{\partial z^2} + \frac{\partial^2 E_\phi}{\partial r^2} + \frac{1}{r}\frac{\partial E_\phi}{\partial r} - \left(\frac{1}{r^2} + j\omega\mu_r\mu_0\sigma\right)E_\phi = 0 \tag{9.27}$$

Region 4 ($z < -t$):

This is similar to region 2 and Equation 9.26 again applies.

The solution of E_ϕ in regions 1–4 is obtained by invoking the Fourier Bessel integral transformation [6]:

$$E* = \int_0^\infty E_\phi(r) r J_1(kr) dr \tag{9.28}$$

and noting that:

$$\int_0^\infty \delta(r-a) J_1(kr) r\, dr = a J_1(ka) \tag{9.29}$$

resulting in the transformed version of Equation 9.27:

$$\frac{d^2 E*}{dz^2} = k^2 E * + j\omega\mu_0 I_\phi\, a\, J_1(ka)\, \delta(z-d_1) \tag{9.30}$$

The solution is of the form:

$$E* = I e^{kz} + A e^{-kz} \tag{9.31}$$

I and A are constants to be determined by the boundary conditions.

$$I \to 0 \text{ since } E* \to 0 \text{ at infinity.}$$

The solution in $E*$ in each region becomes:

$$\text{Region 1} \quad z \ge d_1 \quad E* = A e^{-kz} \tag{9.32}$$

$$\text{Region 2} \quad 0 \leq z < d_1 \quad E* = Be^{kz} + Ce^{-kz} \tag{9.33}$$

$$\text{Region 3} \quad -t \leq z < 0 \quad E* = De^{\Lambda z} + Fe^{-\Lambda z} \tag{9.34}$$

$$\text{Region 4} \quad z < -t \quad E* = Ge^{kz} \tag{9.35}$$

with:

$$\Lambda = \sqrt{k^2 + j\omega\mu_r\mu_0\sigma} \tag{9.36}$$

There are six constants to be established on the basis of the boundary conditions.

The electric field is continuous at the following boundaries: $z = d_1$, $z = 0$ and $z = -t$, giving:

$$Ae^{-kd_1} = Be^{kd_1} + Ce^{-kd_1} \tag{9.37}$$

$$B + C = D + F \tag{9.38}$$

$$De^{-\Lambda t} + Fe^{\Lambda t} = Ge^{-kt} \tag{9.39}$$

The boundary condition imposed on the radial component of the magnetic field intensity is given by:

$$\mathbf{n} \times (\mathbf{H}_+ - \mathbf{H}_-) = \mathbf{K}_f \tag{9.40}$$

where $\mathbf{n}$ is the unit vector normal to the plane at the boundary and $\mathbf{K}_f$ is the surface current density at the boundary.

The radial component of the magnetic field intensity is given by Maxwell's equations:

$$\frac{\partial E_\phi}{\partial z} = j\omega\mu_r\mu_0 H_r \tag{9.41}$$

At $z = 0$ and $z = -t$, there is no surface current, and equating H_r given by Equation 9.41 at either side of the boundary yields:

$$k(B - C) = \frac{\Lambda}{\mu_r}(D - F) \tag{9.42}$$

and:

$$\frac{\Lambda}{\mu_r}(De^{-\Lambda t} - Fe^{\Lambda t}) = kGe^{-kt} \tag{9.43}$$

At $z = d_1$ the surface current density is:

$$K_f = \int_{d_{1-}}^{d_{1+}} I_\phi \delta(r - a)\delta(z - d_1)dz = I_\phi \delta(r - a) \tag{9.44}$$

and in terms of the transformed variable:

$$K_f^* = I_\phi a J_1(ka) \tag{9.45}$$

H_+ and H_- may be found by using Equation 9.42 in Equation 9.33 and 9.34:

$$-Ae^{-kd_1} - Be^{kd_1} + Ce^{-kd_1} = \frac{j\omega\mu_0}{k} I_\phi a J_1(ka) \tag{9.46}$$

It now remains to solve six equations in six unknowns.

Our primary interest is in the mutual impedance between the filaments in Region 1, where the electric field is:

$$E^* = -j\omega\mu_0 I_\phi \frac{aJ_1(ka)}{2k}\left[e^{-k|z-d_1|} + \lambda(t)e^{-k|z+d_1|}\right] \tag{9.47}$$

$\lambda(t)$ is defined as:

$$\lambda(t) = \phi(k)\frac{1 - e^{-2\Lambda t}}{1 - \phi(k)^2 e^{-2\Lambda t}} \tag{9.48}$$

with:

$$\phi(k) = \frac{\mu_r - \dfrac{\Lambda}{k}}{\mu_r + \dfrac{\Lambda}{k}} \tag{9.49}$$

Applying the inverse transform of the Fourier-Bessel integral defined as:

$$E(r) = \int_0^\infty E^*(k)kJ_1(kr)dk \tag{9.50}$$

results in an expression for $E(r,z)$:

$$E(r,z) = -j\omega\mu_0 I_\phi a\frac{1}{2}\int_0^\infty J_1(kr)J_1(ka)\cdot\left[e^{-k|z-d_1|} + \lambda(t)e^{-k|z+d_1|}\right]dk \tag{9.51}$$

The induced voltage in a circular filament at (r, d_2) due to the source at (a, d_1) is $V = ZI_\phi$ and:

$$V = -\int_0^{2\pi} E(r,d_2)\, r d\phi = -2\pi r E(r,d_2) \tag{9.52}$$

It follows that:

$$Z = -\frac{2\pi E(r,d_2)}{I_\phi} \tag{9.53}$$

and:

$$Z = j\omega M + Z_t^f \tag{9.54}$$

$$M = \mu_0 \pi ar \int_0^\infty J_1(kr)J_1(ka)e^{-k|d_2-d_1|}dk \tag{9.55}$$

$$Z_t^f = R_s + j\omega L_s = j\omega\mu_0 \pi ar \int_0^\infty J_1(kr)J_1(ka)\lambda(t)e^{-k(d_1+d_2)}dk \tag{9.56}$$

In practice, if the substrate is at least five skin depths thick, it may be considered infinite in the $-z$ direction. For $t \to \infty$, $\lambda(t) \to \varphi(k)$ and Equation 9.56 becomes:

$$Z_s^f = R_s + j\omega L_s = j\omega\mu_0 \pi ar \int_0^\infty J_1(kr)J_1(ka)\phi(k)e^{-k(d_1+d_2)}dk \tag{9.57}$$

The subscript t in Z refers to a substrate of finite thickness, while subscript s refers to a substrate of infinite thickness.

The parameter $\lambda(t)$ in Equation 9.56 contains four variables of interest: thickness t, relative permeability μ_r, conductivity σ and frequency ω. At low frequencies, $\eta \to k$ as $\omega \to 0$ and $\phi(k) \to \phi_0$ for low frequency operation:

$$\phi_0 = \frac{\mu_r - 1}{\mu_r + 1} \tag{9.58}$$

This factor describes the increase in inductance in a substrate, when $\mu_r \gg 1$, $\phi = 1$; this means that the substrate component is equal to the air component, resulting in a doubling of the overall inductance.

As a check, if there is no substrate, then $\mu_r = 1$, $\eta = k$ and $\phi(k) = 0$, which means that the additional component $L_s = 0$ as expected.

Figure 9.5 shows two circular concentric planar sections on a magnetic substrate with appropriate dimensions. Integrating the filament formula (Equation 9.56) over the coil cross-

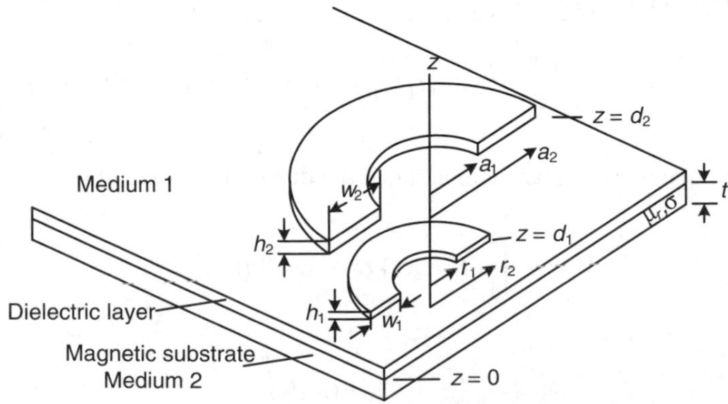

Figure 9.5 Planar coils on a magnetic substrate.

sections yields the impedance formula for coils. The assumption that the current density is inversely proportional to the radius as described in Section 9.0 is also made here.

The total impedance is:

$$Z = j\omega M + Z_t^p \tag{9.59}$$

M is the air term given by Equation 9.16 and the additional component due to the substrate is:

$$Z_t^p = \frac{j\omega\mu_0\pi}{h_1 h_2 \ln\left(\frac{r_2}{r_1}\right)\ln\left(\frac{a_2}{a_1}\right)} \int_0^\infty S(kr_2, kr_1)S(ka_2, ka_1)Q(kh_1, kh_2)\lambda(t)e^{-k(d_1+d_2)}dk \tag{9.60}$$

As before, replace $\lambda(t)$ by $\phi(k)$ for an infinite substrate.

The DC resistance may be combined with the terms given by Equation 9.59 to complete the equivalent circuit model.

Example 9.2

Calculate the self inductance of the two-turn coil in Figure 9.6, with the dimensions shown. Plot the results for μ_r going from 1 to 1000 and for three values of substrate thicknesses $t = 0.05$ mm, 0.1 mm and 0.5 mm. Assume there are no losses in the substrate ($\sigma = 0$).

The self inductance of the device is given by:

$$L = L_{11} + 2M_{12} + L_{22}$$

where L_{11} and L_{22} are the self inductances of sections 1 and 2 in Figure 9.6 and M_{12} is the mutual inductance between sections 1 and 2. The individual terms for the inductance in air are given by Equation 9.16 and the contributions for the substrate are given by Equation 9.60. For a lossless substrate, λ becomes ϕ_0 in Equation 9.58.

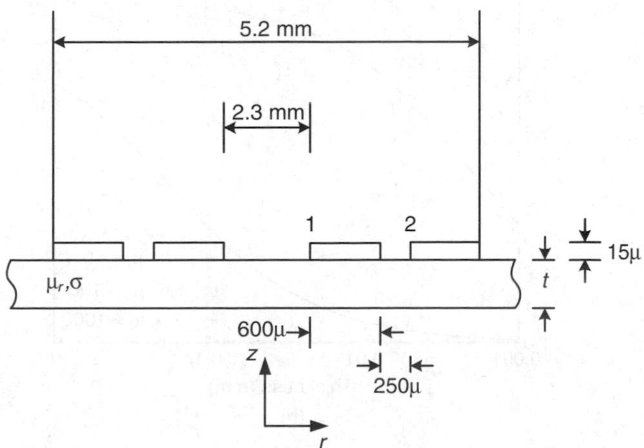

Figure 9.6 Planar sections of a two-turn device on a finite substrate.

The required dimensions for the self inductance calculations are:

$$\text{Coil 1}: r_1 = a_1 = 1.15\,\text{mm},\ r_2 = a_2 = 1.75\,\text{mm},\ h_1 = h_2 = 15\,\mu\text{m},\ d_1 = d_2 = 7.5\,\mu\text{m}$$

$$\text{Coil 2}: r_1 = a_1 = 2.00\,\text{mm},\ r_2 = a_2 = 2.60\,\text{mm},\ h_1 = h_2 = 15\,\mu\text{m},\ d_1 = d_2 = 7.5\,\mu\text{m}$$

The required dimensions for the mutual inductance calculations are:

$$\text{Coils 1 and 2}: r_1 = 1.15\,\text{mm},\ r_2 = 1.75\,\text{mm},\ a_1 = 2.00\,\text{mm},\ a_2 = 2.60\,\text{mm}$$
$$h_1 = h_2 = 15\,\mu\text{m},\ d_2 = d_2 = 7.5\,\mu\text{m}$$

The air terms were calculated in Example 9.1, and the MATLAB program at the end of this chapter gives the calculation for the substrate components.

Figure 9.7(a) shows the inductance enhancement as a function of relative permeability for difference values of substrate thickness t. The enhancement is in the range:

$$1 \leq \frac{L}{L_1} \leq 2$$

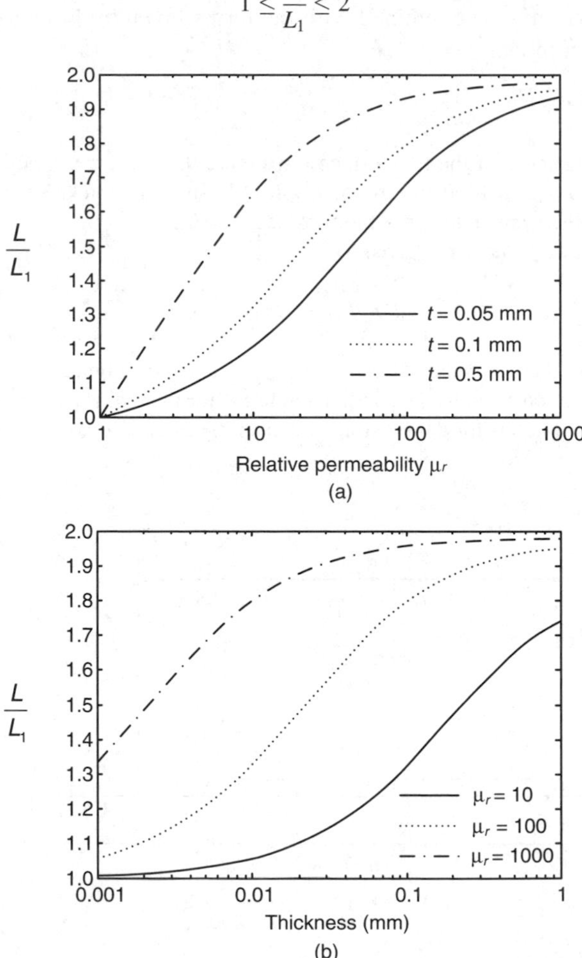

Figure 9.7 Enhancement of inductance with magnetic substrate: (a) as a function of μ_r, (b) as a function of t.

where L_1 is the inductance in air (without the substrate). L_1 for the two-turn coil is 17.14 nH The upper limit is achieved for the product $\mu_r t > 0.1$ m and the lower limit is approached as $\mu_r \to 1$.

Figure 9.7(b) shows the inductance enhancement as a function of substrate thickness for different values of relative permeability μ_r. The enhancement is in the range:

$$1 \leq \frac{L}{L_1} \leq 1 + \phi_0$$

The upper limit is approached for $\mu_r t > 0.1$ m. The lower limit is approached for $\mu_r t < 0.1$ mm.

The calculations are included in the MATLAB program at the end of this chapter.

Example 9.3

Calculate the inductance of the device in Example 9.2 on a ferrite substrate with relative permeability $\mu_r = 1000$ and electrical conductivity $\sigma = 10\ (\Omega - \text{m})^{-1}$. Take the frequency range 1 MHz to 1000 MHz.

The skin depth in the material is given by:

$$\delta = \sqrt{\frac{1}{\pi f \mu_r \mu_0 \sigma}} = \frac{5033}{\sqrt{f}}\ \text{mm}$$

where f is the frequency in Hz.

The self impedance is shown in Figure 9.8(a) for each value of substrate thickness. The frequency dependence due to the eddy currents in the substrate becomes evident when the skin depth becomes comparable to the thickness of the substrate. This is evident at 200 MHz for $t = 0.5$ mm, where the skin depth is 0.36 mm.

The eddy current loss as represented by the resistance in Figure 9.8(b) is pronounced above 100 MHz. The DC resistance of the device in Figure 9.6 is taken as 0.093 Ω for copper turns, and this is included in R. The resistance increases by a factor of 7 at 500 MHz in the 0.5 mm substrate, where the skin depth is less than half the thickness of the substrate. This analysis shows that it is very important that the skin depth in the substrate should be greater than the thickness of the substrate, to ensure that inductance enhancement takes place without incurring the penalty of high eddy current loss.

The calculations are included in the MATLAB program at the end of this chapter.

9.1.3 Spiral Coil in a Sandwich Structure

The addition of a second substrate above the planar coils results in a sandwich structure, as shown in Figure 9.9.

Adopting the dimensions shown in Figure 9.9, the solution of Maxwell's equations proceeds as before. The solution is given by [2]:

$$Z = j\omega M + Z_{s\omega}^{p} \tag{9.61}$$

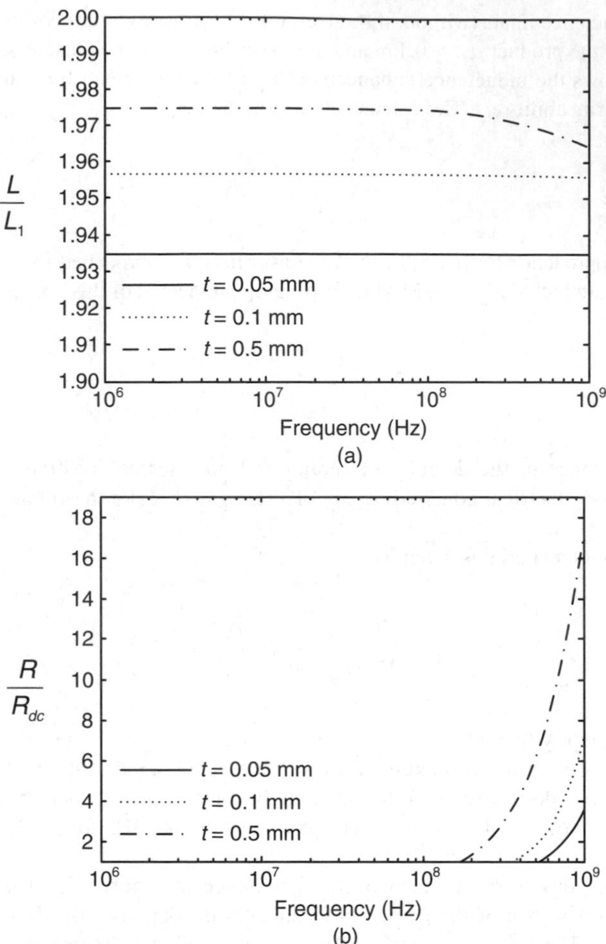

Figure 9.8 Self impedance of a planar coil on a finite substrate: (a) inductance, (b) resistance.

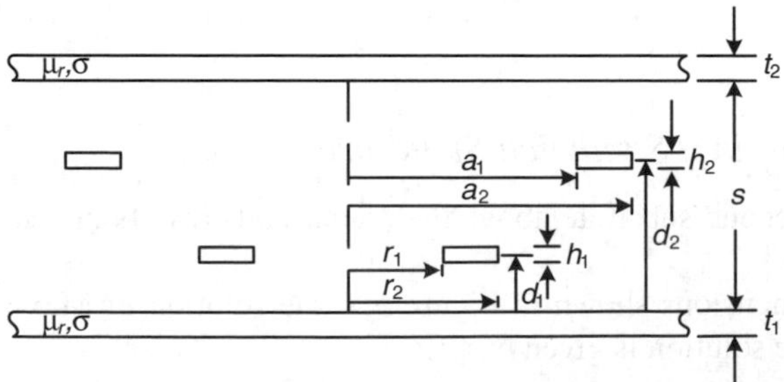

Figure 9.9 Planar coils in a sandwich structure.

M is the air component given by Equation 9.16 and the enhancement due to the substrate is:

$$Z^p_{s\omega} = \frac{j\omega\mu_0\pi}{h_1h_2\ln\left(\frac{r_2}{r_1}\right)\ln\left(\frac{a_2}{a_1}\right)}\int_0^\infty S(kr_2,kr_1)S(ka_2,ka_1)[f(\lambda)+g(\lambda)]Q(kh_1,kh_2)dk \quad (9.62)$$

where:

$$f(\lambda) = \frac{\lambda(t_1)e^{-k(d_1+d_2)} + \lambda(t_2)e^{-k(d_1'+d_2')}}{1-\lambda(t_1)\lambda(t_2)e^{-2ks}} \quad (9.63)$$

and:

$$g(\lambda) = \frac{2\lambda(t_1)\lambda(t_2)e^{-2ks}\cosh[k(d_2-d_1)]}{1-\lambda(t_1)\lambda(t_2)e^{-2ks}} \quad (9.64)$$

In addition to the four parameters (μ_r, σ, t_1, ω) discussed in Section 9.1.2, there are additional parameters s and t_2 to be taken into account in the sandwich inductor. For the purposes of analysis, substrates of equal thickness will be considered ($t_1 = t_2 = t$).

Example 9.4

Calculate the inductance and equivalent resistance of the device in Example 9.2, sandwiched between two ferrite substrates with relative permeability $\mu_r = 1000$ and electrical conductivity $\sigma = 10\ (\Omega\text{-m})^{-1}$. The substrate thickness is 0.5 mm. Plot the inductance for substrate separation ranging from 0.02 mm to 10 mm. Take the frequency range 1–1000 MHz.

The calculations are included in the MATLAB program at the end of this chapter, based on Equation 9.63.

Figure 9.10 shows the variation of inductance with separation s for low frequency (below the frequency where eddy currents in the substrate make a difference). In this case, the range of inductance is between 1 and μ_r.

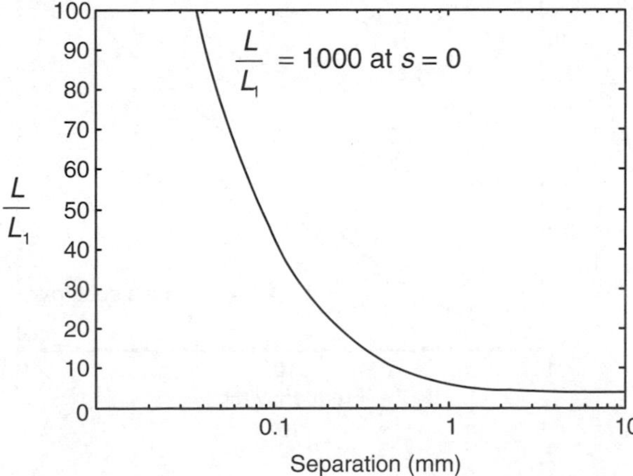

Figure 9.10 Inductance as a function of substrate separation; $L_1 = 17.14$ nH.

The upper limit is approached for $s \to 0$ because the coils are enclosed by a material that has a relative permeability of μ_r.

The lower limit is approached when one of the substrates is removed ($s \to \infty$). In this example, $\mu_r t > 0.1$ m, so that $\frac{L}{L_1} \to 1 + \phi_0$, which is too large for μ_r, as shown by Equation 9.58.

The analysis in Figure 9.10 was carried out for low frequencies where eddy current loss is negligible. The most important feature of Figure 9.10 is that the increase in inductance in a sandwich structure is strongly influenced by the gap between the substrates. The increase drops from a factor of 1000 (μ_r) at $s = 0$ down to 100 at $s = 15\ \mu$m. At $s = 1$ mm, the inductance is just 50% higher than that for a single substrate.

Figure 9.11 shows inductance and resistance as a function of frequency for $s = 15\ \mu$m. The inductance remains relatively flat up to 100 MHz. For comparison purposes, the results of Figure 9.8 with $t = 0.5$ mm for the single substrate are included and, clearly, the presence of the second substrate greatly increases both the overall inductance and the overall loss. At 2 MHz, the loss in the substrates is equal to the DC loss in the conductors.

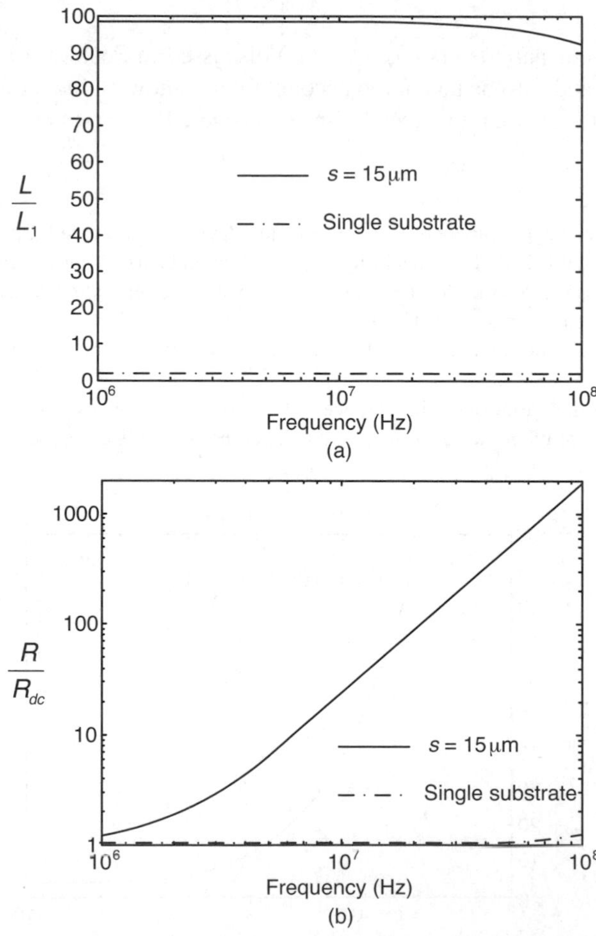

Figure 9.11 Self impedance of a planar sandwich coil as a function of frequency: (a) inductance, (b) resistance; $L_1 = 17.14$ nH, $R_{dc} = 0.093\ \Omega$.

9.2 Fabrication of Spiral Inductors

The fabrication of planar magnetic devices may be broadly classified as printed circuit board (PCB), thick film, low temperature co-fired ceramic (LTCC) and thin film. The technology tends to be aligned to the application; for example, PCB has found applications in microprocessors and computer loads, where etched conductors can carry the load currents. Thick film is normally associated with passive components such as resistors and, to a lesser extent, capacitors [7], and it may be found in biomedical applications, where a ceramic substrate is normally expected to avail of its non-toxic properties. LTCC finds favour in high temperature applications, and thin film has emerged in power supply on a chip (PwrSoC) because of its compatibility with silicon.

9.2.1 PCB Magnetics

Printed circuit board (PCB) technology was developed and adapted to magnetic components to overcome the size and cost problems associated with conventional magnetics. PCB magnetics may be categorized as discrete PCB magnetics and integrated PCB magnetics. The difference lies in the method of fabrication or assembly. PCB magnetics have found application in low power DC-DC converters.

Discrete PCB Magnetics

Discrete PCB magnetics mean that the winding is produced separately from the magnetic core. Low-height magnetic cores are used and the windings do not require winding bobbins. Figure 9.12 shows a typical planar magnetic construction with PCB windings. The windings on printed circuit boards are assembled in layers, along with appropriate insulating layers. Standard ferrite cores are placed around the windings to complete the inductor or transformer in a sandwich structure.

Ferrite cores are available in many shapes and sizes due to their ease of manufacture and, therefore, ferrite is the primary core material for planar magnetics designs. Most planar magnetics assemblies have core materials of manganese-zinc (Mn-Zn) or nickel-zinc (Ni-Zn). Mn-Zn is more popular in applications where the operating frequencies are less than 2 MHz; above 2 MHz, Ni-Zn is more suitable because its resistivity is several orders of magnitude larger than Mn-Zn. The only disadvantage of ferrite material is the relatively low value of saturation flux density compared with other core materials.

One of the main advantages of this planar type construction is that the surface area for heat transfer tends to be higher relative to volume when compared to conventional wire-wound devices. The larger surface area means that more heat can be removed for the same temperature rise and this, in turn, means that higher power densities are achievable. Another advantage is that the parasitics, such as leakage inductance and capacitance, are more consistent for devices built in the same lot, because there is less variation in the winding manufacture. The nature of the mechanical coil winding process in conventional devices leads to more variation from part to part.

New materials are constantly appearing that further enhance the advantages of discrete planar devices, with particular emphasis on extending the operating frequency range with materials that exhibit low core loss.

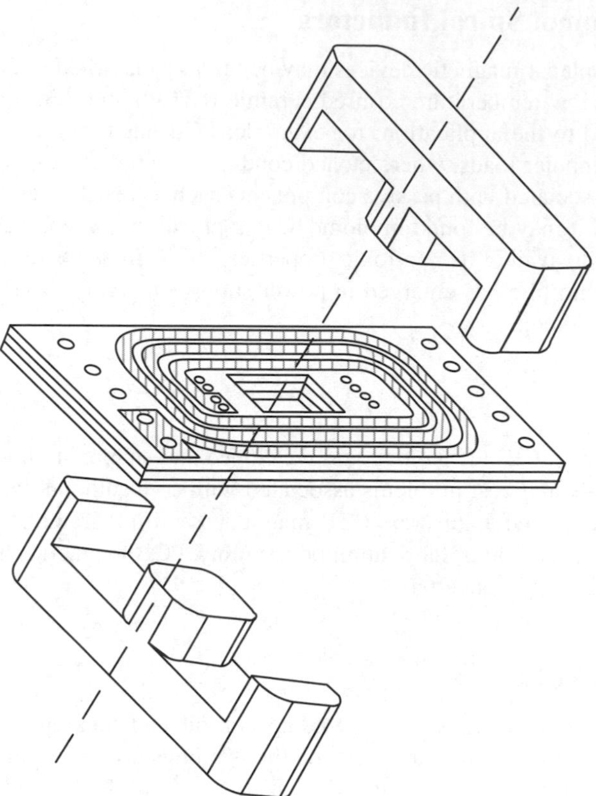

Figure 9.12 A typical planar transformer with an E-I core.

Integrated PCB Magnetics

The automated fabrication of magnetic components, with attendant improvement in reliability, may be achieved by integrating the core and winding so that the magnetic core is formed during the PCB process. Figure 9.13 shows a typical integrated PCB transformer.

The process used to make PCB windings in discrete planar magnetics is also used to integrate the magnetic core into the manufacturing process. The cores are not limited to conventional or off-the-shelf components, and this gives the designer the opportunity to specify the core shape for optimized operation. Due to the low power range addressed to date, the size of the windings is generally smaller in comparison with discrete planar magnetics. PCB track widths down to 150 μm are readily available using standard PCB processing, allowing multi-turn inductors in a small footprint – less than $10 \times 10\,\text{mm}^2$. The electroplating of the core is compatible with the process used for fabricating the PCB winding.

There are two approaches to integrating magnetic material in PCB. One approach involves laminating commercial magnetic foils within multilayer boards. Foils are available in a range of thicknesses and with relative permeability values as high as 100 000. This means that high inductance values may be achieved in a small footprint area. The other approach involves electroplating magnetic materials (e.g. permalloy) onto laminate layers. The process is

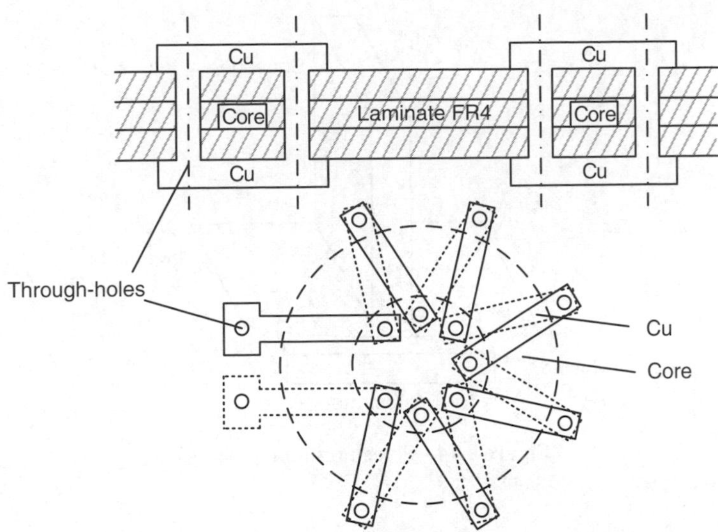

Figure 9.13 PCB integrated magnetic toroidal transformer.

integral to the PCB process, thereby eliminating expensive foils. In addition, the required thickness of core material is controlled, giving greater flexibility in the design.

In many cases, integrated PCB magnetics consist of toroidal structures where the winding is wrapped around the core by the use of etching and conductive vias. However due to the higher resistance of the vias, winding loss is higher. Normally, toroids have a high number of turns, and this adds further to the overall winding loss.

9.2.2 *Thick Film Devices*

A thick film integrated passive RLC circuit offers cost and reliability benefits over those requiring the addition of discrete wire-wound inductor components. Printed conductor through-holes in alumina substrates have been developed that allow double-sided solenoidal-type inductors to be manufactured, with the substrate acting as the interlayer dielectric. Traditionally, thick film materials have been limited in terms of conductivity, line definition and dielectric performance. However, new materials are constantly evolving, with higher conductivity and with track widths as low as ten microns.

Circular spiral inductors may be manufactured on a magnetic substrate with thick film technology [3], thereby introducing a magnetic core. These inductors may consist of multiple layers of several conductor turns, each deposited with thick film conductor paste. These conductor layers are separated by layers of dielectric material. The magnetic substrates are usually ferrite material.

Thick film circuits are produced by a screen printing process, which involves using a mesh screen to produce designs on a suitable substrate. The thickness of film is generally of the order of 10 μm. The mesh screen may be made from stainless steel or from synthetic fibres such as nylon or Dacron. A viscous paste is forced through the screen to deposit a pattern onto the substrate. A typical screen printing set-up is shown in Figure 9.14.

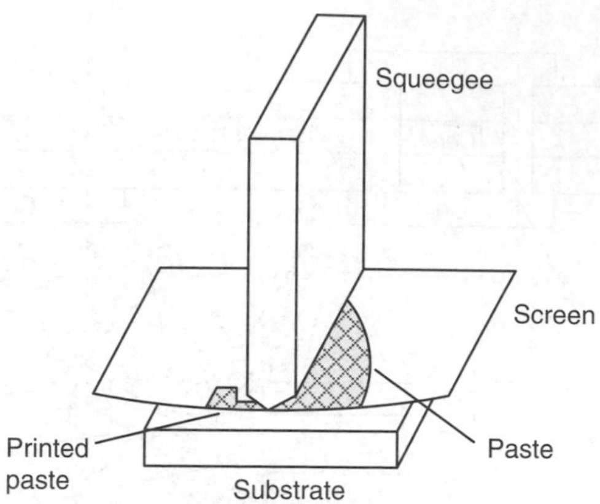

Figure 9.14 Screen printing process.

There are three categories of thick film paste, namely conductive, resistive and insulating pastes.

The fabrication of the spiral inductor involves both conductive paste and insulating paste. Each paste contains:

- a functional material that determines the electrical properties of the paste;
- a solvent/thinner that determines the viscosity of the paste;
- a temporary binder that holds other ingredients together;
- a permanent binder that fuses particles of the functional material together.

There are four main steps in the fabrication process for each layer of printed paste:

- **Screen generation.** Firstly, the desired pattern is laid out using a suitable software layout package. The photoplots for a four-layer inductor are shown in Figure 9.15. The masks for the dielectric layer and a via hole through which the various conductor layers can be interconnected, as shown in Figure 9.16. A photosensitive emulsion is then applied to the entire screen and the mask is placed on the screen and exposed to ultraviolet light. The parts of the emulsion that were not exposed by the mask may be washed away using a spray gun, and a negative image is generated on the screen mesh. This part of the process is illustrated in Figure 9.17.
- **Screen printing.** The required pattern is deposited onto a substrate by forcing a viscous paste through the apertures in the patterned screen.

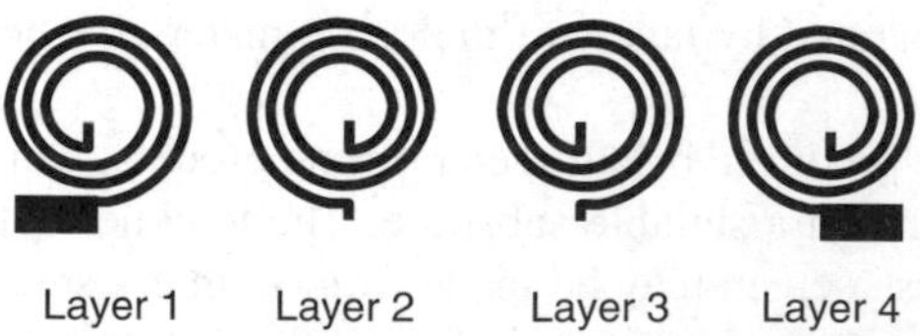

Figure 9.15 Photoplots of conducting layers.

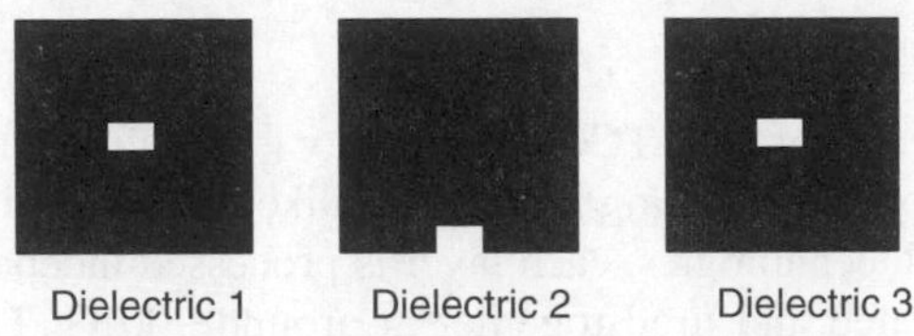

Figure 9.16 Masks for dielectric layers.

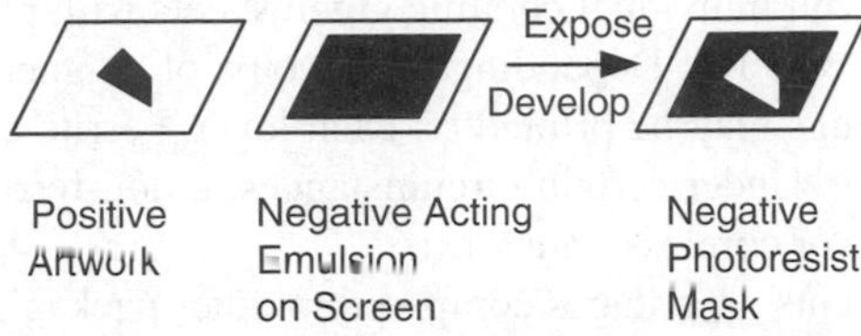

Figure 9.17 Screen generation.

- **Drying/Curing.** The aim of this part of the process is to remove the organic solvents from the screen-printed paste in two steps. At the first step, the substrate with the freshly printed paste is placed in the air for 5–10 minutes in order to settle the paste. At the next step, the substrate is put into an oven at 125°C for 10–20 minutes and the organic solvents are removed by evaporation.
- **Firing.** This is carried out in three steps. First, the temporary organic binder is decomposed by oxidation and removed at temperatures up to 500°C. Next, the permanent binder melts and wets both the surface of the substrate and the particles of the functional material between 500°C and 700°C. In the final step, the functional particles are sintered and become interlocked with the permanent binder and the substrate between 700°C and 850°C.

A microsection of an experimental device is shown in Figure 9.18. Note, the individual conductor layers can be easily recognized. The actual physical dimensions will vary slightly from the ideal rectangular cross-sections because, by its nature, the screen printing process results in tapered walls and the individual layers do not line up exactly in the vertical direction.

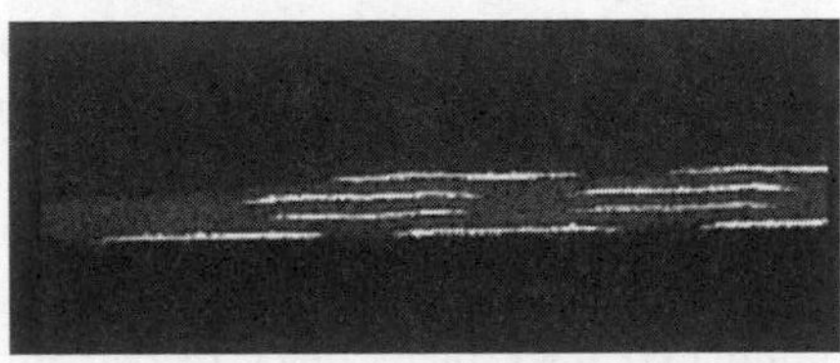

Figure 9.18 Optical photograph of a microsectioned device (scale 30: 1) [3] Reproduced with permission from [3]. Copyright 1999 IEEE.

9.2.3 LTCC Magnetics

Low temperature co-fired ceramic (LTCC) technology is a multilayer technology that has its roots in the microelectronics packaging industry. Unlike the successive build-up of layers in the previously described technologies, each layer is processed independently. The layers are stacked, aligned, laminated and fired together at around 900°C. The constituent ceramic materials must be sintered at the same temperature. LTCC is particularly suitable for high-current switching devices operating in the MHz range.

The key to LTCC is a hybrid multilayer structure that lends itself to integrated magnetics. The layers are prepared with individual ceramic green sheets with punched-through vias and screen-printed conductor patterns. Depending on the type of magnetic components, a partial winding or several windings may be printed on each layer. Ferrite ceramic is typically used to enhance the inductance. Under certain circumstances, a non-ferrite ceramic may be used to form air-core inductors or coreless transformers.

LTCC technology presents advantages compared to other packaging technologies, such as high temperature co-fired ceramic (HTCC) technology: the ceramic is generally fired below 1000°C, due to a special composition of the material. Firing below 1000°C means that highly conductive materials such as silver, copper and gold may be co-fired to form the conductors for low DC resistance and high Q factor.

LTCC technology can incorporate other passive elements, such as resistors and capacitors, minimizing the size of the ceramic package, with better manufacturing yields and lower costs. LTCC technology has been employed in rf and wireless applications that have highly integrated multilayer circuits with resistors, capacitors and inductors, along with active components in a single package.

There are eight main steps in the LTTC process:

- **Tape preparation:** putty-like green sheets are supplied on a roll and are cut by laser or punched. Some tapes are baked for up to 30 minutes at 120°C.
- **Blanking:** individual pieces are cut and orientation marks and tooling holes are formed with a blanking die. Single sheets are rotated to compensate for shrinkage in different directions.
- **Via punching:** vias are formed by punching or laser drilling. The thickness of the tapes determines the via diameter.
- **Via filling:** the vias are filled with thick film screen printer, as described in Section 9.2.2, or with a filler.
- **Printing:** a thick film screen printer is used to print co-fireable conductors onto the green sheets, using standard emulsions. Following printing, the vias and conductors are dried in an oven with a temperature setting between 80°C and 120°C, for up to 30 minutes.
- **Collating and laminating:** each layer is stacked in turn over tooling pins to line them up correctly for lamination. The stacked tapes are pressed together at 20 MPa and 70°C for ten minutes. The stack is rotated through 180° after five minutes. Alternatively, the stacked tapes may be vacuum-sealed in foil. Pressing takes place in hot water for ten minutes at 21MPa.
- **Co-firing:** firing of the laminated stack takes place in two stages. In the first stage, the temperatures increase at a rate of 2–5°C per minute up to 450°C, taking about 1.5–2 hours and causing organic ingredients to burn out. At the next stage, the temperature is increased

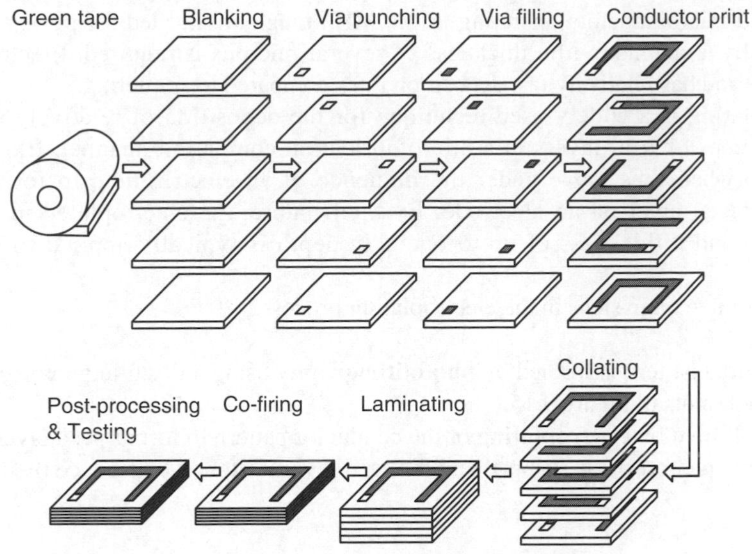

Figure 9.19 LTCC Process flow.

steadily to around 875°C over a period of about 30 minutes and held there for 20 minutes while sintering takes place. The oven is then cooled down over the next hour.

- **Post-firing and singulation:** in post-firing, a paste is applied to the fired parts and fired under conditions dictated by the material used. In singulation, the fired parts are now ready for cutting, trimming with a laser or ultrasonic cutter.

These steps are illustrated in Figure 9.19.

9.2.4 *Thin Film Devices*

Recent trends towards power supply on chip (PwrSoC) technology have placed further demands on the size of the inductor, with typical footprints of 5–10 mm^2. This may be achieved by thin film technology, which involves a microinductor fabricated on silicon using electrochemical deposition techniques. these are sometimes referred to as MEMS (micro-electro-mechanical systems) inductors.

Various approaches to deposit the magnetic core layer have been demonstrated over the last 20 years, such as screen printing, sputtering and electroplating. The screen print technique has been explained earlier in Section 9.2.2.

Sputtering is a widely used microfabrication technique whereby atoms are dislodged from the target material and deposited on a surface at relatively low temperatures; the resulting layers are usually several microns in thickness. Low temperatures rule out ferrites, because they require sintering, which takes place well above the temperatures involved in sputtering. Sputtering has found application in microinductors operating in the GHz frequency range. The main disadvantage of sputtering lies in the fact that it is quite expensive to deposit relatively thick layers up to several microns.

Integrated microinductors operating in the MHz range are needed for power conversion and, typically, a magnetic film thickness of several microns is required. Electroplating is inexpensive and has relatively fast deposition rates compared to sputtering.

Electroplating is a widely used technique for the deposition of relatively thick (several μm to tens of μm) layers in a microfabrication compatible manner. Electroplating takes place when ions move under the influence of an electric field to form a coated conductive seed layer on an electrode. By their nature, the electroplated materials are conductive, and a thickness of up to one skin depth is typically applied to limit eddy current loss.

There are three main steps in the electroplating process:

1. The required pattern is formed by **photolithography** using a dry film photoresist, similar to the photoplots in Figure 9.15.
2. This is followed by **electroplating** of the conductor pattern to form a seed layer.
3. Finally, the photoresist is stripped and the seed layer is **etched** to produce the final circuit pattern.

Inductors formed in this way are limited to the nH range because the thickness of deposited thin film has to be limited to control the eddy current loss. However, this limitation can be overcome by implementing core lamination techniques. Eddy current loss in the core can be substantially reduced by laminating the core in successive layers of magnetic material separated by dielectric.

The top view and cross-section of a microinductor using laminated magnetic core are shown in Figure 9.20 [4,5]. The two-layer magnetic core is effectively laminated to reduce eddy current loss at high frequencies, allowing the device to operate in the tens of MHz range.

As we are dealing with a current-carrying component, low resistance is therefore important, and this can only be achieved by thick conductors. From a performance point of view, it is desirable to increase the conductor thickness to reduce winding loss, provided that the conductor thickness does not exceed the skin depth. In this case, the conductors are deposited by electroplating, which limits the achievable thickness.

There are five layers in a typical microinductor:

1. **Electroplated magnetic layer:** permalloy (Ni Fe) is electroplated to form the bottom magnetic core layer of the inductor.
2. **Dielectric layer:** a dielectric material is used to insulate the bottom core from the conductors.
3. **Electroplated copper layer:** the windings are fabricated, using the electroplating of copper through thick photoresist patterns.
4. **Insulation layer:** thick photoresist is spun on the conductors to provide an insulation layer between the conductors and the top core layer.
5. **Electroplated magnetic layer:** finally, the top layer of permalloy is electroplated and patterned to close the magnetic core by connecting the top core to the bottom core.

These are illustrated in Figure 9.21.

The process flow is shown in Figure 9.17.

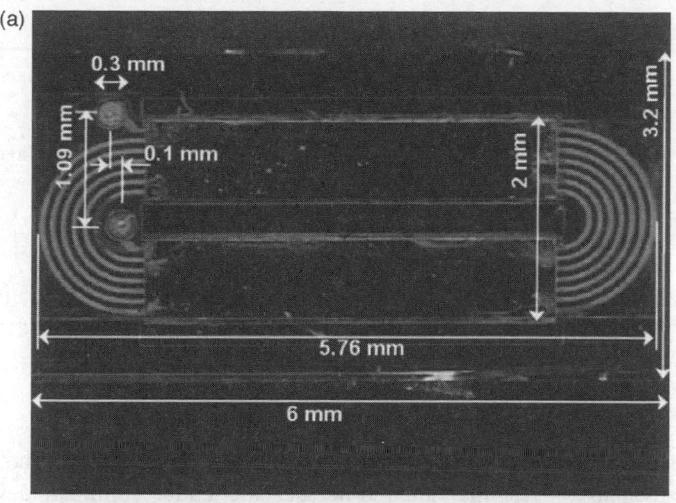

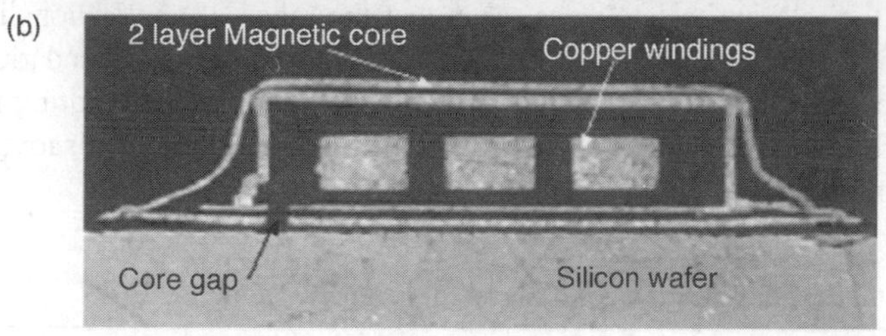

Figure 9.20 Silicon integrated microinductor: (a) Top view. Reproduced with permission from [4]. Copyright 2008 IEEE, (b) Cross-section. Reproduced with permission from [5]. Copyright 2005 IEEE.

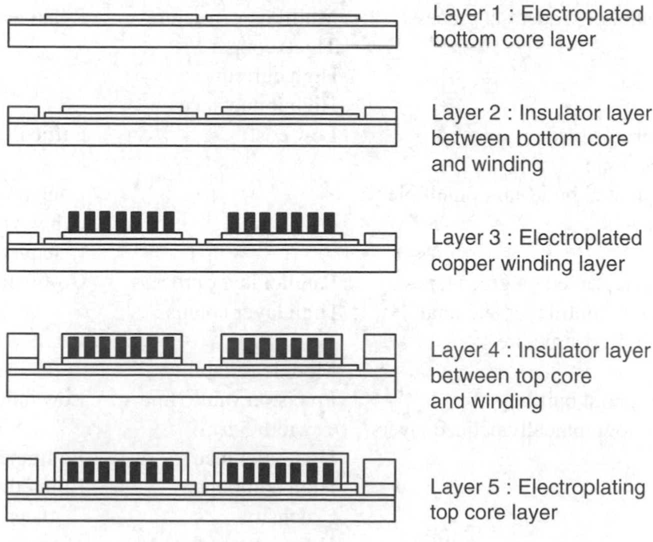

Figure 9.21 Microfabrication process flow for an inductor.

Table 9.1 A comparison on performance of different magnetic technologies

Technology	Frequency (typical)	Power (typical)	Inductance (typical)	Size (typical)
PCB magnetics	20 KHz ~ 2 MHz	1 W ~ 5 kW	10 μH ~ 10 mH	100 mm^2 ~ 100s cm^2
Thick film	< 10 MHz	< 10 W	1 μH ~ 1 mH	< 1 cm^2
LTCC	200 KHz ~ 10 MHz	< 10 W	1 μH ~ 1 mH	< 1 cm^2
Thin film	> 10 MHz	< 1 W	10s ~ 100s nH	< 10 mm^2

9.2.5 Summary

Each of the four technologies described above has its own characteristics and its own advantages and disadvantages. The individual processes are under continual development, and improvements in the relevant technologies mean that the range of applications is increasing in each case. The ongoing research into new materials in pushing out the boundaries of operation.

The achievable performance (inductance and power level), target operating frequency, and size for each technology are listed in Table 9.1. The advantage and disadvantage of each technology is summarized in Table 9.2.

Table 9.2 Advantages and disadvantages of different magnetic technologies

Technology	Integration method	Advantages	Disadvantage
PCB	Discrete core on laminated structure or integrated core in laminated structure	Low cost	Low resolution (line width 100 μm)
	Parallel or sequential process	Multilayer structure Thick copper High current High inductance	Relatively low frequency
Thick film	Screen printed on sintered ceramic	Low cost	Difficult to form
	Sequential build-up of multiple layers		Long process time and low yield due to sequential build-up
LTCC	Screen printed on green tapes	Parallel layer process	Co-fireability of materials
	Parallel multilayer and final co-fired structure	High layer counts	
		Module reliability	
Thin film	Sequential build-up of lithographically defined layers	Precision value (line width 5 μm)	Low inductance
		High tolerance	Equipment costly
		High component density	Limited selection on film materials(material compatibility
		High frequency	

9.3 Problems

9.1 Repeat Example 9.2 for the four-layer device in Figure 9.3.
9.2 Repeat Example 9.3 for the four-layer device in Figure 9.3.
9.3 Repeat Example 9.4 for the four-layer device in Figure 9.3.
9.4 What are the main advantages of planar magnetic as compared to conventional wire-wound magnetic?
9.5 List the main planar magnetic technologies.
9.6 What are the disadvantages and advantages of each planar magnetic technology?

MATLAB Program for Example 9.1

```
%This MATLAB program is used to calculate the self and mutual
inductances
%in Example 9.1
%The parameters are shown in Figure 9.3

rin =[1.15e-3 1.15e-3 2e-3 2e-3];
rout =[1.75e-3 1.75e-3 2.6e-3 2.6e-3];
height =[15e-6 15e-6 15e-6 15e-6];
d=[7.5e-6 62.5e-6 62.5e-6 7.5e-6];

Inductance=ones(4,4);

for i=1:4
    for j=1:4
Inductance(i,j)=air_mutual(rin(i),rout(i),rin(j),rout(j),height
(i),height(j),d(i),d(j));

    end
end

%Export the inductance matrix
Inductance

%Get the total inductance
L_total = sum(sum(Inductance))

%File to define the function air_mutual
function y = air_mutual(r1,r2,a1,a2,h1,h2,d1,d2)

%This function is used to calculate the mutual inductance in air core
%condition.
%r1,r2,h1,d1 are the inside radius, outside radius, height and upright
%position of the cross-section 1.
```

```
%a1,a2,h2,d2 are the inside radius, outside radius, height and upright
%position of the cross-section 2.
%z is the axis separation. z=0 for self-inductance calculation; z=
|d2-d1|
%for mutual inductance calculation.
global uo;
uo=4*pi*1e-7;
g=@(k)aircoremul(r1,r2,a1,a2,h1,h2,d1,d2,k);

for upper=1000:1000:1000000
[integalresult,err] = quadgk(g,0,upper);
if err<0.01*integalresult
   integalresult_real=integalresult;
else
   break;
end
end

y=uo.*pi.*integalresult_real./(h1*log(a2/a1)*h2*log(r2/r1));

end

function y=aircoremul(r1,r2,a1,a2,h1,h2,d1,d2,k)

z=d2-d1;
if z==0
   Q=2.*(h1.*k+exp(-h1.*k)-1)./(k.^2);
else
   Q=2.*(cosh(0.5.*(h1+h2).*k)-cosh(0.5.*(h1-h2).*k))./(k.^2);
end

S1=(besselj(0,r2.*k)-besselj(0,r1.*k))./k;
S2=(besselj(0,a2.*k)-besselj(0,a1.*k))./k;

if k(1)==0
   Q(1)=0;
   S1(1)=0;
   S2(1)=0;
end;

y=S1.*S2.*Q.*exp(-z.*k);
end
```

MATLAB Program for Example 9.2(a)

```
%This MATLAB program is used to calculate the self inductances and plot
the
%results in Example 9.2(a)
%The parameters are shown in Figure 9.6

rin=[1.15e-3 2e-3];
rout=[1.75e-3 2.6e-3];
height=[15e-6 15e-6];
d=[7.5e-6 7.5e-6];

L1_air=air_mutual(rin(1),rout(1),rin(1),rout(1),height(1),height(1),
d(1),d(1));
L2_air=air_mutual(rin(2),rout(2),rin(2),rout(2),height(2),height(2),
d(2),d(2));
L12_air=air_mutual(rin(1),rout(1),rin(2),rout(2),height(1),height
(2),d(1),d(2));
L1=L1_air+L2_air+2*L12_air;

t1=0.05e-3;
t2=0.1e-3;
t3=0.5e-3;

L_t1=L1.*ones(1,1000);
L_t2=L1.*ones(1,1000);
L_t3=L1.*ones(1,1000);

i=2;
for ur=2:1000;

L1_t1=L1_air+inductance_substrate(rin(1),rout(1),rin(1),rout(1),
height(1),height(1),d(1),d(1),t1,ur);

L2_t1=L2_air+inductance_substrate(rin(2),rout(2),rin(2),rout(2),
height(2),height(2),d(2),d(2),t1,ur);

L12_t1=L12_air+inductance_substrate(rin(1),rout(1),rin(2),rout(2),
height(1),height(2),d(1),d(2),t1,ur);
   L_t1(i)=L1_t1+L2_t1+2*L12_t1;

L1_t2=L1_air+inductance_substrate(rin(1),rout(1),rin(1),rout(1),
height(1),height(1),d(1),d(1),t2,ur);

L2_t2=L2_air+inductance_substrate(rin(2),rout(2),rin(2),rout(2),
height(2),height(2),d(2),d(2),t2,ur);
```

```
L12_t2=L12_air+inductance_substrate(rin(1),rout(1),rin(2),rout(2),
height(1),height(2),d(1),d(2),t2,ur);
    L_t2(i)=L1_t2+L2_t2+2*L12_t2;

L1_t3=L1_air+inductance_substrate(rin(1),rout(1),rin(1),rout(1),
height(1),height(1),d(1),d(1),t3,ur);

L2_t3=L2_air+inductance_substrate(rin(2),rout(2),rin(2),rout(2),
height(2),height(2),d(2),d(2),t3,ur);

L12_t3=L12_air+inductance_substrate(rin(1),rout(1),rin(2),rout(2),
height(1),height(2),d(1),d(2),t3,ur);
    L_t3(i)=L1_t3+L2_t3+2*L12_t3;

    i=i+1;
end

x=1:1:1000;
y1=L_t1./L1;
y2=L_t2./L1;
y3=L_t3./L1;

semilogx(x,y1,'-',x,y2,'-',x,y3,'-')
axis([1,1000,1,2])

%File to define the function air_mutual
function y=air_mutual(r1,r2,a1,a2,h1,h2,d1,d2)

%This function is used to calculate the mutual inductance in air core
%condition.
%r1,r2,h1,d1 are the inside radius, outside radius, height and upright
%position of the cross-section 1.
%a1,a2,h2,d2 are the inside radius, outside radius, height and upright
%position of the cross-section 2.
%z is the axis separation. z=0 for self-inductance calculation; z=|d2-d1|
%for mutual inductance calculation.
global uo;
uo=4*pi*1e-7;
g=@(k)aircoremul(r1,r2,a1,a2,h1,h2,d1,d2,k);

for upper=1000:1000:1000000
[integalresult,err]=quadgk(g,0,upper);
if err<0.01*integalresult
    integalresult_real=integalresult;
else
    break;
```

```
end
end

y=uo.*pi.*integalresult_real./(h1*log(a2/a1)*h2*log(r2/r1));

end

function y=aircoremul(r1,r2,a1,a2,h1,h2,d1,d2,k)

z=d2-d1;
if z==0
    Q=2.*(h1.*k+exp(-h1.*k)-1)./(k.^2);
else
    Q=2.*(cosh(0.5.*(h1+h2).*k)-cosh(0.5.*(h1-h2).*k))./(k.^2);
end
S1=(besselj(0,r2.*k)-besselj(0,r1.*k))./k;
S2=(besselj(0,a2.*k)-besselj(0,a1.*k))./k;

if k(1)==0
    Q(1)=0;
    S1(1)=0;
    S2(1)=0;
end;

y=S1.*S2.*Q.*exp(-z.*k);
end

%File to define the function inductance_substrate

function y=inductance_substrate(r1,r2,a1,a2,h1,h2,d1,d2,t,ur)

%This function is used to calculate the additional mutual impedance to the
%presence of the substrate.
%r1,r2,h1,d1 are the insidc radius, outside radius, height and upright
%position of the cross-section 1.
%a1,a2,h2,d2 are the inside radius, outside radius, height and upright
%position of the cross-section 2.
%t is thickness of the substrate
%ur relative permeability of the magnetic substrate

global uo;
uo=4*pi*1e-7;

g=@(k)integrand(r1,r2,a1,a2,h1,h2,d1,d2,t,ur,k);

for upper=1000:1000:1000000
```

```
[integalresult,err]=quadgk(g,0,upper);
if err<0.01*integalresult
    integalresult_real=integalresult;
else
    break;
end
end

y=uo.*pi.*integalresult_real./(h1*log(a2/a1)*h2*log(r2/r1));

end

function y=integrand(r1,r2,a1,a2,h1,h2,d1,d2,t,ur,k)
global uo;

z=d2-d1;
if z==0
    Q=2.*(h1.*k+exp(-h1.*k)-1)./(k.^2);
else
    Q=2.*(cosh(0.5.*(h1+h2).*k)-cosh(0.5.*(h1-h2).*k))./(k.^2);
end

S1=(besselj(0,r2.*k)-besselj(0,r1.*k))./k;
S2=(besselj(0,a2.*k)-besselj(0,a1.*k))./k;

if k(1)==0
    Q(1)=0;
    S1(1)=0;
    S2(1)=0;
end;

phi=(ur-1)./(ur+1);
lambda=phi.*(1-exp(-2*t.*k))./(1-(phi.^2).*exp(-2*t.*k));
y=S1.*S2.*Q.*lambda.*exp(-(d1+d2).*k);
end
```

MATLAB Program for Example 9.2(b)

```
%This MATLAB program is used to calculate the self inductances and plot
the
%results in Example 9.2(b)
%The parameters are shown in Figure 9.6
rin=[1.15e-3 2e-3];
rout=[1.75e-3 2.6e-3];
height=[15e-6 15e-6];
d=[7.5e-6 7.5e-6];
```

```
L1_air=air_mutual(rin(1),rout(1),rin(1),rout(1),height(1),height
(1),d(1),d(1));
L2_air=air_mutual(rin(2),rout(2),rin(2),rout(2),height(2),height
(2),d(2),d(2));
L12_air=air_mutual(rin(1),rout(1),rin(2),rout(2),height(1),height
(2),d(1),d(2));
L1=L1_air+L2_air+2*L12_air;

ur1=10;
ur2=100;
ur3=1000;

L_ur1=L1.*ones(1,1000);
L_ur2=L1.*ones(1,1000);
L_ur3=L1.*ones(1,1000);

i=1;
for t=0.001e-3:0.001e-3:1e-3;

L1_ur1=L1_air+inductance_substrate(rin(1),rout(1),rin(1),rout(1),
height(1),height(1),d(1),d(1),t,ur1);

L2_ur1=L2_air+inductance_substrate(rin(2),rout(2),rin(2),rout(2),
height(2),height(2),d(2),d(2),t,ur1);

L12_ur1=L12_air+inductance_substrate(rin(1),rout(1),rin(2),rout
(2),height(1),height(2),d(1),d(2),t,ur1);
   L_ur1(i)=L1_ur1+L2_ur1+2*L12_ur1;

L1_ur2=L1_air+inductance_substrate(rin(1),rout(1),rin(1),rout(1),
height(1),height(1),d(1),d(1),t,ur2);

L2_ur2=L2_air+inductance_substrate(rin(2),rout(2),rin(2),rout(2),
height(2),height(2),d(2),d(2),t,ur2);

L12_ur2=L12_air+inductance_substrate(rin(1),rout(1),rin(2),rout
(2),height(1),height(2),d(1),d(2),t,ur2);
   L_ur2(i)=L1_ur2+L2_ur2+2*L12_ur2;

L1_ur3=L1_air+inductance_substrate(rin(1),rout(1),rin(1),rout(1),
height(1),height(1),d(1),d(1),t,ur3);

L2_ur3=L2_air+inductance_substrate(rin(2),rout(2),rin(2),rout(2),
height(2),height(2),d(2),d(2),t,ur3);

L12_ur3=L12_air+inductance_substrate(rin(1),rout(1),rin(2),rout
(2),height(1),height(2),d(1),d(2),t,ur3);
```

```
L_ur3(i)=L1_ur3+L2_ur3+2*L12_ur3;
    i=i+1;
end

x=0.001:0.001:1;
y1=L_ur1./L1;
y2=L_ur2./L1;
y3=L_ur3./L1;

semilogx(x,y1,'-',x,y2,'-',x,y3,'-')
axis([0.001,1,1,2])

%File to define the function air_mutual
function y=air_mutual(r1,r2,a1,a2,h1,h2,d1,d2)

%This function is used to calculate the mutual inductance in air core
%condition.
%r1,r2,h1,d1 are the inside radius, outside radius, height and upright
%position of the cross-section 1.
%a1,a2,h2,d2 are the inside radius, outside radius, height and upright
%position of the cross-section 2.
%z is the axis separation. z=0 for self-inductance calculation; z=|d2-d1|
%for mutual inductance calculation.
global uo;
uo=4*pi*1e-7;
g=@(k)aircoremul(r1,r2,a1,a2,h1,h2,d1,d2,k);

for upper=1000:1000:1000000
[integalresult,err] = quadgk(g,0,upper);
if err<0.01*integalresult
    integalresult_real=integalresult;
else
    break;
end
end

y=uo.*pi.*integalresult_real./(h1*log(a2/a1)*h2*log(r2/r1));

end
function y=aircoremul(r1,r2,a1,a2,h1,h2,d1,d2,k)

z=d2-d1;
if z==0
    Q=2.*(h1.*k+exp(-h1.*k)-1)./(k.^2);
else
    Q=2.*(cosh(0.5.*(h1+h2).*k)-cosh(0.5.*(h1-h2).*k))./(k.^2);
```

```
end

S1=(besselj(0,r2.*k)-besselj(0,r1.*k))./k;
S2=(besselj(0,a2.*k)-besselj(0,a1.*k))./k;

if k(1)==0
   Q(1)=0;
   S1(1)=0;
   S2(1)=0;
end;

y=S1.*S2.*Q.*exp(-z.*k);
end

%File to define the function inductance_substrate
function y=inductance_substrate(r1,r2,a1,a2,h1,h2,d1,d2,t,ur)

%This function is used to calculate the additional mutual impedance to the
%presence of the substrate.
%r1,r2,h1,d1 are the inside radius, outside radius, height and upright
%position of the cross-section 1.
%a1,a2,h2,d2 are the inside radius, outside radius, height and upright
%position of the cross-section 2.
%t is thickness of the substrate
%ur relative permeability of the magnetic substrate

global uo;
uo=4*pi*1e-7;

g=@(k)integrand(r1,r2,a1,a2,h1,h2,d1,d2,t,ur,k);

for upper=1000:1000:1000000
[integalresult,err] = quadgk(g,0,upper);
if err<0.01*integalresult
   integalresult_real=integalresult;
else
   break;
end
end

y=uo.*pi.*integalresult_real./(h1*log(a2/a1)*h2*log(r2/r1));

end

function y=integrand(r1,r2,a1,a2,h1,h2,d1,d2,t,ur,k)
global uo;
```

```
z = d2-d1;
if z == 0
    Q = 2.*(h1.*k+exp(-h1.*k)-1)./(k.^2);
else
    Q = 2.*(cosh(0.5.*(h1+h2).*k)-cosh(0.5.*(h1-h2).*k))./(k.^2);
end

S1=(besselj(0,r2.*k)-besselj(0,r1.*k))./k;
S2=(besselj(0,a2.*k)-besselj(0,a1.*k))./k;

if k(1)==0
    Q(1)=0;
    S1(1)=0;
    S2(1)=0;
end;

phi=(ur-1)./(ur+1);
lambda=phi.*(1-exp(-2*t.*k))./(1-(phi.^2).*exp(-2*t.*k));
y=S1.*S2.*Q.*lambda.*exp(-(d1+d2).*k);
end
```

MATLAB Program for Example 9.3

```
%This MATLAB program is used to calculate the self impedance and plot the
%results in Example 9.3
%The parameters are shown in Figure 9.6

rin=[1.15e-3 2e-3];
rout=[1.75e-3 2.6e-3];
height=[15e-6 15e-6];
d=[7.5e-6 7.5e-6];

L1_air=air_mutual(rin(1),rout(1),rin(1),rout(1),height(1),height
(1),d(1),d(1));
L2_air=air_mutual(rin(2),rout(2),rin(2),rout(2),height(2),height
(2),d(2),d(2));
L12_air=air_mutual(rin(1),rout(1),rin(2),rout(2),height(1),height
(2),d(1),d(2));
L1=L1_air+L2_air+2*L12_air;
ur=1000;
sigma=10;
Rdc=0.093;

t1=0.05e-3;
```

```
t2=0.1e-3;
t3=0.5e-3;

L_t1 = L1.*ones(1,1000);
L_t2 = L1.*ones(1,1000);
L_t3 = L1.*ones(1,1000);
Rac_t1 = Rdc.*ones(1,1000);
Rac_t2 = Rdc.*ones(1,1000);
Rac_t3 = Rdc.*ones(1,1000);

i = 1;
for frequency=1e6:1e6:1000e6;

L1_t1=impedance_substrate(rin(1),rout(1),rin(1),rout(1),height(1),
height(1),d(1),d(1),t1,ur,sigma,frequency);
    R1_t1=real(L1_t1);
    L1_t1=(imag(L1_t1))/(2*pi*frequency)+L1_air;

L2_t1=impedance_substrate(rin(2),rout(2),rin(2),rout(2),height(2),
height(2),d(2),d(2),t1,ur,sigma,frequency);
    R2_t1=real(L2_t1);
    L2_t1=(imag(L2_t1))/(2*pi*frequency)+L2_air;

L12_t1=impedance_substrate(rin(1),rout(1),rin(2),rout(2),height
(1),height(2),d(1),d(2),t1,ur,sigma,frequency);
    R12_t1=real(L12_t1);
    L12_t1=(imag(L12_t1))/(2*pi*frequency)+L12_air;
    L_t1(i)=L1_t1+L2_t1+2*L12_t1;
    Rac_t1(i)=R1_t1+R2_t1+2*R12_t1;

L1_t2=impedance_substrate(rin(1),rout(1),rin(1),rout(1),height(1),
height(1),d(1),d(1),t2,ur,sigma,frequency);
    R1_t2=real(L1_t2);
    L1_t2=(imag(L1_t2))/(2*pi*frequency)+L1_air;

L2_t2=impedance_substrate(rin(2),rout(2),rin(2),rout(2),height(2),
height(2),d(2),d(2),t2,ur,sigma,frequency);
    R2_t2=real(L2_t2);
    L2_t2=(imag(L2_t2))/(2*pi*frequency)+L2_air;

L12_t2=impedance_substrate(rin(1),rout(1),rin(2),rout(2),height
(1),height(2),d(1),d(2),t2,ur,sigma,frequency);
    R12_t2=real(L12_t2);
    L12_t2=(imag(L12_t2))/(2*pi*frequency)+L12_air;
    L_t2(i)=L1_t2+L2_t2+2*L12_t2;
    Rac_t2(i)=R1_t2+R2_t2+2*R12_t2;
```

```
L1_t3=impedance_substrate(rin(1),rout(1),rin(1),rout(1),height(1),
height(1),d(1),d(1),t3,ur,sigma,frequency);
    R1_t3=real(L1_t3);
    L1_t3=(imag(L1_t3))/(2*pi*frequency)+L1_air;

L2_t3=impedance_substrate(rin(2),rout(2),rin(2),rout(2),height(2),
height(2),d(2),d(2),t3,ur,sigma,frequency);
    R2_t3=real(L2_t3);
    L2_t3=(imag(L2_t3))/(2*pi*frequency)+L2_air;

L12_t3=impedance_substrate(rin(1),rout(1),rin(2),rout(2),height
(1),height(2),d(1),d(2),t3,ur,sigma,frequency);
    R12_t3=real(L12_t3);
    L12_t3=(imag(L12_t3))/(2*pi*frequency)+L12_air;
    L_t3(i)=L1_t3+L2_t3+2*L12_t3;
    Rac_t3(i)=R1_t3+R2_t3+2*R12_t3;

    i=i+1;
end

x = 1e6:1e6:1000e6;
y1= L_t1./L1;
y2 = L_t2./L1;
y3 = L_t3./L1;
z1= Rac_t1./Rdc+1;
z2 = Rac_t2./Rdc+1;
z3 = Rac_t3./Rdc+1;

semilogx(x,y1,'-',x,y2,'-',x,y3,'-')
axis([1e6,1000e6,1.9,2])

figure;
semilogx(x,z1,'-',x,z2,'-',x,z3,'-')
axis([1e6,1000e6,1,19])

%File to define the function air_mutual
function y=air_mutual(r1,r2,a1,a2,h1,h2,d1,d2)

%This function is used to calculate the mutual inductance in air core
%condition.
%r1,r2,h1,d1 are the inside radius, outside radius, height and upright
%position of the cross-section 1.
%a1,a2,h2,d2 are the inside radius, outside radius, height and upright
%position of the cross-section 2.
%z is the axis separation. z=0 for self-inductance calculation; z=|d2-
d1|
```

```
%for mutual inductance calculation.
global uo;
uo = 4*pi*1e-7;
g = @(k)aircoremul(r1,r2,a1,a2,h1,h2,d1,d2,k);

for upper=1000:1000:1000000
[integalresult,err] = quadgk(g,0,upper);
if err<0.01*integalresult
integalresult_real=integalresult;
else
   break;
end
end

y = uo.*pi.*integalresult_real./(h1*log(a2/a1)*h2*log(r2/r1));

end

function y = aircoremul(r1,r2,a1,a2,h1,h2,d1,d2,k)

z = d2-d1;
if z == 0
   Q = 2.*(h1.*k+exp(-h1.*k)-1)./(k.^2);
else
   Q=2.*(cosh(0.5.*(h1+h2).*k)-cosh(0.5.*(h1-h2).*k))./(k.^2);
end

S1 = (besselj(0,r2.*k)-besselj(0,r1.*k))./k;
S2 = (besselj(0,a2.*k)-besselj(0,a1.*k))./k;

if k(1)==0
   Q(1)=0;
   S1(1)=0;
   S2(1)=0;
end;

y = S1.*S2.*Q.*exp(-z.*k);
end

%File to define the function impedance_substrate
function y=impedance_substrate(r1,r2,a1,a2,h1,h2,d1,d2,t,ur,sigma,
freq)

%This function is used to calculate the additional mutual impedance to
the
%presence of the substrate in sandwich structures.
```

```
%r1,r2,h1,d1 are the inside radius, outside radius, height and upright
%position of the cross-section 1.
%a1,a2,h2,d2 are the inside radius, outside radius, height and upright
%position of the cross-section 2.
%t is the thickness of the substrate
%ur relative permeability of the magnetic substrate
%sigma relative permeability of the magnetic substrate
%freq operation frequency

global uo;
uo = 4*pi*1e-7;
omega = 2*pi*freq;

g = @(k)integrand(r1,r2,a1,a2,h1,h2,d1,d2,t,ur,sigma,freq,k);

for upper=1000:1000:1000000
[integalresult,err] = quadgk(g,0,upper);
if err<0.01*integalresult
    integalresult_real=integalresult;
else
    break;
end
end

y=1j.*omega.*uo.*pi.*integalresult_real./(h1*log(a2/a1)*h2*log(r2/
r1));

end

function y=integrand(r1,r2,a1,a2,h1,h2,d1,d2,t,ur,sigma,freq,k)
global uo;

z=d2-d1;
if z==0
    Q=2.*(h1.*k+exp(-h1.*k)-1)./(k.^2);
else
    Q=2.*(cosh(0.5.*(h1+h2).*k)-cosh(0.5.*(h1-h2).*k))./(k.^2);
end

S1=(besselj(0,r2.*k)-besselj(0,r1.*k))./k;
S2=(besselj(0,a2.*k)-besselj(0,a1.*k))./k;

if k(1)==0
    Q(1)=0;
    S1(1)=0;
    S2(1)=0;
```

```
end;

eta=sqrt((1j*2*pi.*freq*sigma*ur*4*pi*1e-7)+(k.^2));

phi=(ur.*k-eta)./(ur.*k+eta);
lambda=phi.*(1-exp(-2*t.*eta))./(1-(phi.^2).*exp(-2*t.*eta));
y=S1.*S2.*Q.*lambda.*exp(-(d1+d2).*k);
end
```

MATLAB Program for Example 9.4(a)

```
%This MATLAB program is used to calculate the self impedance and plot the
%results in Example 9.4(a)
%The parameters are shown in Example 9.2 sandwiched between two ferrite
%substrates

rin=[1.15e-3 2e-3];
rout=[1.75e-3 2.6e-3];
height=[15e-6 15e-6];
d=[7.5e-6 7.5e-6];

L1_air=air_mutual(rin(1),rout(1),rin(1),rout(1),height(1),height
(1),d(1),d(1));
L2_air=air_mutual(rin(2),rout(2),rin(2),rout(2),height(2),height
(2),d(2),d(2));
L12_air=air_mutual(rin(1),rout(1),rin(2),rout(2),height(1),height
(2),d(1),d(2));
L1=L1_air+L2_air+2*L12_air;
ur=1000;
sigma=10;
t1=0.5e-3;
t2=0.5e-3;
frequency=1e6;

L_total=L1.*ones(1,1000);

i=1;
for s=0.01e-3:0.01e-3:10e-3;

L1=impedance_sandwich(rin(1),rout(1),rin(1),rout(1),height(1),
height(1),d(1),d(1),t1,t2,s,ur,sigma,frequency);
    L1=(imag(L1))/(2*pi*frequency)+L1_air;

L2=impedance_sandwich(rin(2),rout(2),rin(2),rout(2),height(2),
height(2),d(2),d(2),t1,t2,s,ur,sigma,frequency);
```

```
    L2=(imag(L2))/(2*pi*frequency)+L2_air;

L12=impedance_sandwich(rin(1),rout(1),rin(2),rout(2),height(1),
height(2),d(1),d(2),t1,t2,s,ur,sigma,frequency);
    L12=(imag(L12))/(2*pi*frequency)+L12_air;
    L_total(i)=L1+L2+2*L12;

i=i+1;
end

x=0.01e-3:0.01e-3:10e-3;
y=L_total./L1;

semilogx(x,y,'-')
axis([0.01e-3,10e-3,0,100])

%File to define the function air_mutual
function y=air_mutual(r1,r2,a1,a2,h1,h2,d1,d2)

%This function is used to calculate the mutual inductance in air core
%condition.
%r1,r2,h1,d1 are the inside radius, outside radius, height and upright
%position of the cross-section 1.
%a1,a2,h2,d2 are the inside radius, outside radius, height and upright
%position of the cross-section 2.
%z is the axis separation. z=0 for self-inductance calculation; z=|d2-
d1|
%for mutual inductance calculation.
global uo;
uo=4*pi*1e-7;
g=@(k)aircoremul(r1,r2,a1,a2,h1,h2,d1,d2,k);

for upper=1000:1000:1000000
[integalresult,err] = quadgk(g,0,upper);
if err<0.01*integalresult
integalresult_real=integalresult;
else
    break;
end
end

y=uo.*pi.*integalresult_real./(h1*log(a2/a1)*h2*log(r2/r1));

end
```

```
function y=aircoremul(r1,r2,a1,a2,h1,h2,d1,d2,k)

z=d2-d1;
if z==0
   Q=2.*(h1.*k+exp(-h1.*k)-1)./(k.^2);
else
   Q=2.*(cosh(0.5.*(h1+h2).*k)-cosh(0.5.*(h1-h2).*k))./(k.^2);
end

S1=(besselj(0,r2.*k)-besselj(0,r1.*k))./k;
S2=(besselj(0,a2.*k)-besselj(0,a1.*k))./k;

if k(1)==0
   Q(1)=0;
   S1(1)=0;
   S2(1)=0;
end;

y=S1.*S2.*Q.*exp(-z.*k);
end

%File to define the function impedance_sandwich
function y=impedance_sandwich(r1,r2,a1,a2,h1,h2,d1,d2,t1,t2,s,ur,
sigma,freq)

%This function is used to calculate the additional mutual impedance to
the
%presence of the substrate in sandwich structures.
%r1,r2,h1,d1 are the inside radius, outside radius, height and upright
%position of the cross-section 1.
%a1,a2,h2,d2 are the inside radius, outside radius, height and upright
%position of the cross-section 2.
%t1 and t2 are thickness of bottom and upper substrates
%s substrate separation
%ur relative permeability of the magnetic substrate
%sigma relative permeability of the magnetic substrate
%freq operation frequency

global uo;
uo=4*pi*1e-7;
omega=2*pi*freq;

g=@(k)integrand(r1,r2,a1,a2,h1,h2,d1,d2,t1,t2,s,ur,sigma,freq,k);

for upper=1000:1000:1000000
[integalresult,err] = quadgk(g,0,upper);
```

```
if err<0.01*integalresult
    integalresult_real=integalresult;
else
    break;
end
end

y=1j.*omega.*uo.*pi.*integalresult_real./(h1*log(a2/a1)*h2*log(r2/
r1));

end

function y=integrand(r1,r2,a1,a2,h1,h2,d1,d2,t1,t2,s,ur,sigma,
freq,k)
global uo;
z=d2-d1;
d1_s=s-d1;
d2_s=s-d2;
if z==0
    Q=2.*(h1.*k+exp(-h1.*k)-1)./(k.^2);
else
    Q=2.*(cosh(0.5.*(h1+h2).*k)-cosh(0.5.*(h1-h2).*k))./(k.^2);
end

S1=(besselj(0,r2.*k)-besselj(0,r1.*k))./k;
S2=(besselj(0,a2.*k)-besselj(0,a1.*k))./k;

if k(1)==0
    Q(1)=0;
    S1(1)=0;
    S2(1)=0;
end;

eta=sqrt((1j*2*pi.*freq*sigma*ur*4*pi*1e-7)+(k.^2));

phi=(ur.*k-eta)./(ur.*k+eta);
lambda1=phi.*(1-exp(-2*t1.*eta))./(1-(phi.^2).*exp(-2*t1.*eta));
lambda2=phi.*(1-exp(-2*t2.*eta))./(1-(phi.^2).*exp(-2*t2.*eta));
denominator=1-lambda1.*lambda2.*exp(-2*s.*k);
F=(lambda1.*exp(-(d1+d2).*k)+lambda2.*exp(-(d1_s+d2_s).*k))./
denominator;
G=2.*lambda1.*lambda2.*exp(-2*s.*k).*cosh((d2-d1).*k)./
denominator;
y=S1.*S2.*Q.*(F+G);
end
```

MATLAB Program for Example 9.4(b)

```
%This MATLAB program is used to calculate the self impedance and plot the
%results in Example 9.4 (b)
%The parameters are shown in Example 9.2 sandwiched between two ferrite
%substrates
rin=[1.15e-3 2e-3];
rout=[1.75e-3 2.6e-3];
height=[15e-6 15e-6];
d=[7.5e-6 7.5e-6];

L1_air=air_mutual(rin(1),rout(1),rin(1),rout(1),height(1),height
(1),d(1),d(1));
L2_air=air_mutual(rin(2),rout(2),rin(2),rout(2),height(2),height
(2),d(2),d(2));
L12_air=air_mutual(rin(1),rout(1),rin(2),rout(2),height(1),height
(2),d(1),d(2));
L1=L1_air+L2_air+2*L12_air;
ur=1000;
sigma=10;
t=0.5e-3;
s=15e-6;
Rdc=0.093;

L_s=L1.*ones(1,991);
L_single=L1.*ones(1,991);
Rac_s=Rdc.*ones(1,991);
Rac_single=Rdc.*ones(1,991);

i=1;
for frequency=1e6:0.1e6:100e6;

L1_s=impedance_sandwich(rin(1),rout(1),rin(1),rout(1),height(1),
height(1),d(1),d(1),t,t,s,ur,sigma,frequency);
    R1_s=real(L1_s);
    L1_s=(imag(L1_s))/(2*pi*frequency)+L1_air;

L2_s=impedance_sandwich(rin(2),rout(2),rin(2),rout(2),height(2),
height(2),d(2),d(2),t,t,s,ur,sigma,frequency);
    R2_s=real(L2_s);
    L2_s=(imag(L2_s))/(2*pi*frequency)+L2_air;

L12_s=impedance_sandwich(rin(1),rout(1),rin(2),rout(2),height(1),
height(2),d(1),d(2),t,t,s,ur,sigma,frequency);
    R12_s=real(L12_s);
    L12_s=(imag(L12_s))/(2*pi*frequency)+L12_air;
```

```
    L_s(i)=L1_s+L2_s+2*L12_s;
    Rac_s(i)=R1_s+R2_s+2*R12_s;

L1_single=impedance_substrate(rin(1),rout(1),rin(1),rout(1),height
(1),height(1),d(1),d(1),t,ur,sigma,frequency);
    R1_single=real(L1_single);
    L1_single=(imag(L1_single))/(2*pi*frequency)+L1_air;

L2_single=impedance_substrate(rin(1),rout(1),rin(1),rout(1),height
(1),height(1),d(1),d(1),t,ur,sigma,frequency);
    R2_single=real(L2_single);
    L2_single=(imag(L2_single))/(2*pi*frequency)+L2_air;

L12_single=impedance_substrate(rin(1),rout(1),rin(1),rout(1),
height(1),height(1),d(1),d(1),t,ur,sigma,frequency);
    R12_single=real(L12_single);
    L12_single=(imag(L12_single))/(2*pi*frequency)+L12_air;
    L_single(i)=L1_single+L2_single+2*L12_single;
    Rac_single(i)=R1_single+R2_single+2*R12_single;
    i=i+1;
end

x=1e6:0.1e6:100e6;
y1=L_s./L1;
y2=L_single./L1;
z1=Rac_s./Rdc+1;
z2=Rac_single./Rdc+1;

semilogx(x,y1,'-',x,y2,'-')
axis([1e6,100e6,0,100])

figure;
loglog(x,z1,'-',x,z2,'-')
axis([1e6,100e6,1,2e3])

%File to define the function air_mutual
function y=air_mutual(r1,r2,a1,a2,h1,h2,d1,d2)

%This function is used to calculate the mutual inductance in air core
%condition.
%r1,r2,h1,d1 are the inside radius, outside radius, height and upright
%position of the cross-section 1.
%a1,a2,h2,d2 are the inside radius, outside radius, height and upright
%position of the cross-section 2.
%z is the axis separation. z=0 for self-inductance calculation; z=|d2-d1|
%for mutual inductance calculation.
```

```
global uo;
uo=4*pi*1e-7;
g=@(k)aircoremul(r1,r2,a1,a2,h1,h2,d1,d2,k);

for upper=1000:1000:1000000
[integalresult,err] = quadgk(g,0,upper);
if err<0.01*integalresult
    integalresult_real=integalresult;
else
    break;
end
end

y=uo.*pi.*integalresult_real./(h1*log(a2/a1)*h2*log(r2/r1));

end

function y=aircoremul(r1,r2,a1,a2,h1,h2,d1,d2,k)

z=d2-d1;
if z==0
    Q=2.*(h1.*k+exp(-h1.*k)-1)./(k.^2);
else
    Q=2.*(cosh(0.5.*(h1+h2).*k)-cosh(0.5.*(h1-h2).*k))./(k.^2);
end

S1=(besselj(0,r2.*k)-besselj(0,r1.*k))./k;
S2=(besselj(0,a2.*k)-besselj(0,a1.*k))./k;

if k(1)==0
    Q(1)=0;
    S1(1)=0;
    S2(1)=0;
end;

y=S1.*S2.*Q.*exp(-z.*k);
end

%File to define the function impedance_sandwich
function y=impedance_sandwich(r1,r2,a1,a2,h1,h2,d1,d2,t1,t2,s,ur,
sigma,freq)

%This function is used to calculate the additional mutual impedance to the
%presence of the substrate in sandwich structures.
%r1,r2,h1,d1 are the inside radius, outside radius, height and upright
%position of the cross-section 1.
```

```
%a1,a2,h2,d2 are the inside radius, outside radius, height and upright
%position of the cross-section 2.
%t1 and t2 are thickness of bottom and upper substrates
%s substrate separation
%ur relative permeability of the magnetic substrate
%sigma relative permeability of the magnetic substrate
%freq operation frequency

global uo;
uo=4*pi*1e-7;
omega=2*pi*freq;

g=@(k)integrand(r1,r2,a1,a2,h1,h2,d1,d2,t1,t2,s,ur,sigma,freq,k);

for upper=1000:1000:1000000
[integalresult,err] = quadgk(g,0,upper);
if err<0.01*integalresult
    integalresult_real=integalresult;
else
    break;
end
end

y=1j.*omega.*uo.*pi.*integalresult_real./(h1*log(a2/a1)*h2*log(r2/
r1));

end

function y=integrand(r1,r2,a1,a2,h1,h2,d1,d2,t1,t2,s,ur,sigma,freq,
k)
global uo;
z=d2-d1;
d1_s=s-d1;
d2_s=s-d2;
if z==0
    Q=2.*(h1.*k+exp(-h1.*k)-1)./(k.^2);
else
    Q=2.*(cosh(0.5.*(h1+h2).*k)-cosh(0.5.*(h1-h2).*k))./(k.^2);
end

S1=(besselj(0,r2.*k)-besselj(0,r1.*k))./k;
S2=(besselj(0,a2.*k)-besselj(0,a1.*k))./k;

if k(1)==0
    Q(1)=0;
    S1(1)=0;
```

```
    S2(1)=0;
end;

eta=sqrt((1j*2*pi.*freq*sigma*ur*4*pi*1e-7)+(k.^2));

phi=(ur.*k-eta)./(ur.*k+eta);
lambda1=phi.*(1-exp(-2*t1.*eta))./(1-(phi.^2).*exp(-2*t1.*eta));
lambda2=phi.*(1-exp(-2*t2.*eta))./(1-(phi.^2).*exp(-2*t2.*eta));
denominator=1-lambda1.*lambda2.*exp(-2*s.*k);
F=(lambda1.*exp(-(d1+d2).*k)+lambda2.*exp(-(d1_s+d2_s).*k))./
denominator;
G=2.*lambda1.*lambda2.*exp(-2*s.*k).*cosh((d2-d1).*k)./denominator;
y=S1.*S2.*Q.*(F+G);
end

%File to define the function impedance_substrate
function y=impedance_substrate(r1,r2,a1,a2,h1,h2,d1,d2,t,ur,sigma,
freq)

%This function is used to calculate the additional mutual impedance to the
%presence of the substrate in sandwich structures.
%r1,r2,h1,d1 are the inside radius, outside radius, height and upright
%position of the cross-section 1.
%a1,a2,h2,d2 are the inside radius, outside radius, height and upright
%position of the cross-section 2.
%t is the thickness of the substrate
%ur relative permeability of the magnetic substrate
%sigma relative permeability of the magnetic substrate
%freq operation frequency

global uo;
uo=4*pi*1e-7;
omega=2*pi*freq;

g=@(k)integrand(r1,r2,a1,a2,h1,h2,d1,d2,t,ur,sigma,freq,k);

for upper=1000:1000:1000000
[integalresult,err] = quadgk(g,0,upper);
if err<0.01*integalresult
    integalresult_real=integalresult;
else
    break;
end
end

y=1j.*omega.*uo.*pi.*integalresult_real./(h1*log(a2/a1)*h2*log(r2/
```

```
r1));

end

function y=integrand(r1,r2,a1,a2,h1,h2,d1,d2,t,ur,sigma,freq,k)
global uo;

z=d2-d1;
if z==0
    Q=2.*(h1.*k+exp(-h1.*k)-1)./(k.^2);
else
    Q=2.*(cosh(0.5.*(h1+h2).*k)-cosh(0.5.*(h1-h2).*k))./(k.^2);
end

S1=(besselj(0,r2.*k)-besselj(0,r1.*k))./k;
S2=(besselj(0,a2.*k)-besselj(0,a1.*k))./k;

if k(1)==0
    Q(1)=0;
    S1(1)=0;
    S2(1)=0;
end;

eta=sqrt((1j*2*pi.*freq*sigma*ur*4*pi*1e-7)+(k.^2));

phi=(ur.*k-eta)./(ur.*k+eta);
lambda=phi.*(1-exp(-2*t.*eta))./(1-(phi.^2).*exp(-2*t.*eta));
y=S1.*S2.*Q.*lambda.*exp(-(d1+d2).*k);
end
```

References

1. Hurley, W.G. and Duffy, M.C. (1995) Calculation of self and mutual impedances in planar magnetic structures. *IEEE Transactions on Magnetics* **31** (4), 2416–2422.
2. Hurley, W.G. and Duffy, M.C. (1997) Calculation of self- and mutual impedances in planar sandwich inductors. *IEEE Transactions on Magnetics* **33** (3), 2282–2290.
3. Hurley, W.G., Duffy, M.C., O'Reilly, S., and O'Mathuna, S.C. (1999) Impedance formulas for planar magnetic structures with spiral windings. *IEEE Transactions on Industrial Electronics* **46** (2), 271–278.
4. Wang, N., O'Donnell, T., Meere, R. *et al.* (2008) Thin-film-integrated power inductor on Si and its performance in an 8-MHz buck converter. *IEEE Transactions on Magnetics* **44** (11), 4096–4099.
5. Mathuna, S.C.O., O'Donnell, T., Wang, N., and Rinne, K. (2005) Magnetics on silicon: an enabling technology for power supply on chip. *IEEE Transactions on Magnetics* **20** (3), 585–592.
6. Morse, P. and Feshback, H. (1953) *Methods of Theoretical Physics*, McGraw-Hill, New York.
7. O'Reilly, S., O'Mathúna, S.C., Duffy, M.C., and Hurley, W.G. (1998) A comparative analysis of interconnection technologies for integrated multilayer inductors. *Microelectronics International: Journal of the International Society for Hybrid Microelectronics* **15** (1), 6–10.

Further Reading

1. Brunet, M., O'Donnell, T., Connell, A.M., McCloskey, P., Mathuna, S.C.O. (2006). Electrochemical process for the lamination of magnetic cores in thin-film magnetic components. *Journal of Microelectromechanical Systems* **15** (1), 94–100.
2. Brunet, M., O'Donnell, T., O'Brien, J. *et al.* (2001) Design study and fabrication techniques for high power density microtransformers. *Proceedings of the IEEE Applied Power Electronics Conference and Exposition, APEC*, pp. 1189–1195.
3. Chi Kwan, L., Su, Y.P., and Hui, S.Y.R. (2011) Printed spiral winding inductor with wide frequency bandwidth. *IEEE Transactions on Power Electronics* **26** (10), 2936–2945.
4. Di, Y., Levey, C.G., and Sullivan, C.R. (2011) Microfabricated V-groove power inductors using multilayer Co-Zr-O thin films for very-high-frequency DC-DC converters. *Proceedings of the IEEE Energy Conversion Congress and Exposition, ECCE*, pp. 1845–1852.
5. Dwight, H.B. (1919) Some new formulas for reactance coils. *Transactions of the American Institute of Electrical Engineers*, **XXXVIII** (2) 1675–1696.
6. Ferreira, J.A. (1994) Improved analytical modeling of conductive losses in magnetic components. *IEEE Transactions on Power Electronics* **9** (1), 127–131.
7. Gray, A. (1893) *Absolute Measurements in Electricity and Magnetism*, MacMillan, London.
8. Grover, F.W. (2004) *Inductance Calculations: Working Formulas and Tables*, Dover Publications Inc., New York.
9. Hui, S.Y.R. and Ho, W.W.C. (2005) A new generation of universal contactless battery charging platform for portable consumer electronic equipment. *IEEE Transactions on Power Electronics* **20** (3), 620–627.
10. Hui, S.Y.R. and Liu, X. (2010) Semiconductor transformers. *Electronics Letters* **46** (13), 947–949.
11. Jian, L., Hongwei, J., Xuexin, W., Padmanabhan, K., Hurley, W.G. and Shen, Z.J. (2010). Modeling, design, and characterization of multiturn bondwire inductors with ferrite epoxy glob cores for power supply system-on-chip or system-in-package applications. *IEEE Transactions on Power Electronics* **25** (8), 2010–2017.
12. Jiankun, H. and Sullivan, C.R. (2001) AC resistance of planar power inductors and the quasidistributed gap technique. *IEEE Transactions on Power Electronics* **16** (4), 558–567.
13. Laili, W., Yunqing, P., Xu, Y. *et al.* (2010) A class of coupled inductors based on LTCC technology. *Proceedings of the IEEE Applied Power Electronics Conference and Exposition, APEC*, pp. 2042–2049.
14. Liu, X. and Hui, S.Y.R. (2007) Equivalent circuit modeling of a multilayer planar winding array structure for use in a universal contactless battery charging platform. *IEEE Transactions on Power Electronics* **22** (1), 21–29.
15. Ludwig, G.W. and El-Hamamsy, S.A. (1991) Coupled inductance and reluctance models of magnetic components. *IEEE Transactions on Power Electronics* **6** (2), 240–250.
16. Ludwig, M., Duffy, M., O'Donnell, T. *et al.* (2003) PCB integrated inductors for low power DC/DC converter. *IEEE Transactions on Power Electronics* **18** (4), 937–945.
17. Lyle, T.R. (1914) *Philosophical Transactions of the Royal Society of London. Series A, Containing Papers of a Mathematical or Physical Character*, The Royal Society, London.
18. Mathuna, S.C.O., Byrne, P., Duffy, G. *et al.* (2004) Packaging and integration technologies for future high-frequency power supplies. *IEEE Transactions on Industrial Electronics* **51** (6), 1305–1312.
19. Maxwell, J.C. (1881) *A Treatise on Electricity and Magnetism*, Clarendon Press, Oxford.
20. Meere, R., O'Donnell, T., Wang, N. *et al.* (2009) Size and performance tradeoffs in micro-inductors for high frequency DC-DC conversion. *IEEE Transactions on Magnetics* **45** (10), 4234–4237.
21. Mino, M., Yachi, T., Tago, A. *et al.* (1992) A new planar microtransformer for use in micro-switching converters. *IEEE Transactions on Magnetics* **28** (4), 1969–1973.
22. O'Donnell, T., Wang, N., Kulkarni, S. *et al.* (2010) Electrodeposited anisotropic NiFe 45/55 thin films for high-frequency micro-inductor applications. *Journal of Magnetism and Magnetic Materials* **322** (9–12), 1690–1693.
23. O'Reilly, S., Duffy, M., O'Donnell, T. *et al.* (2000) New integrated planar magnetic cores for inductors and transformers fabricated in MCM-L technology. *International Journal of Microcircuits and Electronic Packaging*, **23** (1), 62–69.
24. Ouyang, Z., Sen, G., Thomsen, O., and Andersen, M. (2013) Analysis and design of fully integrated planar magnetics for primary-parallel isolated boost converter. *IEEE Transactions on Industrial Electronics* **60** (20), 494–508.
25. Ouyang, Z., Thomsen, O.C., and Andersen, M.A.E. (2012) Optimal design and tradeoff analysis of planar transformer in high-power DC-DC converters. *IEEE Transactions on Industrial Electronics* **59** (7), 2800–2810.

26. Ouyang, Z., Zhang, Z., Andersen, M.A.E., and Thomsen, O.C. (2012) Four quadrants integrated transformers for dual-input isolated DC-DC converters. *IEEE Transactions on Power Electronics* **27** (6), 2697–2702.
27. Ouyang, Z., Zhang, Z., Thomsen, O.C., and Andersen, M.A.E. (2011) Planar-integrated magnetics (PIM) module in hybrid bidirectional DC-DC converter for fuel cell application. *IEEE Transactions on Power Electronics* **26** (11), 3254–3264.
28. Quinn, C., Rinne, K., O'Donnell, T. *et al.* (2001) A review of planar magnetic techniques and technologies. *Proceedings of the IEEE Applied Power Electronics Conference and Exposition, APEC*, pp. 1175–1183.
29. Rodriguez, R., Dishman, J., Dickens, F., and Whelan, E. (1980) Modeling of two-dimensional spiral inductors. *IEEE Transactions on Components, Hybrids, and Manufacturing Technology*, **3** (4), 535–541.
30. Roshen, W.A. (1990) Effect of finite thickness of magnetic substrate on planar inductors. *IEEE Transactions on Magnetics* **26** (1), 270–275.
31. Roshen, W.A. (1990) Analysis of planar sandwich inductors by current images. *IEEE Transactions on Magnetics* **26** (5), 2880–2887.
32. Roshen, W.A. and Turcotte, D.E. (1988) Planar inductors on magnetic substrates. *IEEE Transactions on Magnetics* **24** (6), 3213–3216.
33. Sato, T., Hasegawa, M., Mizoguchi, T., and Sahashi, M. (1991) Study of high power planar inductor. *IEEE Transactions on Magnetics* **27** (6), 5277–5279.
34. Su, Y.P., Xun, L., and Hui, S.Y. (2008) Extended theory on the inductance calculation of planar spiral windings including the effect of double-layer electromagnetic shield. *IEEE Transactions on Power Electronics* **23** (4), 2052–2061.
35. Su, Y.P., Xun, L., and Hui, S.Y.R. (2009) Mutual inductance calculation of movable planar coils on parallel surfaces. *IEEE Transactions on Power Electronics* **24** (4), 1115–1123.
36. Sullivan, C.R. (2009) Integrating magnetics for on-chip power: Challenges and opportunities. *Proceedings of the Custom Integrated Circuits Conference, CICC*, pp. 291–298.
37. Sullivan, C.R. and Sanders, S.R. (1993) Microfabrication of transformers and inductors for high frequency power conversion. *Proceedings of the IEEE Power Electronics Specialists Conference, PESC*, pp. 33–41.
38. Sullivan, C.R. and Sanders, S.R. (1996) Design of microfabricated transformers and inductors for high-frequency power conversion. *IEEE Transactions on Power Electronics* **11** (2), 228–238.
39. Tang, S.C., Hui, S.Y., and Chung, H.S.H. (1999) Coreless printed circuit board (PCB) transformers with multiple secondary windings for complementary gate drive circuits. *IEEE Transactions on Power Electronics* **14** (3), 431–437.
40. Tang, S.C., Hui, S.Y., and Chung, H.S.H. (2000) Coreless planar printed-circuit-board (PCB) transformers-a fundamental concept for signal and energy transfer. *IEEE Transactions on Power Electronics* **15** (5), 931–941.
41. Tang, S.C., Hui, S.Y., and Chung, H.S.H. (2000) Characterization of coreless printed circuit board (PCB) transformers. *IEEE Transactions on Power Electronics* **15** (6), 1275–1282.
42. Tegopoulos, J.A. and Kriezis, E.E. (1985) *Eddy Currents in Linear Conducting Media*, Elsevier, Amsterdam.
43. Williams, R., Grant, D.A., and Gowar, J. (1993) Multielement transformers for switched-mode power supplies: toroidal designs. *IEE Proceedings-Electric Power Applications B* **140** (2), 152–160.
44. Xun, L. and Hui, S.Y. (2007) Simulation study and experimental verification of a universal contactless battery charging platform with localized charging features. *IEEE Transactions on Power Electronics* **22** (6), 2202–2210.
45. Yamaguchi, K., Ohnuma, S., Imagawa, T. *et al.* (1993) Characteristics of a thin film microtransformer with circular spiral coils. *IEEE Transactions on Magnetics* **29** (5), 2232–2237.
46. Yamaguchi, M., Arakawa, S., Ohzeki, H. *et al.* (1992) Characteristics and analysis of a thin film inductor with closed magnetic circuit structure. *IEEE Transactions on Magnetics* **28** (5), 3015–3017.
47. Yamaguchi, T., Sasada, I., and Harada, K. (1993) Mechanism of conductive losses in a sandwich-structured planar inductor. *IEEE Translation Journal on Magnetics in Japan*, **8** (3), 182–186.
48. Zhang, Z., Ouyang, Z., Thomsen, O.C., and Andersen, M.A.E. (2012) Analysis and design of a bidirectional isolated DC-DC converter for fuel cells and supercapacitors hybrid system. *IEEE Transactions on Power Electronics* **27** (2), 848–859.
49. Ziwei, O., Thomsen, O.C., and Andersen, M. (2009) The analysis and comparison of leakage inductance in different winding arrangements for planar transformer. *Proceedings of the IEEE Power Electronics and Drive Systems, PEDS*, pp. 1143–1148.

10

Variable Inductance[1]

An inductor with an air gap of fixed length, and operated so that the flux density in the core is below the saturation value, has a fixed value of inductance determined by the length of the air gap (the analysis is provided in Section 2.1). When the current is increased to the point where the flux density enters the saturation region, the increased reluctance of the core (due to the reduced value of relative permeability) reduces the overall inductance. When the core material enters saturation, the operation of the inductor becomes non-linear and, while the resulting characteristics have advantages in certain applications, new challenges arise in terms of the circuit analysis. In this chapter, we explore the main physical manifestations of variable inductance, we examine their applications and analyze the approach to circuit simulation.

The flux level in the core is determined by the magnetic field intensity, as dictated by Ampere's law, and therefore by the current level in the coil. In saturation, the inductance decreases with increasing current as a consequence of progressively increasing saturation. In the case of a fixed air gap, the saturation is dictated by the *B-H* characteristic of the magnetic core material, which means that we have no control of the inductance/current (L/i) characteristic once saturation is reached.

A small measure of control may be introduced by using a stepped air gap; this is called a swinging inductor. A logical extension of the stepped air gap is to introduce an infinite number of steps in the form of a sloped air gap (SAG) so that the onset of saturation takes place progressively from the narrow end of the gap to the wide end and, as we shall see later, this gives us a measure of control over the L/i characteristic to suit the intended application.

In a powder iron core, the gap is distributed and the manufacturer supplies a curve of magnetic permeability versus magnetic field intensity, from which an L/i characteristic may be constructed using the analysis of Chapter 2. There are many applications where the required inductance is a function of the load level, and an L/i characteristic where the inductance falls off with increasing current is desirable. Such an L/i characteristic has an

[1] Parts of this chapter are reproduced with permission from [1] Wolfle, W.H. and Hurley, W.G. (2003) Quasi-active power factor correction with a variable inductive filter: theory, design and practice. *IEEE Transactions on Power Electronics* **18** (1), 248–255; [2] Zhang, L., Hurley, W.G., and Wölfle, W.H. (2011) A new approach to achieve maximum power point tracking for PV system with a variable inductor. *IEEE Transactions on Power Electronics* **26** (4), 1031–1037.

Transformers and Inductors for Power Electronics: Theory, Design and Applications, First Edition.
W. G. Hurley and W. H. Wölfle.

advantage in that the largest inductance for the smallest value of current means that the stored energy (in the form of $\frac{1}{2}Li^2$) is lower at the higher current level. This, in turn, means that the overall size of the inductor is reduced compared to the size of an inductor based on the largest L and the largest i.

In this chapter, we will introduce each type of variable inductor along with an application: the swinging inductor for voltage regulation, the SAG inductor for power factor correction and the powder iron core for maximum power point tracking in a solar PV system. We will show that a considerable reduction in the size of the inductor may be achieved while lower energy is stored in a saturated inductor at high current.

The swinging inductor has a stepped air gap. The core may be considered to have two parallel reluctance paths, with each path having two reluctances in series, the core and the gap. As the current increases, the path containing the smaller gap reaches saturation first and the increased reluctance reduces the overall inductance. This device has been employed for voltage regulation in rectifier circuits, as it achieves continuous current even at small values of current.

The sloped air gap (SAG) inductor [1] operates on the same principle as the swinging inductor. The air gap increases from its minimum value to its maximum value in a graded shape rather than in a discrete step. This means that the variation in inductance with current is more gradual, and we shall see later that the gradual characteristic of the SAG inductor is more suitable for power factor correction than the more abrupt transition afforded by a swinging inductor.

The powder iron core is used in high frequency applications and the inductance is normally found from the manufacturer's data sheet. In solar photovoltaic (PV) systems, impedance matching between the output of the solar panels and the load for maximum power transfer is normally achieved by varying the duty cycle of a buck converter [2]. The minimum inductance required for continuous conduction in maximum power point tracking (MPPT) is a function of the solar insolation and, therefore, a variable inductance is perfectly suited to this application.

The voltage across an inductor is related to its flux linkage and this, in turn is related to the current. The dependence of the inductance on its current must be taken into account. Recall from Chapter 2:

$$\lambda = L(i)i \tag{10.1}$$

Invoking Faraday's law:

$$V = \frac{d\lambda}{dt} = L(i)\frac{di}{dt} + i\frac{dL(i)}{dt} = \left(L(i) + i\frac{dL(i)}{di}\right)\frac{di}{dt} = L_{\text{eff}}\frac{di}{dt} \tag{10.2}$$

$$L_{\text{eff}} = L(i) + i\frac{dL(i)}{di} \tag{10.3}$$

L_{eff} in Equation 10.2 is readily found from the L/i characteristic of the inductor. The L_{eff} versus current characteristic is more insightful, since most simulation models automatically assume this relationship. Most simulation packages assume that the inductance value is fixed for variable inductance.

We will need to take the variation with current into account. The ElectroMagnetic Transient Program (EMTP) [3] is particularly suited for this purpose, and it will be introduced later in the chapter.

10.1 Saturated Core Inductor

The inductance of a gapped core is normally dominated by the air gap. The overall inductance consists of the core and the air gap, with the dimensions shown in Figure 10.1.

Including the reluctance of the core, the overall inductance is:

$$L = \frac{N^2}{\mathcal{R}_c + \mathcal{R}_g} = \frac{N^2}{\mathcal{R}_g} \frac{1}{\frac{\mathcal{R}_c}{\mathcal{R}_g} + 1} \tag{10.4}$$

The reluctance of the core is:

$$\mathcal{R}_c = \frac{l_c}{\mu_r \mu_0 A_c} \tag{10.5}$$

and the reluctance of the gap is:

$$\mathcal{R}_g = \frac{g}{\mu_0 A_g} \tag{10.6}$$

The overall inductance may be expressed in terms of the gap. Fringing may be neglected if the gap length is much smaller than the dimensions of the core cross-section:

$$L = \frac{N^2}{\mathcal{R}_g} \frac{1}{1 + \frac{\mu_{\text{eff}}}{\mu_r}} \tag{10.7}$$

where the effective relative permeability is $\mu_{\text{eff}} = \frac{l_c}{g}$, as described in Chapter 2.

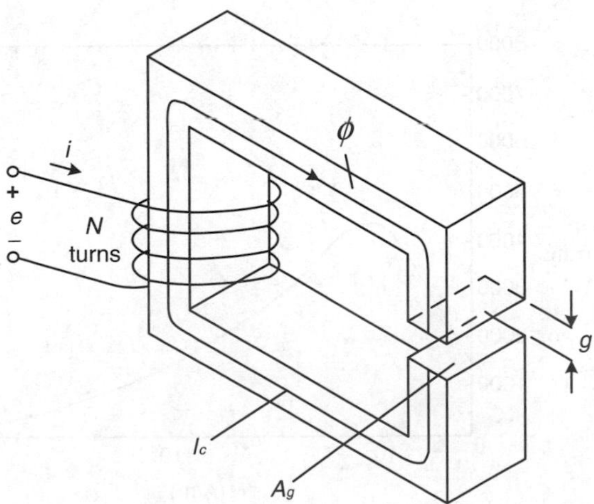

Figure 10.1 Fixed air gap inductor.

When $\mu_{\text{eff}} \ll \mu_r$ the inductance is dominated by the gap. Increasing the current leads to increased magnetic field intensity and a drop in μ_r. In saturation, μ_r decreases towards μ_{eff} and the value of inductance falls. The manufacturer normally supplies the variation of relative permeability with magnetic field intensity in graphical form. An empirical relationship may be established in the following form, which is amenable to analysis in predicting inductance as a function of current:

$$\mu_r = \frac{H_m}{H + H_0} \tag{10.8}$$

H_m and H_0 are constants.

Example 10.1

Establish the constants H_0 and H_m for M530-50A material whose permeability versus magnetic field intensity is shown in Figure 10.2.

Take two data points at:

$$H = 400, \mu_r = 2600 \text{ A/m}$$
$$H = 5000, \mu_r = 255 \text{ A/m}$$

Inserting both values into Equation 10.8:

$$2,600 = \frac{H_m}{400 + H_0}$$
$$255 \quad = \frac{H_m}{5,000 + H_0}$$

Solving yields $H_0 = 100$ A/m and $H_m = 1.3 \times 10^6$ A/m.

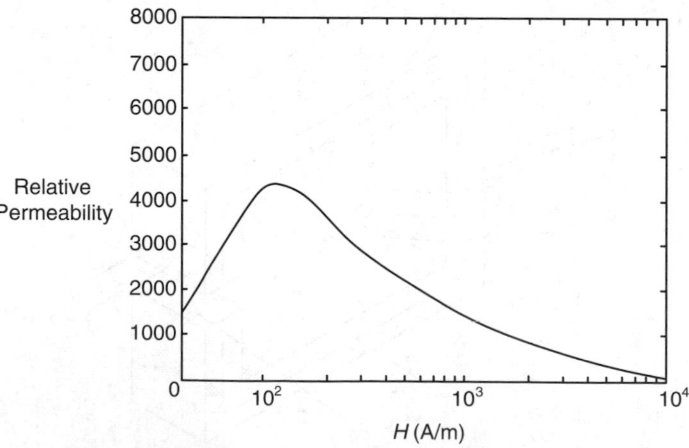

Figure 10.2 Relative permeability data for M530-50A laminated steel.

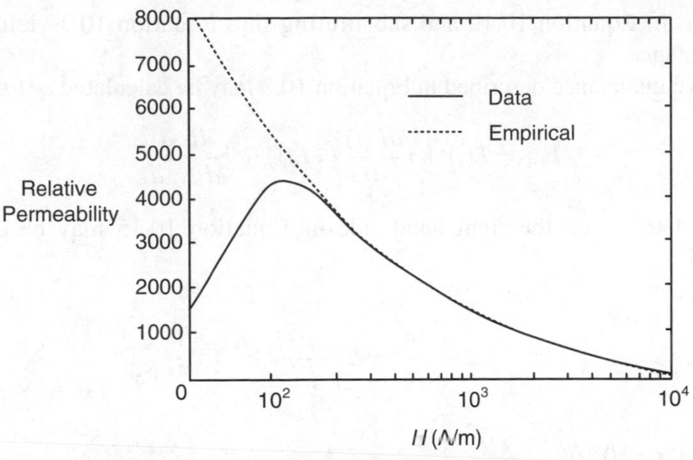

Figure 10.3 Empirical prediction of relative permeability.

The empirical curve in Example 10.1 is compared with the actual data in Figure 10.3. At very low values of H (and current), μ_r is smaller than the value given by the empirical formula. However, this is not important, since the calculations of interest will be in the range of H above 200 A/m, where the agreement is very good. Typically, μ_{eff} is of the order of 100, so that the region of the curve above 2000 A/m would have an impact on the inductance, as indicated by Equation 10.7.

Invoking Ampere's law for a gapped inductor with current i:

$$Ni = H_c l_c + \frac{B}{\mu_0} g \tag{10.9}$$

Noting that $B = \mu_r \mu_0 H_c$ and using the empirical relationship in Equation 10.8 yields a quadratic equation for μ_r:

$$H_0 \mu_r^2 + \left(\frac{Ni}{g} + H_0 \mu_{\text{eff}} - H_m \right) \mu_r - H_m \mu_{\text{eff}} = 0 \tag{10.10}$$

Equation 10.10 is a quadratic equation of the form $ax^2 + bx + c = 0$ with:

$$a = H_0 \tag{10.11}$$

$$b = \left(\frac{Ni}{g} + H_0 \mu_{\text{eff}} - H_m \right) \tag{10.12}$$

$$c = -H_m \mu_{\text{eff}} \tag{10.13}$$

μ_r is now

$$\mu_r = \frac{-b \pm \sqrt{b^2 - 4ac}}{2a} \tag{10.14}$$

Solving for μ_r in Equation 10.10 and substituting into Equation 10.7 yields the desired value of inductance.

The effective inductance described in Equation 10.3 may be calculated as follows:

$$L_{\text{eff}} = L(i) + i\frac{dL(i)}{di} = L(i) + i\frac{dL}{d\mu_r}\frac{d\mu_r}{di} \tag{10.15}$$

The individual terms on the right hand side of Equation 10.15 may be deduced from Equation 10.7:

$$\frac{dL}{d\mu_r} = L\frac{\dfrac{\mu_{\text{eff}}}{\mu_r^2}}{1 + \dfrac{\mu_{\text{eff}}}{\mu_r}} \tag{10.16}$$

and from Equation 10.10:

$$\frac{d\mu_r}{di} = \left[-\frac{1}{2a} + \frac{b}{2a\sqrt{b^2 - 4ac}}\right]\frac{N}{g} \tag{10.17}$$

Example 10.2

An inductor constructed with an EI assembly EI42 (corresponding to IEC YEI 1-14) with M530-50A material, as described in Example 10.1, was constructed with $g = 0.5$ mm, core length $l_c = 8.4$ cm, core cross-sectional area $A_g = 2.072\ \text{cm}^2$ and with $N = 365$ turns. Calculate the inductance L and the effective inductance L_{eff} as a function of current for the range 0.25 to 4 A.

For low values of current before the onset of saturation, the inductance is a function of the gap and is:

$$L_{\max} = \frac{N^2}{\mathcal{R}_g} = \frac{\mu_0 N^2 A_g}{g} = \frac{(4\pi \times 10^{-7})(365)^2(2.072 \times 10^{-4})}{0.5 \times 10^{-3}} \times 10^3 = 69.4\ \text{mH}$$

L is found from Equation 10.7:

$$\mu_{\text{eff}} = \frac{l_c}{g} = \frac{84}{0.5} = 168$$

And for $i = 1.5$ A:

$$a = H_0 = 100\ \text{A/m}$$

$$b = \frac{Ni}{g} + H_0\mu_{\text{eff}} - H_m = \frac{(365)(1.5)}{0.5 \times 10^{-3}} + (100)(168) - 1.3 \times 10^6 = -18.82 \times 10^4\ \text{A/m}$$

$$c = -H_m\mu_{\text{eff}} = -(1.3 \times 10^6)(168) = -2.184 \times 10^8\ \text{A/m}$$

$$\mu_r = \frac{-b + \sqrt{b^2 - 4ac}}{2a} = \frac{(18.82 \times 10^4) + \sqrt{(-18.82 \times 10^4)^2 - (4)(100)(-2.184 \times 10^8)}}{(2)(100)} = 2693$$

$$L = \frac{N^2}{\mathcal{R}_g}\frac{1}{1 + \dfrac{\mu_{\text{eff}}}{\mu_r}} = (69.4 \times 10^{-3})\frac{1}{1 + \dfrac{168}{2693}} = 65.3\ \text{mH}$$

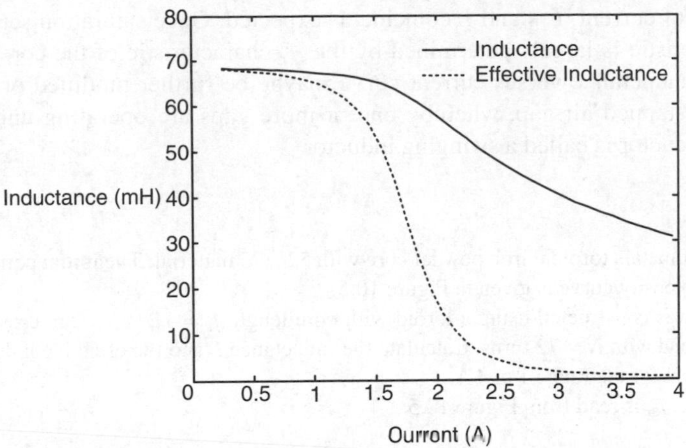

Figure 10.4 Inductance and effective inductance of an inductor with a constant air gap.

L_{eff} is found from Equation 10.15:

$$\frac{dL}{d\mu_r} = L\frac{\mu_{eff}/\mu_r^2}{1+\mu_{eff}/\mu_r} = (0.0653)\frac{168/(2693)^2}{1+168/2693} = 1.424\times10^{-6}\,\mathrm{H}$$

$$\frac{d\mu_r}{di} = \left[-\frac{1}{2a}+\frac{b}{2a\sqrt{b^2-4ac}}\right]\frac{N}{g}$$

$$= \left[-\frac{1}{(2)(100)}+\frac{-18.82\times10^4}{(2)(100)\sqrt{(18.82\times10^4)^2-(4)(100)(-2.184\times10^8)}}\right]\times\frac{365}{0.5\times10^{-3}}$$

$$= -5.61\times10^3\,\mathrm{A}^{-1}$$

$$L_{eff} = L(i)+i\frac{dL}{du_r}\frac{d\mu_r}{di} = 0.0653+(1.5)(1.424\times10^{-6})(-5.61\times10^3) = 53.3\,\mathrm{mH}$$

This process is repeated for the range of current and the plots of L and L_{eff} are shown in Figure 10.4.

The low current asymptotic value of inductance (L_{max} in Figure 10.4) is determined solely by the air gap, as indicated by Equation 10.7 with $\mu_r \gg \mu_{eff}$, The inductance is reduced to 80% of L_{max} where $\mu_r = 4\mu_{eff}$. Substituting this result into Equation 10.10 with $H_m \gg H_0$ yields the roll-off current:

$$I_{80} = \frac{5H_m g}{4N}. \tag{10.18}$$

In Example 10.2:

$$I_{80} = \frac{5H_m g}{4N} = \frac{(5)(1.3\times10^6)(0.5\times10^{-3})}{(4)(365)} = 2.2\,A.$$

At low values of current, L_{eff} and L coincide as expected. Once saturation sets in, the shape of the characteristic is largely determined by the μ_r characteristic of the core material. The characteristic inductance versus current curve maybe be further modified or controlled by introducing a stepped air gap, whereby one or more gaps are operating under saturation. This type of inductor is called a swinging inductor.

Example 10.3

Consider a Micrometals toroidal iron powder core with 52/μ75 material. The initial permeability versus magnetic field intensity curve is given in Figure 10.5.

An inductor was constructed using a toroid with core length $l_c = 4.23$ cm, core cross-sectional area $A_c = 0.179$ cm^2 and with $N = 72$ turns. Calculate the inductance L and the effective inductance L_{eff} as a function of current for the range 1 to 4 A.

For $H = 30$ Oe, μ_r is read from Figure 10.5:

$$i = \frac{Hl_c}{N} = \frac{(2387)(4.23 \times 10^{-2})}{72} = 1.403 \text{ A}$$

$$L(i) = \mu_r(i) \times \mu_0 \times N^2 \times \frac{A_c}{l_c} = (56.1)(4\pi \times 10^{-7})(72)^2 \left(\frac{0.179 \times 10^{-4}}{4.23 \times 10^{-2}}\right) = 154.7 \ \mu\text{H}$$

$$L_{eff} = L(i) + i\frac{\Delta L}{\Delta i} = (154.7 \times 10^{-6}) + (1.403)\frac{(154.7 - 176.7) \times 10^{-6}}{(1.403 - 0.935)} = 88.7 \ \mu\text{H}$$

The full set of calculations are summarized in Table 10.1, and L and L_{eff} are plotted in Figure 10.6.

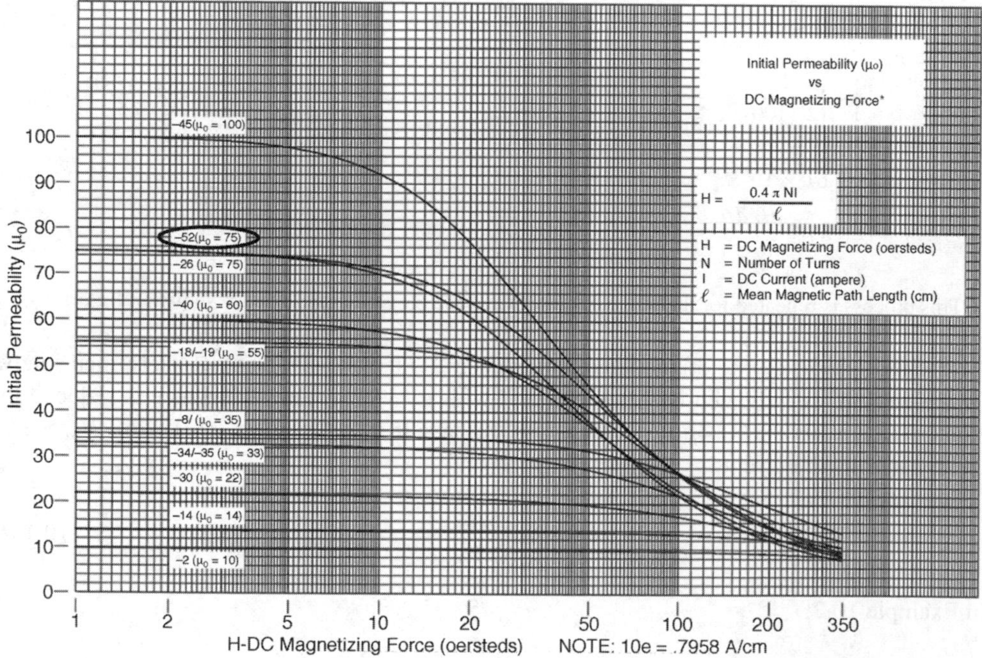

Figure 10.5 Initial permeability curve for iron powder material. Reproduced with permission from Micrometals, Inc. Copyright 2007 Micrometals, Inc.

Table 10.1 Calculations for inductance and effective inductance

H (Oe)	H (A/m)	μ_r	i (A)	L (μH)	L_{eff} (μH)
1	80	75	0.047	206.8	206.8
1.4	111	75	0.065	206.8	206.8
2	159	74.9	0.094	206.5	205.6
3	239	74	0.140	204.0	196.6
5	398	73.5	0.234	202.6	199.2
7	557	72.5	0.327	199.9	190.2
10	796	71.1	0.468	196.0	183.1
14	1114	68.4	0.655	188.6	162.5
20	1592	64.1	0.935	176.7	137.2
30	2387	56.1	1.403	154.7	88.7
50	3979	44	2.338	121.3	37.9
70	5571	35	3.273	96.5	9.6
100	7958	27	4.675	74.4	0.9

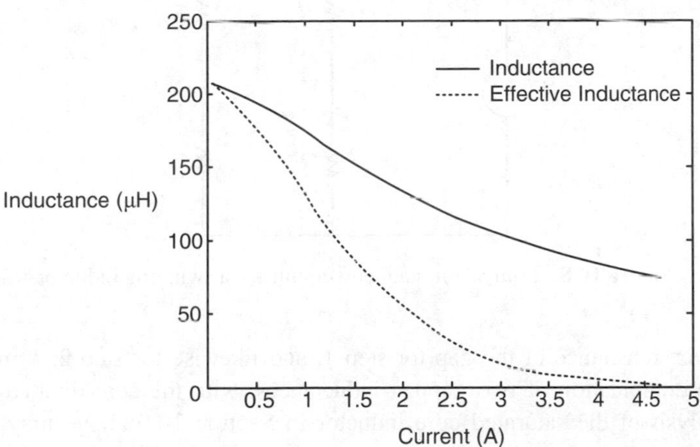

Figure 10.6 Inductance and effective inductance of an inductor with an iron powder core.

10.2 Swinging Inductor

The swinging inductor has a two-stepped air gap, as shown in Figure 10.7. The equivalent magnetic circuit may be obtained by dividing the core into two sections corresponding to the steps in the gap, as shown in Figure 10.8.

Each branch represents one step and the overall inductance is found from the equivalent reluctance:

$$L = \frac{N^2}{\mathcal{R}_{g_1}} \frac{1}{1 + \dfrac{\mu_{\text{eff}_1}}{\mu_{r_1}}} + \frac{N^2}{\mathcal{R}_{g_2}} \frac{1}{1 + \dfrac{\mu_{\text{eff}_2}}{\mu_{r_2}}} \tag{10.19}$$

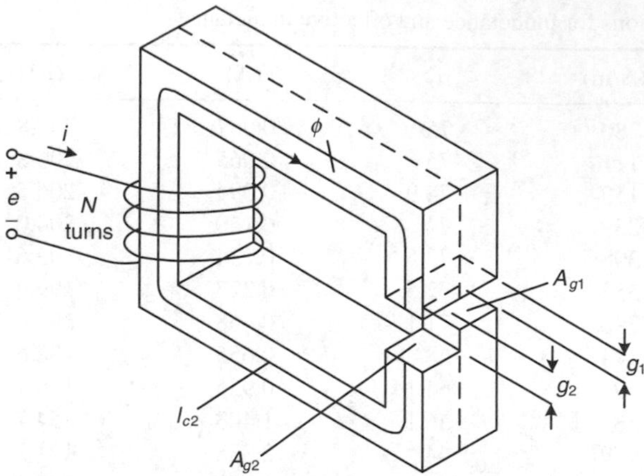

Figure 10.7 Swinging inductor.

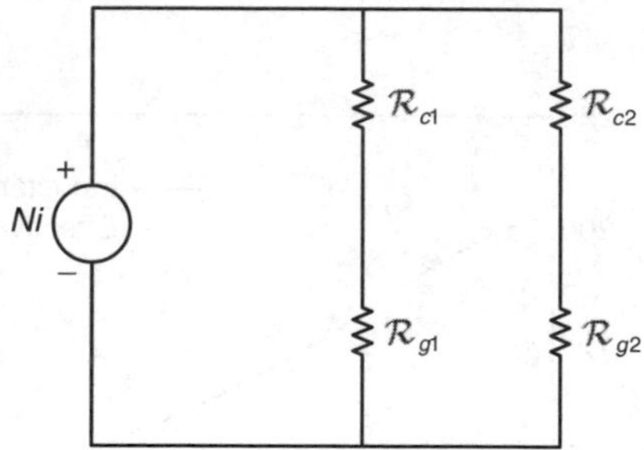

Figure 10.8 Equivalent magnetic circuit for a swinging inductor.

where $\mathcal{R}_{g_1}$ is the reluctance of the gap for step 1, and likewise for step 2. L may be considered the series combination of two gapped inductances, with the core divided into two sections. The analysis of the saturated core inductor in Section 10.0 above may be applied to each step.

Example 10.4

An inductor constructed with an EI assembly EI42 (corresponding to IEC YEI 1-14) with M530-50A material described in Example 10.1 was constructed with $g_1 = 0.39$ mm, $A_{g_1} = 1.036\,\text{cm}^2$, $g_2 = 0.69$ mm, $A_{g_2} = 1.036\,\text{cm}^2$, core length $l_c = 8.4$ cm and with $N = 365$ turns. Calculate the inductance L and the effective inductance L_{eff} as a function of current for the range 0.25 to 4 A.

For low values of current before the onset of saturation, the inductance is a function of the gaps. We can treat the swinging inductor as two inductors in series.

For $g_1 = 0.39$ mm:

$$L_{\max_1} = \frac{N^2}{\mathcal{R}_{g_1}} = \frac{\mu_0 N^2 A_c}{g_1} = \frac{(4\pi \times 10^{-7})(365)^2(1.036 \times 10^{-4})}{0.39 \times 10^{-3}} \times 10^3 = 44.47\ \text{mH}$$

$$\mu_{\text{eff}_1} = \frac{l_{c_1}}{g_1} = \frac{84}{0.39} - 215$$

And for $i = 2.0$ A:

$$\begin{aligned}
a &= H_0 = 100\ \text{A/m} \\
b_1 &= \frac{Ni}{g_1} + H_0\mu_{\text{eff}} - H_m = \frac{(365)(2.0)}{0.39 \times 10^{-3}} + (100)(215) - (1.3 \times 10^6) = 5.93 \times 10^5\ \text{A/m} \\
c_1 &= -H_m\mu_{\text{eff}_1} = -(1.3 \times 10^6)(215) = -2.80 \times 10^8\ \text{A/m} \\
\mu_{r_1} &= \frac{-b + \sqrt{b^2 - 4ac}}{2a} = \frac{-(5.93 \times 10^5) + \sqrt{(5.93 \times 10^5)^2 - (4)(100)(-2.80 \times 10^8)}}{(2)(100)} = 439 \\
L_1 &= L_{\max_1}\frac{1}{1 + \frac{\mu_{\text{eff}_1}}{\mu_{r_1}}} = (44.47)\frac{1}{1 + \frac{215}{439}} = 29.8\ \text{mH} \\
\frac{dL}{d\mu_r} &= L\frac{\mu_{\text{eff}}/\mu_r^2}{1 + \mu_{\text{eff}}/\mu_r} = (0.0298)\frac{215/(439)^2}{1 + 215/439} = 2.23 \times 10^{-5}\ \text{H} \\
\frac{d\mu_r}{di} &= \left[-\frac{1}{2a} + \frac{b}{2a\sqrt{b^2 - 4ac}}\right]\frac{N}{g} \\
&= \left[-\frac{1}{(2)(100)} + \frac{5.93 \times 10^5}{(2)(100)\sqrt{(5.93 \times 10^5)^2 - (4)(100)(-2.80 \times 10^8)}}\right]\left(\frac{365}{0.39 \times 10^{-3}}\right) \\
&= -604/\text{A} \\
L_{\text{eff}_1} &= L(i) + i\frac{dL}{du_r}\frac{d\mu_r}{di} = 0.0298 + (2.0)(2.23 \times 10^{-5})(-604) = 2.9\ \text{mH}
\end{aligned}$$

For: $g_2 = 0.69$ mm:

$$\begin{aligned}
L_{\max_2} &= 25.14\ \text{mH} \\
\mu_{\text{eff}_2} &= 122
\end{aligned}$$

And for $i = 2.0$ A:

$$\begin{aligned}
a &= H_0 = 100\ \text{A/m} \\
b_2 &= -2.299 \times 10^5\ \text{A/m} \\
c_2 &= -1.583 \times 10^8\ \text{A/m} \\
\mu_{r_2} &= 2853 \\
L_2 &= 24.1\ \text{mH} \\
\frac{dL}{d\mu_r} &= 3.458 \times 10^{-7}\ \text{H} \\
\frac{d\mu_r}{di} &= -4.429 \times 10^3\ /\text{A} \\
L_{\text{eff}_2} &= 21\ \text{mH}
\end{aligned}$$

The low current asymptotic value of the inductance in the swinging inductor is the sum of these two values:

$$\begin{aligned}
L &= L_1 + L_2 = 29.8 + 24.1 = 53.9\ \text{mH} \\
L_{\text{eff}} &= 2.9 + 21 = 23.9\ \text{mH}
\end{aligned}$$

This process is repeated for the range of current, and the plot of L and L_{eff} is shown in Figure 10.9.

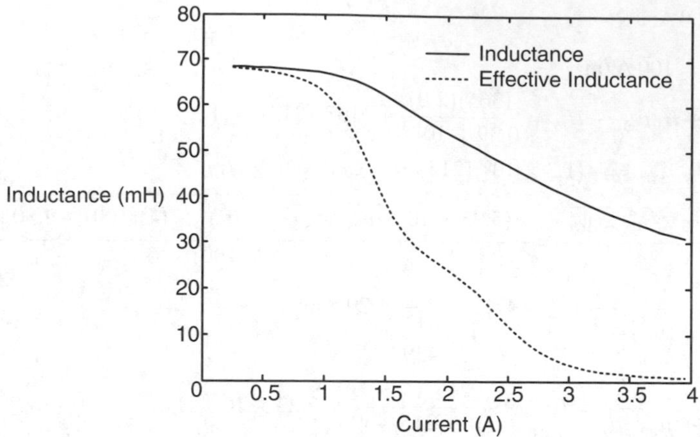

Figure 10.9 Inductance and effective inductance of a swinging inductor.

The low current asymptotic value of inductance (L_{max} in Figure 10.7) is given by Equation 10.19 with $\mu_{r_1} \gg \mu_{eff_1}$ and $\mu_{r_2} \gg \mu_{eff_2}$. It is straightforward to establish the roll-off current in this case, by adopting the approach in Section 10.0. I_{80} is given by Equation 10.18 with [5/4] replaced by [5/(4 – g_1/g_2)] and $g = g_1$, yielding:

$$I_{80} = \frac{5H_m g_1}{\left(4 - \frac{g_1}{g_2}\right)N} \tag{10.20}$$

In Example 10.4:

$$I_{80} = \frac{(5)(1.3 \times 10^6)(0.39 \times 10^{-3})}{\left(4 - \frac{0.39}{0.69}\right)(365)} = 2.0\,A$$

Before the onset of saturation in the larger gap, the overall inductance is determined by the relative permeability of the material (μ_{r1}) and the larger air gap (g_2). The designer has additional control over the *L*/*i* characteristic by judicious selection of g_1 and g_2, so it seems logical to introduce on infinite number of gaps or a sloped air gap.

10.3 Sloped Air Gap Inductor

The inductor with a sloped air gap is illustrated in Figure 10.10, along with its dimensions.

Neglecting fringing, the flux linkage $d\lambda$ in the element $dx \times D$ (assuming $\mu_r \gg \mu_{eff}$) is:

$$d\lambda = \frac{N^2 i}{dR_{g(x)}} = \frac{\mu_0 N^2 i D}{g(x)} dx \tag{10.21}$$

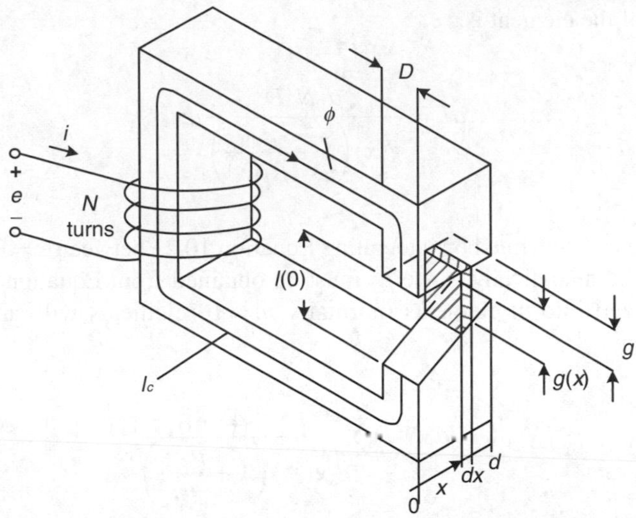

Figure 10.10 Sloped air gap (SAG) inductor.

where $g(x)$ is the length of the air gap at x, given by:

$$g(x) = G - \frac{(G-g)x}{d} \tag{10.22}$$

G and g are the maximum and minimum dimensions of the sloped gap.

The flux linkage between $x = 0$ and x is:

$$\lambda(x) = \int_0^x \frac{\mu_0 N^2 iD}{g(x)} dx \tag{10.23}$$

The total flux linkage is found, by performing the integration Equation 10.17 with $x = d$, the total inductance is $\lambda(d)/i$:

$$L_d = \frac{\mu_0 N^2 A_g}{G-g} \ln\left(\frac{G}{g}\right) \tag{10.24}$$

This is the total inductance before any part of the core becomes saturated. Saturation effects may be treated in the same manner described in Section 10.1. Consider again the element dx and apply Ampere's law (Equation 10.9), with g replaced by $g(x)$. Equation 10.10 applies with $\mu_{\text{eff}} = l_c/g(x)$.

$$H_0\mu_r(x)^2 + \left(\frac{Ni}{g(x)} + H_0\mu_{\text{eff}} - H_m\right)\mu_r(x) - H_m\mu_{\text{eff}} = 0 \tag{10.25}$$

The inductance of the element is:

$$dL = \frac{\mu_0 N^2 D}{g(x)\left(1+\dfrac{\mu_{\text{eff}}}{\mu_r(x)}\right)}dx \tag{10.26}$$

The total inductance is obtained by integrating Equation 10.21 between $x=0$ and $x=d$. This cannot be achieved analytically, since μ_r must be obtained from Equation 10.25. The core must be discretized into m elements (normally $m=10$ elements will suffice). The total inductance is then:

$$L = \mu_0 N^2 D \sum_{i=0}^{m-1} \frac{\Delta x}{g(x_i)\left(1+\dfrac{\mu_{\text{eff}_i}}{\mu_{r_i}}\right)} \tag{10.27}$$

For m equal elements:

$$\Delta x = \frac{d}{m} \tag{10.28}$$

$$x_i = i\Delta x + \frac{\Delta(x)}{2} \tag{10.29}$$

$$\mu_{\text{eff}_i} = \frac{l_c}{g(x_i)} \tag{10.30}$$

μ_{r_i} is obtained from Equation 10.25.

Example 10.5

An inductor constructed with an EI assembly EI42 (corresponding to IEC YEI 1-14), with M530-50A material as described in Example 10.1, was constructed with $G=1.0$ mm, $g=0.2$ mm, $l_c=8.4$ cm, $A_g=2.027\ \text{cm}^2$, d=1.4394 cm and $N=365$ turns. Calculate the inductance L and the effective inductance L_{eff} as a function of current for the range 0.25 to 4 A.

For low values of current before the onset of saturation, the inductance is a function of the gap and is given by Equation 10.24:

$$L_d = \frac{\mu_0 \cdot N^2 \cdot A_g}{G-g}\ln\left(\frac{G}{g}\right) = \frac{(4\pi \times 10^{-7})(365)^2(2.072\times 10^{-4})}{(1.0-0.2)\times 10^{-3}}\ln\left(\frac{1.0}{0.2}\right) \times 10^3 = 69.8\ \text{mH}$$

We may divide the core into ten discrete elements, so that $\Delta x = 0.1\,d$.

The MATLAB program to perform these calculations is included at the end of this chapter.

The plot of L and L_{eff} is shown in Figure 10.11.

The shape of the graph, in this case, is controlled by the sloped gap, which means that we can achieve the desired characteristic by controlling the geometry of the gap.

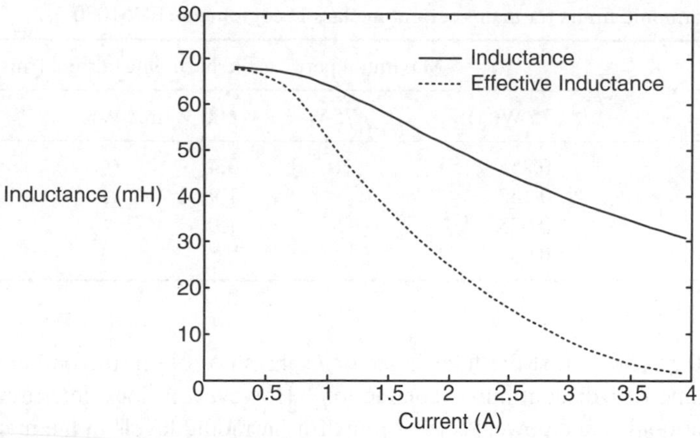

Figure 10.11 Inductance and effective inductance of a SAG inductor.

10.4 Applications

10.4.1 Power Factor Correction

A typical AC/DC converter with an output buffer capacitor and a passive inductor for power factor correction is shown in Figure 10.12. P represents the input power to a second DC-DC converter stage. The output voltage ripple and the hold-up time of the circuit are determined by the time constant of the capacitor and load resistance.

When the output voltage ripple is less than 10% of the peak DC output voltage, approximate analysis [1] shows that:

$$C = \frac{P_{\text{out}} T}{2V \cdot \Delta V} \tag{10.31}$$

where T is the mains period, V is the peak value of the input voltage and ΔV is the output voltage ripple. The value of the capacitance is proportional to the output power. For $\Delta V/V = 10\%$, the hold-up time for a drop of 20% in the output voltage is approximately equal to T or 20 ms in the case of a 50 Hz supply.

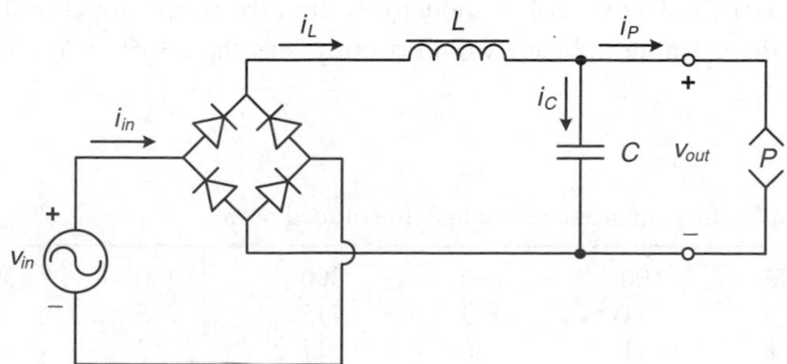

Figure 10.12 Rectifier circuit with passive power factor correction.

Table 10.2 Harmonic limits for mains current in class D equipment (EN61000-3-2)

Harmonic order	Maximum permissible harmonic current (rms)		
	At 75 W(A)	75 W < P < 600 W(mA/W)	At 600 W(A)
3	0.255	3.4	2.04
5	0.142	1.9	1.14
7	0.075	1.0	0.60
9	0.037	0.5	0.30

Equation 10.31 shows that the filter inductor L plays no role in the output voltage ripple and hold-up time (for discontinuous conduction). However, it does determine the rectifier line current harmonics and power factor. Limits for harmonic levels in the mains current are specified in the international standard EN61000-3-2 and are listed in Table 10.2 for class D equipment, which covers rectifier circuits used in TVs and computer equipment. The standard does not require the power supply to meet the limits over the entire power range, but every customer will operate a standard power supply at a different power level. The nominal power of the end users equipment can be anywhere from 25% to 100% of the maximum rating of the power supply. It is reasonable, then, to assume that the power supply is designed to meet the limits between 75 W and the maximum power rating.

The circuit in Figure 10.12 was simulated using the ElectroMagnetic Transient Program [3]. The input was 220 V at 50 Hz and the capacitance C was chosen for 10% output ripple. L was selected so that the limits in Table 10.2 were not exceeded. The simulations where carried out for the output load range from 75 W to 600 W. The values of L and C and the peak value of the input current are listed in Table 10.3.

The capacitance values, predicted by the approximate formula in Equation 10.31, are higher than the exact values in Table 10.3. In practice, L is largely independent of C. The required value of L is a function of the current flowing through it. The variation of filter inductance with load is illustrated in Figure 10.13 for the data in Table 10.3.

A conventional inductor, which would ensure compliance to the harmonic limits for a power range of 100 to 600 W, would have a value of 51 mH. At 600 W, the 51 mH inductor would carry a peak current of 6 A and the stored energy in the inductor would be 918 mJ.

An inductor with an inductance versus current characteristics as shown in Figure 10.13 would have its maximum stored energy at 600 W (i.e. 9 mH at 8.3 A), which represents a stored energy of 310 mJ. The size of an inductor is directly proportional to its energy storage capacity, so that the variable inductor would occupy less than 35% of the volume of a conventional inductor.

Table 10.3 Minimum filter inductance as a function of load power

Load (W)	75	100	200	300	400	500	600
C (μF)	54	71	143	215	285	359	430
L (mH)	68	51	26	17	13	11	9
I_{peak} (A)	1	1.4	2.8	10.2	5.6	6.9	8.3

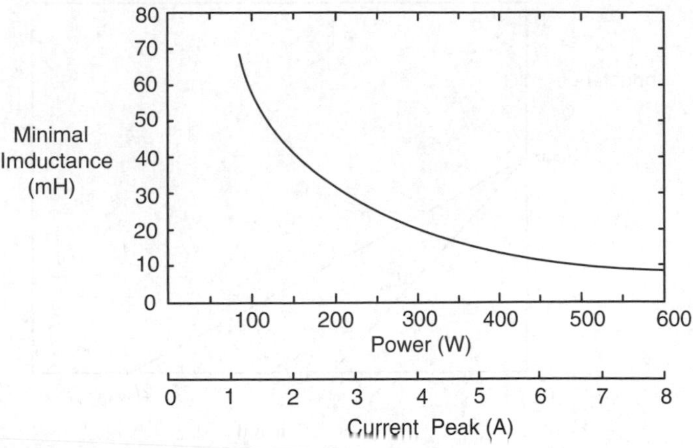

Figure 10.13 Variation of filter inductance with load power and input peak current.

10.4.2 Harmonic Control with Variable Inductance

The circuit in Figure 10.14 has a variable inductance driven by a sinusoidal source of voltage. The variable inductance characteristic is shown in Figure 10.15 and may be described as:

$$L = L_{max}\left(1 - \frac{1}{2}\frac{|i|}{I_{max}}\right) \tag{10.32}$$

This inductance described by Equation 10.32 and illustrated in Figure 10.15 is an idealized form of variable inductance. The inductance falls to zero at $i = 2I_{max}$. The effective inductance is obtained from Equation 10.3:

$$L_{eff} = L_{max}\left(1 - \frac{|i|}{I_{max}}\right). \tag{10.33}$$

In this case, the effective inductance falls to zero at $i = I_{max}$, as shown in Figure 10.15.

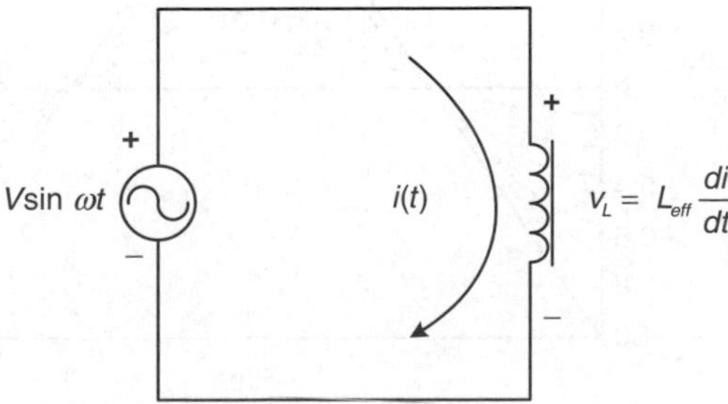

Figure 10.14 Circuit with variable inductance.

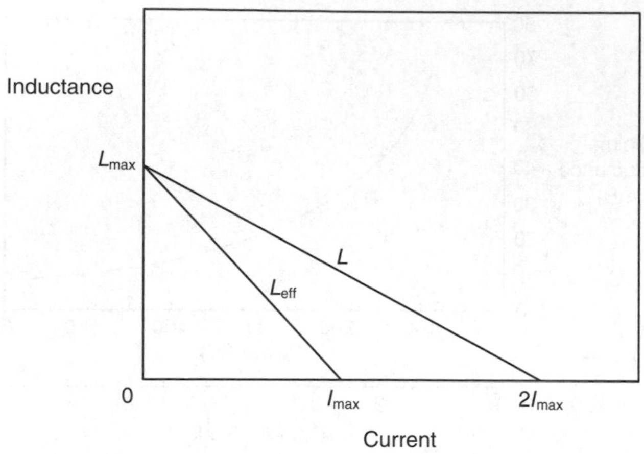

Figure 10.15 Variable inductance as a function of current.

The solution for the current $i(t)$ in Figure 10.14 is straightforward and is given by:

$$i(t) = -sgn(\cos(t))I_{max}\left[1 - \sqrt{1 - k|\cos(\omega t)|}\right] \tag{10.34}$$

where $k = \dfrac{2I_0}{I_{max}}$ and $I_0 = \dfrac{V}{\omega L_{max}}$ is the peak current for constant inductance L_{max}.

$i(t)/I_{max}$ is illustrated in Figure 10.16 for several values of k. For small values of k, the current is almost sinusoidal and therefore would have very low harmonics.

The Fourier series for $i(t)$ is given by [1]:

$$i(t) = \sum_{n=0}^{\infty} a_{2n+1}\cos(2n+1)\omega t \tag{10.35}$$

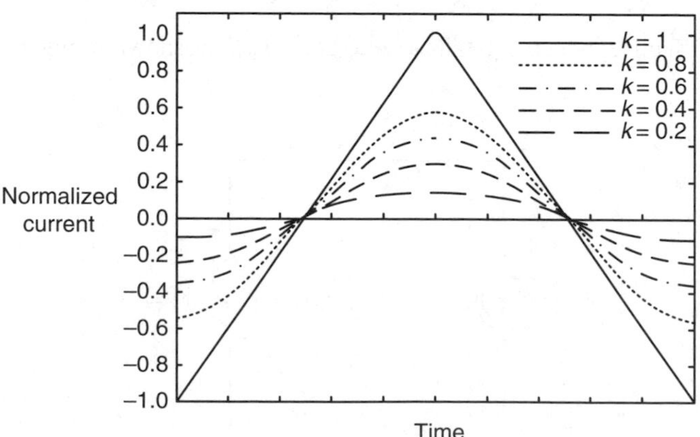

Figure 10.16 Current in the variable inductor.

where:

$$a_{2n+1} = \frac{4}{\pi}\frac{(-1)^n}{2n+1}\left[-1 + {}_3F_2\left(-\frac{1}{4},\frac{1}{4},1;n+\frac{3}{2},\frac{1}{2}-n;k^2\right)\right]$$
$$-2\begin{pmatrix}\frac{1}{2}\\ 2n+1\end{pmatrix}\left(\frac{k}{2}\right)^{2n+1} {}_2F_1\left(n+\frac{1}{4},n+\frac{3}{4};2n+2;k^2\right) \quad (10.36)$$

${}_2F_1$ and ${}_3F_2$ are hypergeometric functions of type (2, 1) and (3, 2) respectively, and $\binom{n}{m}$ is a binomial coefficient. This may be simplified for $k > 0.5$ to yield:

$$i(t) \approx -\frac{\pi}{4}\left[k^{\frac{3}{2}}\cos\omega t + \sum_{n=1}^{\infty}\frac{k^{2n+1}}{(2n+1)^2}\cos(2n+1)\omega t\right] \quad (10.37)$$

which clearly show the dependency of the harmonics on k.

As the peak current in the circuit approaches $I_{\max}$ ($k \rightarrow 1$), the harmonic content increases as illustrated in Figure 10.17 (values shown are normalized to the fundamental in each case).

Adopting the conventional definition of total harmonic distortion (THD):

$$THD = \sqrt{\frac{\sum_{n\neq 1}^{\infty} I_n^2}{I_1^2}} \quad (10.38)$$

where I_n is the amplitude of the nth harmonic.

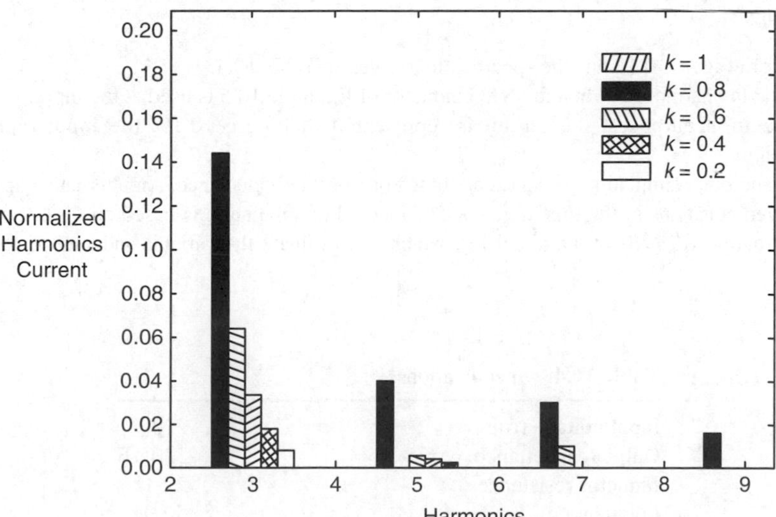

Figure 10.17 Harmonics of the current.

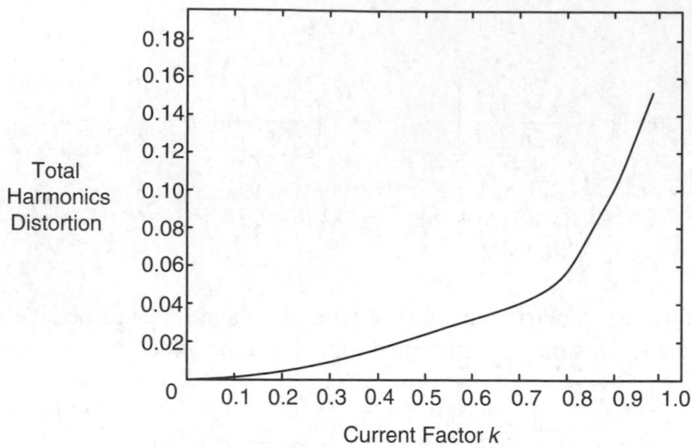

Figure 10.18 Total harmonic distortion.

The severity of the harmonic problem is illustrated in Figure 10.18, particularly as the peak current in the circuit, $\left[I_{peak} = I_{max}\left(1 - \sqrt{1-k}\right)\right]$, approaches I_{max}, that is $k \to 1$, where I_{max} is interpreted as the value of current where L_{eff} becomes zero. In the practical inductors of Examples 10.2, 10.4 and 10.5, I_{max} could be taken as 2.2 A, 3.0 A and 3.5 A for the saturated core inductor (Figure 10.4), the swinging inductor (Figure 10.9) and the SAG inductor (Figure 10.11) respectively.

This analysis shows that the effectiveness of the variable inductance in reducing unwanted harmonics is determined by the relationship between the peak current in the circuit and the roll-off of the inductance as current increases.

Example 10.6

The rectifier in Figure 10.12 has the specifications given in Table 10.4.

Determine the harmonics when the SAG inductor of Example 10.5 is used at the input.

For simulation purposes, the circuit is represented in Figure 10.19; the input represents the rectifier output.

PSPICE and other simulation packages are based on constant inductance. An alternative approach is to use discretized equations in the time domain. The method of simulation is based on the ElectroMagnetic Transient Program (EMTP) by Dommel [3], which can evaluate the variable inductance at each value of current.

Table 10.4 Specifications

Input voltage (rms)	220 V
Output capacitance	330 μF
Inductor resistance	2.5 Ω
Frequency	50 Hz
Output power	200 W

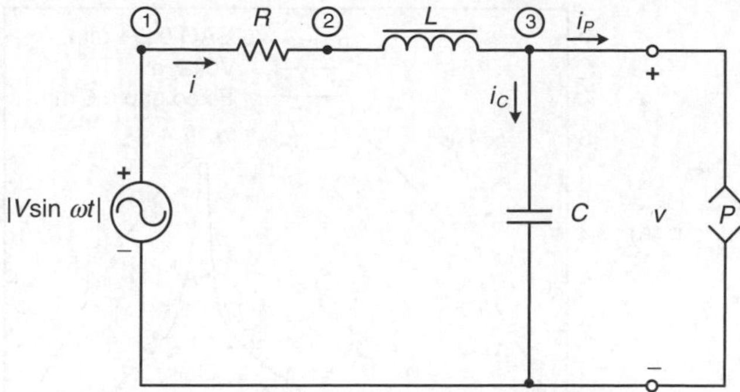

Figure 10.19 Simulated rectifier circuit.

The general solution for the three nodes in Figure 10.19, recognizing that there are no input currents at nodes 2 and 3, is:

$$\begin{bmatrix} i_1(t_n) \\ 0 \\ 0 \end{bmatrix} = \begin{bmatrix} Y_{11} & Y_{12} & Y_{13} \\ Y_{21} & Y_{22} & Y_{23} \\ Y_{31} & Y_{32} & Y_{33} \end{bmatrix} \cdot \begin{bmatrix} v_1(t_n) \\ v_2(t_n) \\ v_3(t_n) \end{bmatrix} + \begin{bmatrix} I_1(t_{n-1}) \\ I_2(t_{n-1}) \\ I_3(t_{n-1}) \end{bmatrix}$$

The terms in the admittance matrix are:

$$Y_{11} = \frac{1}{R} \qquad Y_{12} = -\frac{1}{R} \qquad Y_{13} = 0$$

$$Y_{21} = -\frac{1}{R} \qquad Y_{22} = \frac{1}{R} + \frac{\Delta t}{2L_{\text{eff}}} \qquad Y_{23} = -\frac{\Delta t}{2L_{\text{eff}}}$$

$$Y_{31} = 0 \qquad Y_{32} = -\frac{t}{2L_{\text{eff}}} \qquad Y_{33} = \frac{\Delta t}{2L_{\text{eff}}} + \frac{2C}{\Delta t}$$

$$t_n = n\Delta t$$

L_{eff} must be calculated at $i_1(t_{n-1})$ according to Equation 10.15.
The current sources $I_i(t_{n-1})$ are known from past history:

$$I_1(t_{n-1}) = 0$$

$$I_2(t_{n-1}) = I_2(t_{n-2}) + \frac{\Delta t}{L_{\text{eff}}}[v_2(t_{n-1}) - v_3(t_{n-1})] \quad n = 2, 3\ldots$$

$I_3(t_{n-1})$ includes contributions from L_{eff} and C and, since we are dealing with a constant power output rather than a load resistor, this effect may be included:

$$I_3(t_{n-1}) = I_{30}(t_{n-1}) - I_2(t_{n-1}) + \frac{P_0}{v_3(t_{n-1})} \quad n = 2, 3\ldots$$

$$I_{30}(t_{n-1}) = -I_{30}(t_{n-2}) - \frac{4C}{\Delta t}v_3(t_{n-1}) \quad n = 2, 3\ldots$$

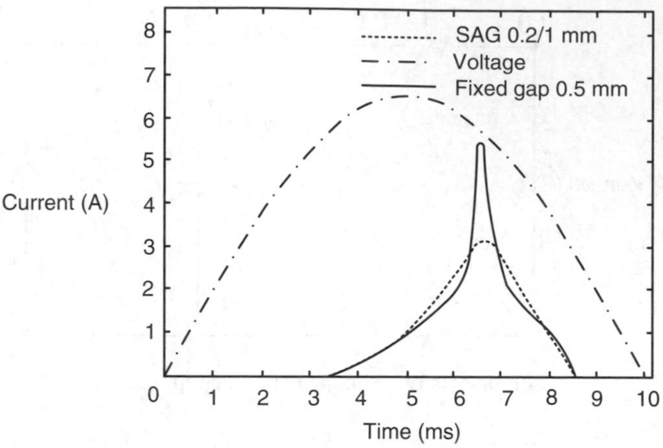

Figure 10.20 Simulation results for 200 W power supply.

The initial conditions are:

$$v_2(0) = v_3(0) = i_i(0) = 0, I_2(0) = 0, I_{30}(0) = 0 I_3(0) = 0.$$

If i_1 results in a negative value, it is reset to zero and both I_2 and I_3 are adjusted accordingly, since in reality the diodes block negative current at the output of the rectifier.

The results are shown in Figure 10.20 for one half cycle. The peak current in the circuit (3.0 A) is less than I_{max} (3.5 A in Figure 10.11). The harmonic spectrum of the current waveform is obtained by the Fast Fourier Transform and is shown in Figure 10.21. The harmonic currents are normalized to the fundamental current in each case. The limits from Table 10.2 are also shown, and the harmonics are within the allowable limits

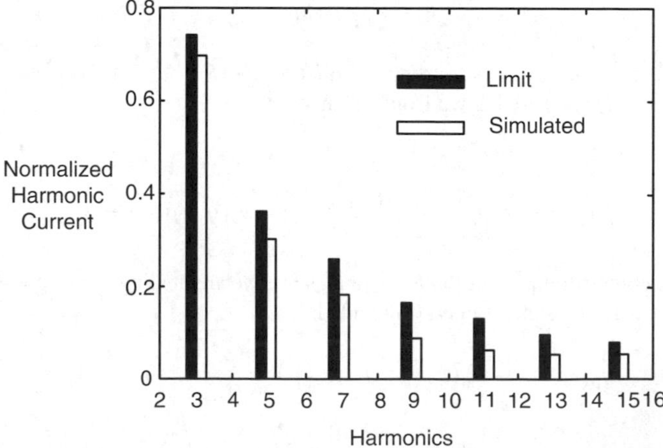

Figure 10.21 Harmonics with the variable inductor.

10.4.3 Maximum Power Point Tracking

Maximum power point tracking (MPPT) [2] is implemented in solar photovoltaic (PV) systems to achieve maximum power output as the ambient conditions, such as incident solar radiation and temperature, change. MPPT is normally achieved by either the perturb and observe method (P&O) or by the incremental conductance method (ICM). In the ICM approach, the output resistance of the PV panel is equal to the load resistance, as expected from the celebrated maximum power transfer theorem; this may be shown by linearizing the *I-V* output characteristic of a PV panel about the operating point, as illustrated in Figure 10.22.

Thus, the equivalent resistance r at the maximum power point should meet the following equation:

$$-r = -\frac{\Delta V}{\Delta I} = R_{LR} = \frac{V_P}{I_P} \tag{10.39}$$

Where R_{LR} is the regulated resistance in order to achieve MPPT, and V_P and I_P are the PV voltage and current at maximum power.

The actual load resistance R_L is matched to R_{LR} by a buck converter, as shown in Figure 10.23. The regulated resistance is related to the load resistance by:

$$R_{LR} = \frac{1}{D^2} R_L \tag{10.40}$$

where D is the duty cycle of the buck converter. The value of D is between 0 and 1, therefore R_{LR} has a value between R_L and infinity.

Consider two levels of illumination intensity at points (1) and (2) in Figure 10.22. The current at the MPP decreases going from (1) to point (2), which changes the value of the PV resistance at the MPP. In order to achieve MPPT, the regulated resistance R_{LR} should be adjusted by changing the duty cycle D in Equation 10.40.

The minimum inductance in a buck converter in continuous conduction mode (CMM) is given by:

$$L_{\min} = \frac{R_L(1 - D)}{2f_s} \tag{10.41}$$

where f_s is the switching frequency.

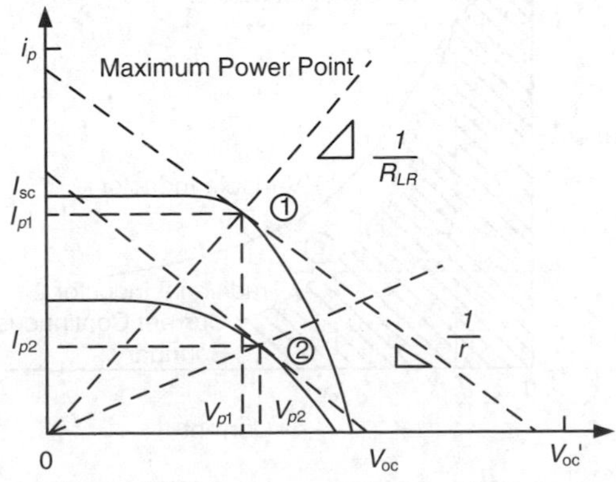

Figure 10.22 MPPT based on impedance matching.

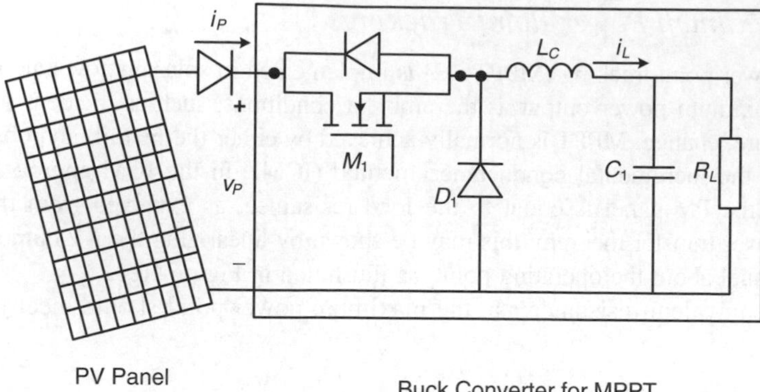

Figure 10.23 Maximum power transfer in a PV module.

Combine Equations 10.39, 10.40 and 10.41 to yield:

$$L_{\min} = \frac{D^2(1-D)V_P}{2f_s I_P} \tag{10.42}$$

The PV voltage is relatively constant over the full range of solar intensity [2], so Equation 10.42 shows that the minimum inductance is a function of duty cycle D and the output current of the PV panel I_p under constant switching frequency. The minimum inductance to achieve CCM falls off with increasing PV current as the solar intensity increases. Conversely, the higher value of inductance required at light loads may be achieved without increasing the volume of the inductor.

The role of the variable inductor in the stable operation of the buck converter is explained by reference to Figure 10.24. Continuous conduction can only be achieved with inductance

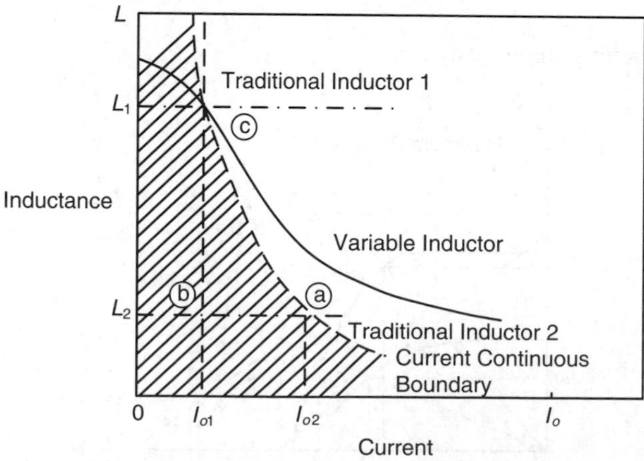

Figure 10.24 Comparisons of CCM conditions in an MPPT DC/DC converter with a variable inductance.

Table 10.5 Parameters under different load conditions

Insolation (W/m^2)	V_P (V)	I_P (A)	Maximum output power (W)
800	41.3	4.1	169
600	41.4	3.1	128
400	41.6	2.0	83.2
200	41.6	1.0	41.6

values above the dashed line in Figure 10.24 (the shaded area is off limits). The lower limit of load current (corresponding to low solar insolation) is given by I_{O1} as long as the inductance is greater than L_1. At higher currents (and higher insolation levels), say, I_{O2}, a smaller inductor L_2 would suffice, with the added advantage of a reduced volume occupied by the inductor. Conversely, setting the inductance at L_2 would limit the lower load range to values of current (and solar insolation) greater than I_{O2}.

Example 10.7

The voltage and current at the maximum power point in a 200 W solar panel are listed in Table 10.5, along with the maximum power for various level of solar insolation.

Calculate the internal resistance of the panel for each level of solar power, the duty cycle of the buck converter for impedance matching with a load resistance of 8 Ω operating at 20 kHz. Calculate the minimum inductance for continuous conduction.

The internal resistance is given by Equation 10.39:

$$r = \frac{V_P}{I_P} = \frac{41.3}{4.1} = 10.07\,\Omega$$

The remaining values are 13.35, 20.80, and 41.60 Ω.

The duty cycle is given by Equation 10.40:

$$D = \sqrt{\frac{R_L}{r}} = \sqrt{\frac{8.0}{10.07}} = 0.89$$

The remaining values are 0.77, 0.62, and 0.44.

The minimum value of inductance to ensure continuous conduction is given by Equation 10.42:

$$L_{\min} = \frac{D^2(1-D)V_P}{2f_s I_P} = \frac{(0.89)^2(1-0.89)(41.3)}{(2)(20\times 10^3)(4.1)} \times 10^6 = 21.9\,\mu\text{H}$$

The remaining values are 45.5, 76.0, and 112.8 μH.

Example 10.8

The inductance characteristic of the inductor in Example 10.3 satisfies the criteria for the inductance-current pairs in Example 10.7.

Table 10.6 Specifications

Input voltage	41.6 V
Output capacitance	80 μF
Inductor resistance	1 Ω
Load resistance	8 Ω
Frequency	20 kHz

The buck converter in Figure 10.23 has the specifications given in Table 10.6.

Simulate the circuit and plot the inductor current for each of the following values of inductance:

1. constant inductance of 21.9 μH at solar insolation of 800 W/m^2
2. constant inductance of 21.9 μH at solar insolation of 200 W/m^2
3. variable inductance of Figure 10.6 at 200 W/m^2

The circuit of Figure 10.23 has been simulated for the purposes of evaluating the response for the variable inductor. The method of simulation is based on the electromagnetic transient program (EMPT). The simulated circuit is shown in Figure 10.25.

The series coil resistance R, which has a damping effect on the inductor current, is included and, for convenience, the solar panels and MOSFET voltage are represented by a rectangular PWM waveform with a variable duty cycle. The appropriate value of the duty cycle was calculated in Example 10.7.

The general solution for the three nodes in Figure 10.25, recognizing that there are no input currents at node 2 and 3, is:

$$\begin{bmatrix} i_1(t_n) \\ i_2(t_n) \\ i_3(t_n) \end{bmatrix} = \begin{bmatrix} Y_{11} & Y_{12} & Y_{13} \\ Y_{21} & Y_{22} & Y_{23} \\ Y_{31} & Y_{32} & Y_{33} \end{bmatrix} \cdot \begin{bmatrix} v_1(t_n) \\ v_2(t_n) \\ v_3(t_n) \end{bmatrix} + \begin{bmatrix} I_1(t_{n-1}) \\ I_2(t_{n-1}) \\ I_3(t_{n-1}) \end{bmatrix}$$

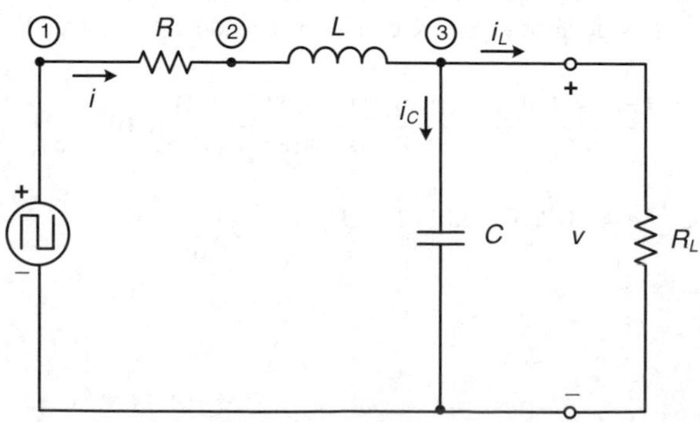

Figure 10.25 Circuit for simulation.

The terms in the admittance matrix are:

$$Y_{11} = \frac{1}{R} \quad Y_{12} = -\frac{1}{R} \quad Y_{13} = 0$$

$$Y_{21} = -\frac{1}{R} \; Y_{22} = \frac{1}{R} + \frac{\Delta t}{2L_{\text{eff}}} \; Y_{23} = -\frac{\Delta t}{2L_{\text{eff}}}$$

$$Y_{31} = 0 \quad Y_{32} = -\frac{\Delta t}{2L_{\text{eff}}} \quad Y_{33} = \frac{\Delta t}{2L_{\text{eff}}} + \frac{2C}{\Delta t}$$

$$t_n = n\Delta t$$

L_{eff} must be calculated at $i_1(t_{n-1})$ according to Equation 10.15.
The current sources $I_1(t_{n-1})$ are known from past history:

$$I_1(t_{n-1}) = 0$$

$$I_2(t_{n-1}) = I_2(t_{n-2}) + \frac{\Delta t}{L_{\text{eff}}}[v_2(t_{n-1}) - v_3(t_{n-1})] \quad n = 2, 3 \ldots$$

$$I_3(t_{n-1}) = I_{30}(t_{n-1}) - I_2(t_{n-1}) + \frac{v_3(t_{n-1})}{R_L} \quad n = 2, 3 \ldots$$

$$I_{30}(t_{n-1}) = -I_{30}(t_{n-2}) - \frac{4C}{\Delta t} v_3(t_{n-1}) \quad n = 2, 3 \ldots$$

We define the initial conditions of voltages and currents equal to 0.

If i_1 results in a negative value, it is reset to zero and I_2 and I_3 are adjusted. This is because, in reality, the diode D is in series with the PV panel, as shown in Figure 10.23, and will block the current flowing into the PV panel.

1. Figure 10.26 shows the inductor current for 21.9 μH, at 800 W/m^2 (this corresponds to point *a* in Figure 10.24 and the current is continuous).

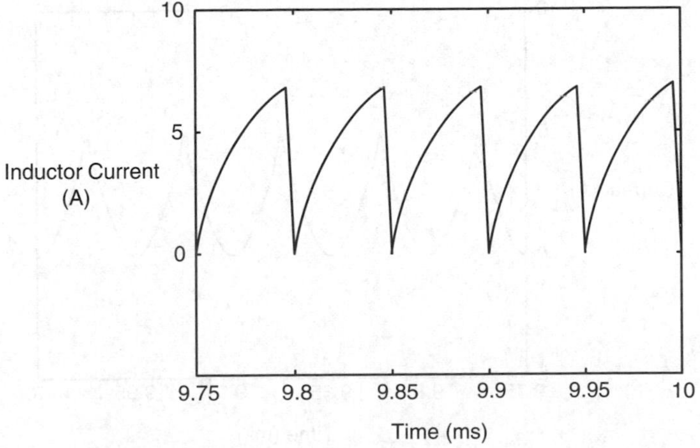

Figure 10.26 Inductor current for 21.9 μH at 800 W/m^2.

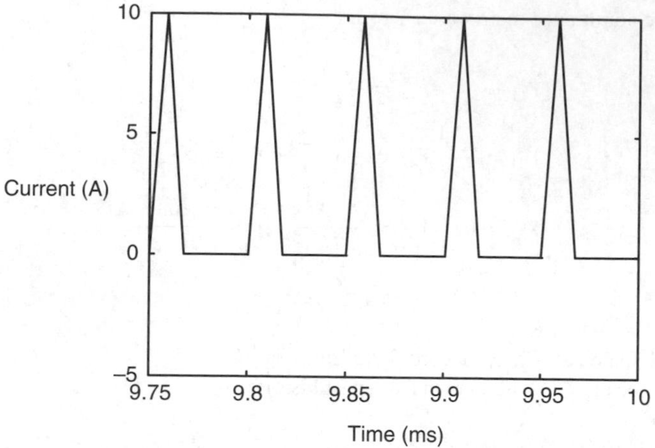

Figure 10.27 Inductor current for 21.9 μH at 200 W/m^2.

2. Figure 10.27 shows the inductor current for 21.9 μH, at 200 W/m^2 (point b in Figure 10.24) and, as expected, the converter is operating in discontinuous conduction mode.
3. Figure 10.28 shows that the variable inductor whose characteristic is shown in Figure 10.6 restores continuous conduction. The current waveform has sharp peaks and wide valleys that arise because of the lower value of inductance at higher current.

The inductor current range is 2.27 A to 4.61 A, corresponding to a solar insolation level from 200 W/m^2 to 800 W/m^2. In this range, the minimum inductor falls from 111 μH to 21.9 μH. A conventional inductor would have 111 μH at 4.61 A, corresponding to a stored energy of 1.2 mJ. With a variable inductor, the stored energy at 111 μH and 2.27 A is 0.29 mJ and, at 21.9 μH with 4.61 A, the stored energy is 0.25 mJ. The size of an inductor is directly proportional to its stored energy, so that the variable inductor would occupy 25% of the volume of a conventional fixed value inductor.

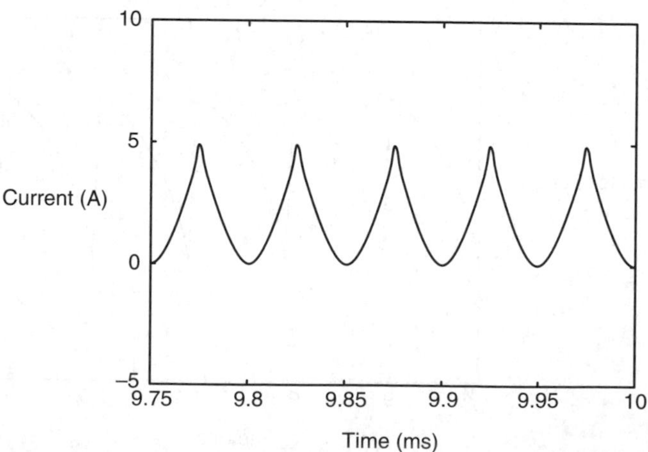

Figure 10.28 Inductor current for variable inductance at 200 W/m^2.

10.4.4 Voltage Regulation

In the AC/DC converter with an output buffer capacitor in Figure 10.12, we showed that the output voltage ripple is determined by the time constant of the capacitor and load resistance. We also showed, in Equation 10.31, that the filter inductor L plays no role in the output voltage ripple for discontinuous conduction of the input current; in this case, the average output voltage is approximately equal to the peak value of the input voltage (adjusted for the voltage ripple). All of the analysis of Section 10.4.1 is based on the assumption of discontinuous conduction.

However, it is possible to introduce continuous conduction with a sufficiently large filter inductor. The advantage of this mode of operation is that the average output voltage is now independent of the load current and is simply given by the average of the rectified voltage waveform. This is possible because the input voltage is always connected to the output; in discontinuous conduction, when the load is disconnected from the input, the output voltage is determined by the characteristics of the load. The high peak charging current typical of a circuit with a buffer capacitor is avoided, the harmonics associated with the peak current are greatly reduced and this leads to an improvement in the input power factor.

This inductor is sized according to the expected rated load current. However, at light loads, this inductance would be too small to maintain continuous conduction; the operation would revert to the discontinuous mode and the output voltage would rise towards the peak value of the input voltage. In the past, this situation was avoided by shunting the buffer capacitor with a bleeding resistor, with a consequential loss of efficiency. Another approach is to have the inductance value 'swing' to a higher value at low current, which led to the advent of the swinging inductor. The maximum inductance at low current ensured continuous conduction while improving voltage regulation over a wider current range. Traditionally, since these are 50 Hz or 60 Hz applications, laminated cores are used. This result would apply in the case of a purely resistive load and for an inductive load as long as continuous conduction is maintained.

The Fourier series for the output voltage of a full wave rectifier is:

$$v = \frac{2V}{\pi} - \frac{4V}{\pi}\sum_{n=1}^{\infty}\frac{\cos(2n\omega t)}{1-4n^2} \tag{10.43}$$

where V is the peak value of the input voltage waveform to the full wave rectifier.

The average output voltage across the output filter inductor is zero, so therefore the average output voltage of a full wave rectifier is:

$$V_{\mathrm{dc}} = \frac{2V}{\pi} \tag{10.44}$$

– thus ensuring good output voltage regulation.

The average current through the load resistance R is simply:

$$I_{\mathrm{dc}} = \frac{2V}{\pi R} \tag{10.45}$$

However, harmonic currents will arise from the harmonic voltages in Equation 10.43, and the sum of these harmonics must not exceed the DC current if continuous conduction is to be maintained. The amplitude of the harmonics of voltage decrease in the order 1/3 ($n = 1$), 1/15 ($n = 2$) and so on. The lowest frequency of the current is $2f$, due to rectifier action, and therefore it is a reasonable assumption to ensure that this current harmonic should be less than the DC component to ensure continuous conduction.

The impedance of the output circuit consisting of the inductor, capacitor and load resistance in Figure 10.12 is:

$$Z = R\frac{\sqrt{1 + x^2\left[1 - \frac{y}{x} + y^2\right]}}{1 + y^2} \tag{10.46}$$

where:

$$x = \frac{\omega L}{R} = \frac{X_L}{R} \tag{10.47}$$

$$y = \omega RC = \frac{R}{X_C} \tag{10.48}$$

At frequency $2f$, this becomes:

$$Z_2 = R\frac{\sqrt{1 + 4x^2\left[1 - \frac{y}{x} + 4y^2\right]}}{1 + 4y^2} \tag{10.49}$$

The amplitude of this harmonic of current is found from Equation 10.43:

$$I_2 = \frac{4V}{3\pi Z_2} \tag{10.50}$$

Taking I_{dc} in Equation 10.45 and I_2 with Z_2 given by Equation 10.49 yields:

$$5 - 16y^2 - 72xy + 36x^2(1 + 4y^2) > 0 \tag{10.51}$$

A good rule of thumb [4] for the design of a choke to give continuous conduction with good regulation, based on this inequality, is:

$$\frac{X_L}{R} \geq 0.4 \tag{10.52}$$

and:

$$\frac{R}{X_C} \geq 4.0 \tag{10.53}$$

These conditions ensure that the AC component of the current waveform is less than the DC component, which thereby maintains continuous conduction.

Example 10.9

Design the filter inductor for a 25 W load at 25 V rms input at 50 Hz that will maintain conduction down to 10% of rated load.

The DC output voltage is:

$$V_{dc} = \frac{(2)\sqrt{2}(25)}{\pi} = 22.5\ \text{V}$$

The DC output resistance is:

$$R = \frac{(22.5)^2}{25} = 20.25\ \Omega$$

The inductance required by Equation 10.55 is:

$$L = \frac{0.4R}{2\pi f} = \frac{(0.4)(20.25)}{(2\pi)(50)} \times 10^3 = 25.8\ \text{mH}$$

The DC current is:

$$I_{dc} = \frac{V_{dc}}{R} = \frac{22.5}{20.25} = 1.11\ \text{A}$$

At 10% of rated load, the current is 0.111 A and the load resistance is:

$$R = \frac{(22.5)^2}{2.5} = 202.5\ \Omega$$

The inductance required by Equation 10.55 is:

$$L = \frac{0.4R}{2\pi f} = \frac{(0.4)(202.5)}{(2\pi)(50)} \times 10^3 = 258\ \text{mH}$$

10.5 Problems

10.1 Repeat Example 10.2 for a fixed air gap with $g = 1.1$ mm and comment on the results.

10.2 Repeat Example 10.3 for Micrometals −45 material with the characteristic data in Figure 10.5.

10.3 Repeat Example 10.4 for a swinging inductor with $g_1 = 0.6$ mm and $g_2 = 0.12$ mm.

10.4 Repeat Example 10.5 for a sloped air gap inductor with $G = 1.2$ mm and $g = 0.6$ mm.

10.5 Calculate the value of capacitance in Example 10.9 to ensure continuous conduction.

10.6 Design a swinging inductor to meet the requirements of Example 10.9, using the characteristics of the inductor in Figure 10.9, by changing the number of turns.

MATLAB Program for Example 10.2

```
% example 10.2 : Inductance and effective inductance of an inductor with
a constant air-gap

%constants
muo = 4*pi*10^-7;
lc = 84e-3;
g = 0.5e-3;
mueff = lc/g;
N = 365;
Ag = 2.072e-4;
Ho = 100;
Hm = 1.3e6;
Lmax = muo*N^2*Ag/g;
u=0;

%loop
for i = [0.25:0.1:4];
a = Ho;
b = N*i/g+Ho*mueff-Hm;
c = -Hm*mueff;
mur = (-b+sqrt(b^2-4*a*c))/(2*a);
L = Lmax*1/(1+mueff/mur);
dL = L*(mueff/mur^2)/(1+mueff/mur);
dmur = ((-1/(2*a))+b/(2*a*sqrt(b^2-4*a*c)))*N/g;
Leff = L+i*dL*dmur;
u = u+1;
A(u) = i;
B(u) = L;
C(u) = Leff;
end

% plot
plot(A,B,'k',A,C,'b','LineWidth',2)
title('Inductance and effective inductance of an inductor with a
constant air-gap')
xlabel('i')
ylabel('L')
axis([0 4 0 80e-3])
grid off
hold on
```

MATLAB Program for Example 10.4

```
% example 10.4 : Inductance and effective inductance of a swinging
inductor

%constants
muo = 4*pi*10^-7;
lc = 84e-3;
g1 = 0.39e-3;
g2 = 0.69e-3;
mueff1 = lc/g1;
mueff2 = lc/g2;
N = 365;
Ag1 = 1.036e-4;
Ag2 = 1.036e-4;
Ho = 100;
Hm = 1.3e6;
Lmax1 = muo*N^2*Ag1/g1;
Lmax2 = muo*N^2*Ag2/g2;
u=0;

%loop
for i = [0.25:0.1:4];
a = Ho;
b1 = N*i/g1+Ho*mueff1-Hm;
b2 = N*i/g2+Ho*mueff2-Hm;
c1 = -Hm*mueff1;
c2 = -Hm*mueff2;
mur1 = (-b1+sqrt(b1^2-4*a*c1))/(2*a);
mur2 = (-b2+sqrt(b2^2-4*a*c2))/(2*a);
L1 = Lmax1*1/(1+mueff1/mur1);
L2 = Lmax2*1/(1+mueff2/mur2);
dL1 = L1*(mueff1/mur1^2)/(1+mueff1/mur1);
dL2 = L2*(mueff2/mur2^2)/(1+mueff2/mur2);
dmur1 = ((-1/(2*a))+b1/(2*a*sqrt(b1^2-4*a*c1)))*N/g1;
dmur2 = ((-1/(2*a))+b2/(2*a*sqrt(b2^2-4*a*c2)))*N/g2;
Leff1 = L1+i*dL1*dmur1;
Leff2 = L2+i*dL2*dmur2;
L = L1+L2;
Leff = Leff1+Leff2;
u = u+1;
A(u) = i;
B(u) = L;
C(u) = Leff;
end
```

```
%plot
plot(A,B,'k',A,C,'b','LineWidth',2)
title('Inductance and effective inductance of a swinging inductor')
xlabel('i')
ylabel('L')
axis([0 4 0 80e-3])
grid off
hold on
```

MATLAB Program for Example 10.5

```
% example 10.5 : Inductance and effective inductance of a SAG inductor

clear all
close all

%constants
muo = 4*pi*10^-7;
lc = 84e-3;
G = 1e-3;
g = 0.2e-3;
N = 365;
Ag = 2.072e-4;
D = 1.4394e-2;
d = 1.4394e-2;
Ho = 100;
Hm = 1.3e6;
Ld = (muo*N^2*Ag)/(G-g)*log(G/g);
m = 10;
deltax = d/m;
%loop for different current
u=0;
for i = [0.25:0.1:4];
L = 0;
Leff = 0;
  %loop of gap division
  for j = [0:1:m-1];
    xj = j*deltax+deltax/2;
    gxj = G-((G-g)*xj)/d;
    mueffj = lc/gxj;
    a = Ho;
    b = N*i/gxj+Ho*mueffj-Hm;
    c = -Hm*mueffj;
    murj = (-b+sqrt(b^2-4*a*c))/(2*a);
    T = muo*N^2*D*deltax/(gxj*(1+mueffj/murj));
```

```
    L = L+T;
    %-----------------------------------------------------------%
    dLj = T*(mueffj/murj^2)/(1+mueffj/murj);
    dmurj = ((-1/(2*a))+b/(2*a*sqrt(b^2-4*a*c)))*N/gxj;
    Leffj = T+i*dLj*dmurj;
    Leff = Leff+Leffj;
    %-----------------------------------------------------------%
  end

u = u+1;
A(u) = i;
B(u) = L;
C(u) = Leff;
end

%plot
plot(A,B,'k',A,C,'b','LineWidth',2)
title('Inductance and effective inductance of a SAG inductor')
xlabel('i')
ylabel('L')
axis([0 4 0 80e-3])
grid off
hold on
```

References

1. Wolfle, W.H. and Hurley, W.G. (2003) Quasi-active power factor correction with a variable inductive filter: theory, design and practice. *IEEE Transactions on Power Electronics* **18** (1), 248–255.
2. Zhang, L., Hurley, W.G., and Wölfle, W.H. (2011) A new approach to achieve maximum power point tracking for PV system with a variable inductor. *IEEE Transactions on Power Electronics* **26** (4), 1031–1037.
3. Dommel, H.W. (1969) Digital computer solution of electromagnetic transients in single-and multiphase networks. *IEEE Transactions on Power Apparatus and Systems* **PAS-88** (4), 388–399.
4. Dunham, C.R. (1934) Some considerations in the design of hot-cathode mercury-vapour rectifier circuits. *Proceedings of the Institution of Electrical Engineers, Wireless Section* **9** (27), 275–285.

Further Reading

1. Benavides, N.D. and Chapman, P.L. (2007) Boost converter with a reconfigurable inductor. *Proceedings of the IEEE Power Electronics Specialists Conference, PESC*, pp. 1695–1700.
2. Dishman, J.M., Kressler, D.R., and Rodriguez, R. (1981) Characterization, modeling and design of swinging inductors. *Proceedings of the Power Conversion Conference, Powercon* 8, pp. B3.1-B3.13.
3. Jain, S. and Agarwal, V. (2007) A single-stage grid connected inverter topology for solar PV systems with maximum power point tracking. *IEEE Transactions on Power Electronics* **22** (5), 1928–1940.
4. Kelley, A.W., Nance, J.L., and Moore, M.D. (1991) Interactive analysis and design program for phase-controlled rectifiers. *Proceedings of the IEEE Applied Power Electronics Conference and Exposition, APEC*, pp. 271–277.
5. Liserre, M., Teodorescu, R., and Blaabjerg, F. (2006) Stability of photovoltaic and wind turbine grid-connected inverters for a large set of grid impedance values. *IEEE Transactions on Power Electronics* **21** (1), 263–272.

6. Medini, D. and Ben-Yaakov, S. (1994) A current-controlled variable-inductor for high frequency resonant power circuits. *Proceedings of the IEEE Applied Power Electronics Conference and Exposition, APEC*, pp. 219–225.
7. Patel, H. and Agarwal, V. (2008) Maximum power point tracking scheme for PV systems operating under partially shaded conditions. *IEEE Transactions on Industrial Electronics* **55** (4), 1689–1698.
8. Redl, R. (1994) Power-factor correction in single phase switching mode power supplies - An overview. *International Journal of Electronics* **77** (5), 555–582.
9. Redl, R., Balogh, L., and Sokal, N.O. (1994) A new family of single-stage isolated power-factor correctors with fast regulation of the output voltage. *Proceedings of the IEEE Power Electronics Specialists Conference, PESC*, pp. 1137–1144.
10. Wolfle, W., Hurley, W.G., and Arnold, S. (2000) Power factor correction for AC-DC converters with cost effective inductive filtering. *Proceedings of the IEEE Power Electronics Specialists Conference, PESC*, pp. 332–337.
11. Wolfle, W., Hurley, W.G., and Lambert, S. (2001) Quasi-active power factor correction: the role of variable inductance. *Proceedings of the IEEE Power Electronics Specialists Conference, PESC*, pp. 2078–2083.

Appendix A

Table A.1 Wire data

AWG Number	IEC Bare Diameter (mm)	AWG Bare Diameter (mm)	Resistance @ 20 °C (mΩ/m)	Weight (g/m)	Overall[a] Diameter (mm)	Current @ 5 A/mm² (A)	Turns per cm²
10		2.588	3.270	46.76	2.721	26.30	12
	2.5		3.480	43.64	2.631	24.54	12
11		2.308	4.111	37.19	2.435	20.92	14
	2.24		4.340	38.14	2.366	19.70	14
12		2.05	5.211	29.34	2.171	16.50	20
	2.0		5.440	27.93	2.120	15.71	20
13		1.83	6.539	23.38	1.947	13.15	23
	1.8		6.720	22.62	1.916	12.72	23
14		1.63	8.243	18.55	1.742	10.43	30
	1.6		8.500	17.87	1.711	10.38	30
15		1.45	10.42	14.68	1.557	8.256	39
	1.4		11.10	13.69	1.506	7.700	42
16		1.29	13.16	11.62	1.392	6.535	52
	1.25		13.90	10.91	1.351	6.140	52
17		1.15	16.56	9.234	1.248	5.193	68
	1.12		17.40	8.758	1.217	4.930	68
18		1.02	21.05	7.264	1.114	4.086	80
	1.00		21.80	6.982	1.093	3.930	85
19		0.912	26.33	5.807	1.002	3.266	99
	0.9		26.90	5.656	0.9900	3.180	105
20		0.813	33.13	4.615	0.8985	2.596	126
	0.8		34.00	4.469	0.8850	2.510	126
21		0.724	41.78	3.660	0.8012	2.058	161
	0.71		43.20	3.520	0.7900	1.980	168
22		0.643	52.97	2.887	0.7197	1.624	195

(continued)

Transformers and Inductors for Power Electronics: Theory, Design and Applications, First Edition.
W. G. Hurley and W. H. Wölfle.

Table A.1 (*Continued*)

AWG Number	IEC Bare Diameter (mm)	AWG Bare Diameter (mm)	Resistance @ 20 °C (mΩ/m)	Weight (g/m)	Overall[a] Diameter (mm)	Current @ 5 A/mm^2 (A)	Turns per cm^2
	0.63		54.80	2.771	0.7060	1.559	216
23		0.574	66.47	2.300	0.6468	1.294	247
	0.56		69.40	2.190	0.6320	1.232	270
24		0.511	83.87	1.823	0.5806	1.025	314
	0.5		87.10	1.746	0.5690	0.982	340
25		0.455	105.8	1.445	0.5213	0.813	389
	0.45		108.0	1.414	0.5160	0.795	407
26		0.404	134.2	1.114	0.4663	0.6409	492
	0.4		136.0	1.117	0.4620	0.6280	504
27		0.361	168.0	0.9099	0.4204	0.5118	621
	0.355		173.0	0.8800	0.4140	0.4950	635
28		0.32	213.9	0.7150	0.3764	0.4021	780
	0.315		219.0	0.6930	0.3710	0.3900	780
29		0.287	265.9	0.5751	0.3494	0.3235	896
	0.28		278.0	0.5470	0.3340	0.3080	941
30		0.254	339.4	0.4505	0.3054	0.2534	1184
	0.25		348.0	0.4360	0.3010	0.2450	1235
31		0.226	428.8	0.3566	0.2742	0.2006	1456
	0.224		434.0	0.3500	0.2720	0.1970	1512
32		0.203	531.4	0.2877	0.2484	0.1618	1817
	0.2		554.0	0.2790	0.2450	0.1570	1840
33		0.18	675.9	0.2262	0.2220	0.1272	2314
	0.18		672.0	0.2270	0.2220	0.1270	2314
34		0.16	855.5	0.1787	0.1990	0.1005	2822
	0.16		850.0	0.1790	0.1990	0.1010	2822
35		0.142	1086.0	0.1408	0.1783	0.0792	3552
	0.14		1110.0	0.1370	0.1760	0.0770	3640
36		0.127	1358.0	0.1126	0.1613	0.0633	4331
	0.125		1393.0	0.1100	0.1590	0.0610	4645
37		0.114	1685.0	0.0907	0.1455	0.0510	5372
	0.112		1735.0	0.0880	0.1430	0.0490	5520
38		0.102	2105.0	0.0726	0.1313	0.0409	6569
	0.1		2176.0	0.0700	0.1290	0.0390	6853
39		0.0889	2771.0	0.0552	0.1160	0.0310	8465
	0.08		3401.0	0.0450	0.1050	0.0250	10300
40		0.0787	3536.0	0.0433	0.1036	0.0243	10660
41		0.0711	4328	0.0353	0.091	0.0199	13734
42		0.0635	5443	0.0282	0.0855	0.0158	15544
43		0.0559	7016	0.0218	0.0739	0.0123	20982

[a]Grade 2 or medium insulation.

Table A.2 List of manufacturers

ACME	http://www.acme-ferrite.com.tw
CERAMINC MAGNETICS	http://www.cmi-ferrite.com
DMEGC	http://www.chinadmegc.com
EILOR	http://www.magmet.com
EPCOS	http://www.epcos.com
FAIR-RITE	http://www.fair-rite.com
FDK	http://www.fdk.com
FERRITE INT	http://www.tscinternational.com
FERRONICS	http://www.ferronics.com
FERROXCUBE	http://www.ferroxcube.com
HIMAG	http://www.himag.co.uk
HITACHI	http://www.hitachimetals.com
ISKRA	http://www.iskra-ferrites.com
ISU	http://www.isu.co.kr
JFE(KAWTATETSU)	http://www.jfe-frt.com
KASCHKE	http://www.kaschke.de
KRVSTINEL	http://www.mmgna.com
MAGNETICS	http://www.mag-inc.com
MAGNETICS METALS	http://www.magmet.com
MICROMETALS	http://www.micrometals.com
MK MAGNETICS	http://www.mkmagnetics.com
NEOSID	http://www.neosid.com.au
NICERA	http://www.nicera.co.jp
ORB ELECTRICAL STEELS	http://www.orb.gb.com
PAYTON	http://www.paytongroup.com
SAILCREST	http://www.sailcrestmagnetics.com
SAMWHA	http://www.samwha.com
STEWARD	http://www.steward.com
TAKRON (TOHO)	http://www.toho-zinc.co.jp
TDG	http://www.tdgcore.com
TDK	http://www.tdk.com
THOMSON	http://www.avx.com
TOKIN	http://www.nec-tokin.com
TOMITA	http://www.tomita-electric.com
TRANSTEK MAGNETICS	http://www.transtekmagnetics.com

Index

Transformers and Inductors for Power Electronics: Theory, Design and Applications, First Edition.
W. G. Hurley and W. H. Wölfle.